W9-AGQ-281

FOUNDATIONS
OF ANALYSIS
IN THE
COMPLEX
PLANE

FOUNDATIONS OF ANALYSIS IN THE COMPLEX PLANE

THOMAS McCULLOUGH
California State University

KEITH PHILLIPS
New Mexico State University

HOLT, RINEHART AND WINSTON, INC.

New York · Chicago · San Francisco · Atlanta · Dallas · Montreal · Toronto · London · Sydney

Preface

The book is written for a beginning elementary, one-semester or one-quarter course in complex analysis, typically attended by a mixture of undergraduate mathematics students and both undergraduate and graduate physics and engineering students. The exposition and arrangement are intended to make the book usable for minimally prepared students (a good background in elementary calculus suffices). However, more discriminating students will find that the discussion of each topic treated is logically complete. We have not strived for mathematical elegance or complete generality, but rather we have attempted to be simple, direct, and useful. Most concepts introduced appear in various contexts throughout the book, not as isolated topics.

The subject proper begins in Chapter 1 where we first discuss mapping properties and continuity of complex-valued functions of a complex variable and then present the topological prerequisites necessary for much of the book. Chapter 1 concludes with the definition and elementary properties of derivatives of functions of a complex variable. Functions which have a complex derivative are called analytic. The study of such functions is the main topic of the book.

Chapter 0 is a review of material which most students will have seen although perhaps not all of it in the context of complex numbers. We urge all readers to at least scan this chapter for use as a reference for the rest of the book. Readers unfamiliar with the material should go through it carefully.

We consider complex power series early (Chapter 2) for three reasons: it requires the smallest possible conceptual jump from elementary real analysis; power series play a central role in both pure and applied complex analysis; and we can then give complete definitions of the elementary functions at an early point in the development.

The theory of integration of complex-valued functions with respect to complex weight functions, that is integration along paths, is introduced in Chapter 3. Included in this introduction is a discussion of the index (or winding number) of a closed path in the context of several examples. More information and examples are provided in Chapter 4 where the index is extensively used. When amply illustrated with examples, our experience has been that the index is not a difficult concept (though it is a brilliant and beautiful one) for students to understand. The inclusion of a treatment of the index makes for more precise statements or proofs of several results (the residue and local mapping theorems, for example), rendering them easier to understand. It also enables the reader to better distinguish difficult topological problems (for example, the Jordan curve theorem) from elementary aspects of complex analysis.

The central result of complex analysis is Cauchy's theorem, stating that the integral of an analytic function on a closed path is zero provided that the path and its interior are included in the domain of analyticity of the function. We follow the example of Walter Rudin in *Real and Complex Analysis* by first proving this result for closed paths in convex regions. The result is very well motivated in this way, and it is not terribly difficult to prove. This version of Cauchy's theorem suffices for most of our subsequent development. By splitting regions into convex subsets and modifying curves in ad hoc ways, the theorem suffices for the great majority of regions encountered in applied work. More general, and considerably more difficult, versions of Cauchy's theorem are proved in the last part of Chapter 4 in which simple connectedness instead of convexity is assumed.

Chapter 5 covers the Laurent series, the argument principle, and the residue calculus. All these are essential topics for both pure and applied complex analysis. Chapter 6 is a brief introduction to harmonic functions and the Dirichlet problem.

Chapter 7 is devoted to a study of mapping properties of one-to-one analytic functions. Elementary general results are proved and many examples are given. The theory has important applications in the solution of boundary value problems. The chapter ends with a discussion of the Schwartz-Christoffel formula, in which it is shown that the upper half-plane (or unit disk) can be mapped onto the interior of a closed polygon with preassigned exterior angles.

The exercises range from the routine and/or computational to the very challenging. We have split them into Type I and Type II exercises. Type I are straightforward illustrations of the material preceding them. Our idea is that most students should be able to work most of these as a check on their surface understanding of the material. Type II exercises are either conceptually or computationally more difficult or in some cases develop small theories of related topics. Applications are mentioned mainly in carefully modulated exercises. This method fits in with our plan of making the book adaptable. The text proper is primarily mathematical; individual instructors and students will choose which applications they want to emphasize. We believe that working on the exercises is the most important single aspect of learning this material. To this extent one may say that the function of the text is to provide a mathematical context for the exercises. This is especially true in the early parts of the text. As the material and the student become more sophisticated, the emphasis changes.

We provide a notational index in Appendix A. Readers are urged to consult it when they come upon an unfamiliar symbol or notation. In Appendix B we prove basic facts about the real numbers.

For most classes, there is probably too much material in the text to be covered in one quarter or even one semester. When teaching from preliminary versions, the authors have usually covered the unstarred sections of Chapters 1–5, but only scattered topics in Chapters 6 and 7. Starred sections can be omitted without disrupting the logical development. For example, we feel that most classes will want to skip 4.23–4.33 in which a very elementary (but long and difficult) proof of Cauchy's theorem for simply connected regions and related topics are presented.

The notes from which the book evolved were written during the spring and summer terms of 1969 for the undergraduate course in complex analysis which is given regularly at New Mexico State University. Edited versions were used in the fall and spring semesters of 1969–1970 at New Mexico State University and during the spring and summer of 1970 at California State University at Long Beach.

The entire manuscript was read by Professor Edwin Hewitt and by Dr. John Higgins. Each offered many valuable suggestions, pointed out several awkward constructions, and found a few errors. We greatly appreciate their efforts and interest. The major portion of the final manuscript was typed with real professional expertise by Mrs. Tess Pearce of the secretarial staff at New Mexico State University. Eva Bazan, Joy Krog, Jackie Hancock, and Mona Sailer all did excellent typing at various stages of the work. Our thanks are extended to all these people.

Finally, Holt, Rinehart and Winston has been very cooperative. In particular, Lois Wernick of the Holt, Rinehart and Winston editorial staff saw the book through various editorial and production stages with patience, thoroughness, and efficiency, complementing in each case our own characteristics.

THOMAS McCULLOUGH
KEITH PHILLIPS

Long Beach, California
Las Cruces, New Mexico

December 1972

Contents

FOUNDATIONS OF ANALYSIS IN THE COMPLEX PLANE

0

Review

0.0

The topics discussed in this chapter are generally covered either in pre-calculus mathematics courses or in the usual calculus sequence. The information is here for review and reference and to establish various notational conventions. Almost all of the action in this book takes place in a plane equipped with Cartesian coordinates. As usual, a point in a plane is located by specifying its x and y coordinates in that order. We will let the symbol $\mathbf{C}$ denote such a plane. Hence we write

$$\mathbf{C} = \{(x, y) \mid x \text{ and } y \text{ are real numbers}\}.$$

Two points in $\mathbf{C}$ are said to be equal if and only if both coordinates are equal; that is, $(a, b) = (c, d)$ means $a = c$ and $b = d$.

0.1 ADDITION

There are various operations on the points of $\mathbf{C}$. One of the simplest is addition. If $z_1 = (a, b)$ and $z_2 = (c, d)$ are two points of $\mathbf{C}$, then $z_1 + z_2 = (a, b) + (c, d) = (a + c, b + d)$; that is, points are added by adding corresponding coordinates. Note that this is exactly the way vectors are added. We let 0 denote the origin; that is, $0 = (0, 0)$. Thus the symbol 0 has two possible meanings: the number zero and the point $(0, 0)$. If $z_1 + z_2 = 0$ and $z_1 = (a, b)$, then it is evident that $z_2 = (-a, -b)$. In this case we call z_2 the *additive inverse* or *negative* of z_1 and write $-z_1 = z_2$. Subtraction is then defined by $z_1 - z_2 = z_1 + (-z_2)$.

1

0.2 DISTANCE

Another useful operation that can be performed is the computation of distance. If $z \in \mathbf{C}$ then, of course, $z - 0 = z$ and we let $|z - 0| = |z|$ denote the distance from z to 0. Recall that for a real number x, $|x|$ denotes the distance from x to 0. The distance $|z|$ has several names, a few of which are: (1) the length of z, (2) the absolute value of z, and (3) the modulus of z. Because we have chosen Cartesian coordinates, if $z = (a, b)$ then $|z| = (a^2 + b^2)^{1/2}$. If z_1 and z_2 are points of $\mathbf{C}$, then $|z_1 - z_2|$ is the distance from z_1 to z_2. Writing $z_1 = (a, b)$ and $z_2 = (c, d)$, we have

$$|z_1 - z_2| = [(a - c)^2 + (b - d)^2]^{1/2}.$$

Figure 0.1, a picture of a parallelogram, should be kept in mind.

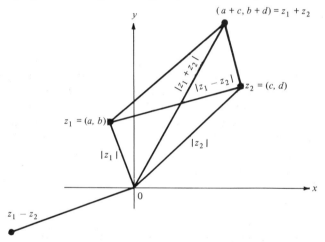

Figure 0.1

0.3 COMPLEX NUMBERS

We introduce a new notation for the points of $\mathbf{C}$ by denoting the point $(0, 1)$ by the letter i, a point $(0, b)$ by ib, and a point $(a, 0)$ by simply a. For $z = (a, b)$, we thus obtain

$$z = (a, 0) + (0, b) = a + ib.$$

If $z_1 = a + ib$ and $z_2 = c + id$, notice that

$$z_1 - z_2 = (a - c) + i(b - d)$$

and

$$|z_1 - z_2| = [(a - c)^2 + (b - d)^2]^{1/2}.$$

We will usually use the $a + ib$ notation. In this notation, the points of $\mathbf{C}$ are called *complex numbers*.

0.4 MULTIPLICATION OF COMPLEX NUMBERS

Multiplication of complex numbers is defined by first defining the product of i with itself; that is, i^2. We let $i^2 = -1 \; [=(-1, 0)]$. To see what $(a + ib) \cdot (c + id)$ should be, we treat the factors as first-degree polynomials in i and follow the rules of algebra to write

$$(a + ib)(c + id) = ac + aid + ibc + ibid$$
$$= ac + i(ad + bc) + i^2bd$$
$$= (ac - bd) + i(ad + bc).$$

We are thus led to the definition

$$(a + ib)(c + id) = (ac - bd) + i(ad + bc).$$

We now make several observations about multiplication of complex numbers.

First, i satisfies the polynomial equation $x^2 + 1 = 0$. In truth, i^2 was defined to be -1 just so that this equation, which has no real number as a solution, would have i as a solution. Notice that $-i$ also solves the equation.

For two points on the horizontal axis, say $z_1 = a + i0$ and $z_2 = c + i0$, we have $z_1z_2 = ac$. If b is real, then the product of b and i is $b \bullet i = (b + i0)(0 + i1) = ib$; similarly, $i \cdot b = ib$. Also note that for any $z = a + ib$, we have $0 \cdot z = 0$ and $1 \cdot z = (1 + i0)(a + ib) = z$. Finally, suppose that $z = a + ib \neq 0$. If we wish to find an element $x + iy$ of $\mathbf{C}$ such that $(x + iy)(a + ib) = 1$, then we can solve the system

$$ax - by = 1$$
$$ay + bx = 0$$

to obtain $x = a/(a^2 + b^2)$ and $y = -b/(a^2 + b^2)$. It follows that the product

$$\left\{ \frac{a}{a^2 + b^2} - i\left(\frac{b}{a^2 + b^2}\right) \right\} (a + ib) = 1$$

and we call the number $a/(a^2 + b^2) - i[b/(a^2 + b^2)]$ the multiplicative *inverse* (or simply the inverse) of z. We often write $1/z$ or z^{-1} for the inverse of z. Division of z by w is defined by $z/w = zw^{-1}$ if $w \neq 0$.

As an example, we have

$$(1 + 2i)(3 + 4i) = (1 \cdot 3 - 2 \cdot 4) + (2 \cdot 3 + 1 \cdot 4)i = -5 + 10i$$

and

$$\frac{3 + 4i}{1 + 2i} = (3 + 4i) \cdot \left(\frac{1}{5} - \frac{2}{5}i\right) = \frac{11}{5} - \frac{2}{5}i.$$

The reader should verify that $|zw| = |z||w|$ for any complex numbers z and w. In particular, the number $(1/|z|)z$ has length one for any $z \neq 0$.

The multiplication and addition just defined gives to **C** the same rules of arithmetic as have the real numbers. That is, **C** is a field. [See (B.1) for the definition of a field.] Later we shall prove that every polynomial equation having coefficients in **C** has a solution in **C** (for example, $x^2 + 2x + 2 = 0$ has no real solutions but has two solutions in **C**).

0.5   ORDER

We have seen that the complex numbers, like the real numbers, have a natural addition, multiplication, and absolute value. They do not, however, have a natural order. To see this, consider the number i. Clearly, $i \neq 0$. If $i > 0$ held, then we would have $-1 = i^2 > 0$; if $i < 0$ held, then $-i > 0$ would hold, and hence $(-i)^2 = -1 > 0$. Thus if "$<$" is to satisfy its familiar properties for complex numbers, i can be neither positive, negative, nor zero. The reader can easily construct other examples in **C** where the usual requirement for order (that is, $z_1 z_2 > 0$ whenever $z_1 > 0$ and $z_2 > 0$) fails. The absence of order (a useful tool in calculus) forces distance (absolute value) to assume a very important role in complex analysis.

0.6   CONJUGATE

If $z = a + ib$, then the reflection of z in the x-axis is $a - ib$. The reflection of z in the x-axis is so important that it is given a special name. If $z = a + ib$, then $\bar{z} = a - ib$ is called the *conjugate* of z. Notice that $z\bar{z} = a^2 + b^2 = |z|^2$ and that $z^{-1} = \bar{z}/|z|^2$ (Figure 0.2). It is illuminating to compare the

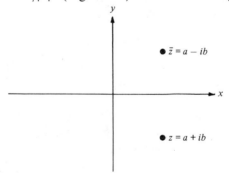

Figure 0.2

following two sets of computations.

(i) $\dfrac{1}{1 + \sqrt{2}} = \dfrac{1}{1 + \sqrt{2}} \dfrac{1 - \sqrt{2}}{1 - \sqrt{2}} = \dfrac{1 - \sqrt{2}}{-1} = -1 + \sqrt{2}, \quad [(\sqrt{2})^2 = 2].$

(ii) $\dfrac{1}{1 + i} = \dfrac{1}{1 + i} \dfrac{1 - i}{1 - i} = \dfrac{1 - i}{2} = \dfrac{1}{2} - \dfrac{1}{2} i, \quad [(i)^2 = -1].$

Thus the formula $z^{-1} = \bar{z}/|z|^2$ is similar to the technique of "rationalizing the denominator." In analogy, one could say that multiplying $1/z$ by $\bar{z}/\bar{z}$ "realizes the denominator."

For $z = a + ib$ it is customary to call a the real part of z and b the imaginary part of z and to write Re $(z) = a$ and Im $(z) = b$. Thus we have Re $(z) = (z + \bar{z})/2$ and Im $(z) = (z - \bar{z})/2i$. For example,

$$\frac{(1 + i) - (1 - i)}{2i} = 1 = \text{Im} (1 + i).$$

The equalities $\overline{z + w} = \bar{z} + \bar{w}$, $\overline{zw} = \bar{z}\bar{w}$, $\overline{z}\overline{w} = \bar{z}w$, $|z| = |\bar{z}|$, and the inequalities $|\text{Re} (z)| \le |z|$ and $|\text{Im} (z)| \le |z|$ are all easily established; the reader is advised to verify them.

0.7 TRIANGLE INEQUALITY

Perhaps the most important inequality in complex analysis is the so-called triangle inequality. From Figure 0.1 it is apparent that

$$|z_1 + z_2| \le |z_1| + |z_2|.$$

This inequality is so useful, however, that we will emphasize its truth by presenting another proof. We have

$$
\begin{aligned}
|z_1 + z_2|^2 &= (z_1 + z_2)(\overline{z_1 + z_2}) = (z_1 + z_2)(\bar{z}_1 + \bar{z}_2) \\
&= z_1\bar{z}_1 + z_2\bar{z}_1 + \bar{z}_2 z_1 + \bar{z}_2 z_2 \\
&= |z_1|^2 + z_2\bar{z}_1 + \overline{z_2\bar{z}_1} + |z_2|^2 \\
&= |z_1|^2 + 2 \text{ Re } (z_2\bar{z}_1) + |z_2|^2 \\
&\le |z_1|^2 + 2|\text{Re } (z_2\bar{z}_1)| + |z_2|^2 \\
&\le |z_1|^2 + 2|z_2\bar{z}_1| + |z_2|^2 \\
&= |z_1|^2 + 2|z_1||z_2| + |z_2|^2 \\
&= (|z_1| + |z_2|)^2.
\end{aligned}
$$

Since $|z_1 + z_2| \ge 0$ and $|z_1| + |z_2| \ge 0$, it follows that

$$|z_1 + z_2| \le |z_1| + |z_2|.$$

As an exercise in technique the reader should verify each step in the above proof.

0.8 ARGUMENT AND POLAR FORM

If $z = a + ib \ne 0$, then the line from 0 through z and the x-axis form an angle. We can choose the measure of this angle, t, to satisfy $-\pi < t \le \pi$. From trigonometry we have

$$\cos (t) = \frac{a}{\sqrt{a^2 + b^2}} \quad \text{and} \quad \sin (t) = \frac{b}{\sqrt{a^2 + b^2}}.$$

Evidently (see Figure 0.3)

$$z = \sqrt{a^2 + b^2}\,(\cos t + i \sin t)$$
$$= |z|\,(\cos t + i \sin t).$$

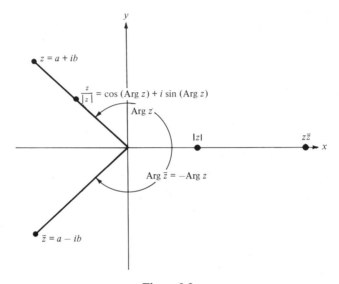

Figure 0.3

The number t is called the *principal value* of the *argument* of $z = a + ib$ and is written $t = \text{Arg}\,(z)$. That is, Arg (z) denotes the unique real number (angle measured in radians) between $-\pi$ and π $(-\pi < \text{Arg}\,(z) \le \pi)$ such that $z = |z|\,(\cos [\text{Arg}\,(z)] + i \sin [\text{Arg}\,(z)])$. The term arg (z) (observe the lower-case a here) is used to denote *any* θ such that $z = |z|\,(\cos \theta + i \sin \theta)$. The distinction here between Arg (z) and arg (z) is similar to the distinction made in trigonometry between the arctangent and the function called the principal value of the arctangent. The arctangent of a real number u is any real number θ satisfying tan $(\theta) = u$, while the principal value of the arctangent is usually required to satisfy $-\pi/2 < \theta \le \pi/2$ and is denoted Arctan (u). It is apparent from Figure 0.3 that Arg $(z) = $ Arctan (b/a) provided that z lies in the right half-plane. If z is in the upper left quadrant, then Arg $(z) = $ Arctan $(b/a) + \pi$ holds; if z is in the lower left quadrant, then Arg $(z) = $ Arctan $(b/a) - \pi$ holds. As examples, the reader should check that Arg $(1) = 0$, Arg $(i) = \pi/2$, Arg $(1 + i) = \pi/4$, Arg $(-1) = \pi$, and Arg $(-1 - i) = -3\pi/4$.

Thus, every $z \ne 0$ has a *polar form*

(i) $z = |z|\,(\cos (t) + i \sin (t)) = |z|e^{it}.$

In (i), $e^{it} = \cos(t) + i\sin(t)$ is taken as the definition of e^{it}, for any real number t. Presumably, the reader is familiar with the definition of e^x for x a real number, so now the exponential function is defined for real and for imaginary numbers. In Chapter 2, e^z will be defined for all complex numbers z, and we will see that the definitions agree. We summarize below some important properties of the function which takes t into e^{it}.

(ii) $|e^{it}| = 1$, all real numbers t.

(iii) $e^{it}e^{is} = e^{i(t+s)}$, all real t and s.

(iv) $(e^{it})^n = e^{int}$, all real t and integers n.

(v) $e^{it} = 1$ if and only if $t = 2\pi j$ for some integer j.

Equality (iii) is proved by a straightforward calculation depending upon trigonometric identities. The equalities (iv) follow from (iii) by induction for $n \geq 0$. In case $n < 0$, use the definition $z^{-n} = 1/z^n$. Equality (v) is easy to establish using properties of cos and sin. It is worth emphasizing that the unit circle, $\mathbf{T} = \{z \mid |z| = 1\}$, is equal to $\{e^{it} \mid -\pi < t \leq \pi\}$. In the latter form, the mental picture of $\mathbf{T}$ being traced out counterclockwise as t moves from $-\pi$ to π is useful.

The following example illustrates some of the basic notions discussed in Sections 0.1 through 0.8.

Example. Suppose that $z = -\sqrt{2} + i\sqrt{2}$. We have

$$|z| = \sqrt{2 + 2} = 2,$$

$$\bar{z} = -\sqrt{2} - i\sqrt{2},$$

$$|z|^2 = z\bar{z} = 4 = |z^2|,$$

$$\frac{z}{|z|} = \frac{1}{2}(-\sqrt{2} + i\sqrt{2}) = -\frac{\sqrt{2}}{2} + i\frac{\sqrt{2}}{2},$$

$$\text{Arg}(z) = \frac{3\pi}{4},$$

$$z = 2\left(\cos\left(\frac{3\pi}{4}\right) + i\sin\left(\frac{3\pi}{4}\right)\right) = 2e^{i(3\pi/4)},$$

$$\frac{z}{|z|} = \cos\left(\frac{3\pi}{4}\right) + i\sin\left(\frac{3\pi}{4}\right) = e^{i(3\pi/4)},$$

and

$$z^{-1} = \frac{1}{z} = \frac{\bar{z}}{|z|^2} = \frac{-\sqrt{2} - i\sqrt{2}}{4} = \frac{1}{2}e^{-i(3\pi/4)}.$$

Of course, the reader should verify all these equalities for himself.

The next property of Arg (z) that we shall discuss can be a bit tricky. Suppose that we have two complex numbers z and w with Arg $(z) = t$ and Arg $(w) = s$. What does Arg (zw) equal? To compute it we write $z = |z|e^{it}$ and $w = |w|e^{is}$. Then $zw = |z||w|e^{it}e^{is} = |z||w|e^{i(t+s)}$. Evidently $|zw| = |z||w|$, but what about Arg (zw)? A first guess is that Arg $(zw) = t + s$. However, the sum $t + s$ may not satisfy $-\pi < t + s \leq \pi$. In fact, all we can say is that $-2\pi < t + s \leq 2\pi$. This is not a serious problem. If $t + s$ lies outside the range $-\pi$ to π simply add, as appropriate, 2π or -2π to $s + t$ to obtain a number u such that $-\pi < u \leq \pi$. By equalities (iii) and (v) above, we have

$$zw = |z||w|e^{i(t+s)} = |zw|e^{i(t+s)}e^{\pm 2\pi i} = |zw|e^{i[(t+s)\pm 2\pi]} = |zw|e^{iu}.$$

Thus $u = $ Arg $(zw) = $ Arg $(z) + $ Arg $(w) + \delta 2\pi$ where δ is 0, 1, or -1. Since we have

$$\frac{z}{w} = zw^{-1} = \frac{|z|}{|w|}e^{i(t-s)},$$

the equality Arg $(z/w) = $ Arg $(z) - $ Arg (w) also holds modulo 2π; that is,

$$\text{Arg}\left(\frac{z}{w}\right) = \text{Arg }(z) - \text{Arg }(w) + \delta 2\pi$$

where as before δ is 0, 1, or -1.

Example. If $z = 3e^{i(2\pi/3)}$ and $w = 2e^{i(3\pi/4)}$, then $zw = 6e^{i(17/12)\pi} = 6e^{-i(7/12)\pi}$.

0.9 ROOTS OF COMPLEX NUMBERS

The polar form of a complex number provides an easy way to find the roots of the number. Suppose that $z \neq 0$ and that n is a positive integer. An nth root of z is a complex number w satisfying $w^n = z$. To find the nth roots of z, write z in the form $z = |z|e^{it}$ and write w in polar form also, say $w = |w|e^{i\theta}$ The equality $w^n = z$ then becomes

$$|w|^n e^{in\theta} = |z|e^{it}, \tag{1}$$

by (iv) above. Using (ii) and taking the absolute value of each side we see that $|w|^n = |z|$ must hold; that is, $|w| = |z|^{1/n}$. We thus cancel $|w|^n$ and $|z|$ in (1) and obtain $e^{in\theta} = e^{it}$. By (v) we must have $n\theta = t + 2\pi j$ for some integer j, and so we must have

$$w = |z|^{1/n}e^{i[(t+2\pi j)/n]} \tag{2}$$

for some integer j. Thus every n^{th} root of z must be of the form (2) and conversely for each integer j, the number in (2) is an nth root of $z = |z|e^{it}$. When do distinct j give the same w? To answer this question, we seek conditions under which

$$e^{i[(t+2\pi j)/n]} = e^{i[(t+2\pi k)/n]}. \tag{3}$$

Multiplying both sides of (3) by $e^{-i[(t+2\pi j)/n]}$, we find that $e^{i\{2\pi[(k-j)/n]\}} = 1$ and hence by (v) that $2\pi(k - j)/n = 2\pi m$ for some integer m. Thus (3) holds if and only if

$$k = mn + j \tag{4}$$

for some integer m. Every integer k can be written uniquely in the form (4) for integers m and j where j satisfies $0 \le j < n$ (divide k by n). Thus the n numbers

$$|z|^{1/n}e^{i[(t+2\pi j)/n]} \qquad 0 \le j < n \tag{5}$$

are distinct and are all the nth roots of z.

As an example, we find the fifth roots of i. Write $i = e^{i(\pi/2)}$; then, the fifth roots are

$$e^{i[(\pi/2+2\pi j)/5]} = e^{i(\pi/5)(1/2+2j)} \qquad 0 \le j < 5.$$

The roots are thus $e^{i(\pi/10)}$, $e^{i(5\pi/10)}$, $e^{i(9\pi/10)}$, $e^{i(13\pi/10)}$, and $e^{i(17\pi/10)}$. They are diagrammed in Figure 0.4.

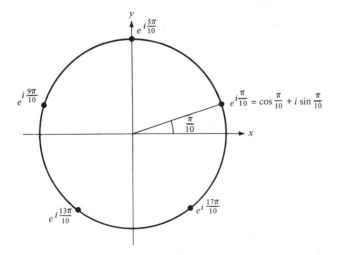

Figure 0.4

EXERCISES

TYPE I

1. Perform the calculations required to put the given complex number in the form $a + ib$.
   (a) $(2 + 3i) + (1 + i)$
   (b) $(1 + i) \cdot 5$
   (c) $(1 + 2i)(7 + 3i)$
   (d) $(1 + 2i)/(7 + 3i)$
   (e) $i/(1 + i)$
   (f) $(1 + i)/i$
   (g) $1/i$

2. Express the given complex number in the form $a + ib$.
   (a) e^i
   (b) $e^{i(\pi/2)}$
   (c) $e^{i(\pi/3)}$
   (d) $e^{i(\pi/6)}$
   (e) $2e^{i(3\pi/2)}$
   (f) $e^{i(\pi/2)} \cdot e^{-i(\pi/8)}$
   (g) $8e^{i300}$
   (h) $1/e^{i(\pi/5)}$

3. Find $|z|$, $\bar{z}$, $z\bar{z}$, and $z/|z|$ for the following complex numbers. Locate each of your answers in the plane.
   (a) $2 + 3i$
   (b) i
   (c) $-i$
   (d) $1/i$
   (e) $1 + i$
   (f) 1
   (g) $(1/\sqrt{2})(1 + i)$
   (h) $1 - i$
   (i) $1/(2 + 3i)$
   (j) $5i$

4. Find Arg (z) for each of the following complex numbers. Find an angle $\theta \neq$ Arg z such that $z = |z|(\cos \theta + i \sin \theta)$ in each case. In each case, express z and $z/|z|$ in polar form.
   (a) $z = 1 + i$
   (b) $z = 1 - i$
   (c) $z = 1 + i\sqrt{3}$
   (d) $z = \sqrt{3} + i$
   (e) $z = i$
   (f) $z = -1 - i$

5. Write each number in Exercise 4 in the form $re^{i\theta}$ with $0 \leq \theta < 2\pi$ and as $re^{i\psi}$ with $-2\pi < \psi \leq 0$.

6. Verify that $|zw| = |z| \cdot |w|$ for any complex z and w. Use the equality to find $|(2 + 3i)(1 + i)|$ and $|i(1 + i)|$.

7. If $z = a + ib$ and $w = c + id$ are distinct complex numbers, show that the line segment joining z and w is the set of points

$$\{sz + (1 - s)w \mid 0 \leq s \leq 1\}.$$

   Find the Cartesian equation of the line determined by z and w.

8. Denote the line segment joining z to w (z and w complex numbers) by $[z, w]$. If z and w are distinct, show that $[0, z]$, $[0, w]$, $[z, z + w]$, $[w, z + w]$ are the sides of a parallelogram. Show that the area of the parallelogram is equal to $||z| \cdot |w| \sin (\text{Arg } z/w)|$.

9. Find and diagram the four 4th roots of 1. The eight 8th roots.

10. Find the 4th roots of $1 + i$. Diagram.

• 11. Prove the binomial expansion for complex numbers.

12. Prove the formula for the sum of a finite geometric series; that is,

$$\sum_{n=0}^{N} z^n = \frac{1 - z^{N+1}}{1 - z} \qquad (z \neq 1).$$

13. The formula $(a + bi)(c + di) = (ac - bd) + i(bc + ad)$ stated as a product of ordered pairs is

$$(a, b) \cdot (c, d) = (ac - bd, bc + ad).$$

Another reasonable product for **C**, call it $*$ to distinguish it from $\cdot$, for order pairs is given by

$$(a, b) * (c, d) = (ac, bd).$$

Show that **C** with this product and the usual addition is not a field.

14. Use the triangle inequality to show that

$$||z| - |w|| \leq |z - w|$$

for any complex numbers z and w. Hence show that $1/|z - w| \leq 1/||z| - |w||$ if $|z| \neq |w|$.

15. What can you say about the statement $1 + i > 1 - i$?

16. Referring if necessary to Appendix B, make a list of the axioms for the real number and a list of the axioms for the complex numbers. Convince yourself that a theorem about real numbers with a proof which involves *only* those axioms common to both the real numbers and the complex numbers is also a theorem about complex numbers.

17. Identify complex numbers $z = x + iy$ and $w = a + ib$ with vectors in the plane. As vectors in the plane, z and w have a scalar (or "dot") product $z \cdot w$. Show that $z \cdot w = \text{Re}\,(z\bar{w})$.

TYPE II

1. Let $\zeta = e^{i(2\pi/n)}$, so that ζ is an nth root of 1. (n^{th} roots of 1 are called *roots of unity*.) Show that the n n^{th} roots of unity are $1, \zeta, \ldots, \zeta^{n-1}$. Prove that $\sum_{k=0}^{n-1} \zeta^k = 0.$ or 1.

2. Let ξ be an nth root of unity.
   (a) Show that ξ^k is an nth root of unity for all $k \in \mathbf{Z}$.
   (b) ξ is called a *primitive n^{th} root* if $\{\xi^k\}_{k=0}^{n-1}$ exhausts all n^{th} roots of unity. Find all primitive n^{th} roots of unity, $n \in \mathbf{Z}^{++}$.

3. A *group* is a nonempty set G with a binary operation $*$ on G satisfying
   (a) $a, b \in G \Rightarrow a * b \in G$.
   (b) There is an $e \in G$ such that $a * e = e * a = a$ for all $a \in G$.
   (c) Every $a \in G$ has a unique inverse $a^{-1} \in G$ satisfying $a * a^{-1} = a^{-1} * a = e$.
   (d) $(a * b) * c = a * (b * c)$ for all a, b, c, in G.

   Examples of groups we have encountered are $\mathbf{Z}$ with $* = +$ and $e = 0$; $\mathbf{R}$ with $* = +$; $\mathbf{R} \backslash \{0\}$ with $* = \cdot$ and $e = 1$. Prove that $\mathbf{T} = \{z \mid |z| = 1\}$ is a group under $\cdot$. Is it a group under $+$?   cyclic

4. Let G_n be the set of n n^{th} roots of unity. Prove that G_n is a group under multiplication.

5. Let α be a fixed real number. Prove that $\{e^{i\alpha n} \mid n \in \mathbf{Z}\}$ is a group under multiplication. For what α is the group finite? Infinite?

6. Show that z_1, z_2, and z_3 are vertices of an equilateral triangle if and only if $z_1^2 + z_2^2 + z_3^2 = z_1 z_2 + z_2 z_3 + z_3 z_1$.

7. Identify the following curves:
   (a) $|z - a| + |z + a| = b^2$ $(b^2 \geq 2|a|)$ $|z - a| + |z - (-a)| = b^2$
   (b) $|z| - \text{Re}(z) = |z|^2$
   (c) $|z - a| - |z + a| = b^2$ $(b^2 \leq 2|a|)$
   (d) $|z - a| = \text{Re}(z)$.

8. For any complex numbers $z_1 z_2, \ldots, z_n$ and $w_1, w_2, \ldots, w_n$, show that

$$\left| \sum_{i=1}^{n} z_i w_i \right|^2 \leq \left(\sum_{i=1}^{n} |z_i|^2 \right) \left(\sum_{i=1}^{n} |w_i|^2 \right).$$

   This inequality often goes by the name of Schwarz's inequality.

9. Show that equality holds in the inequality in Exercise 8 if and only if $\arg(z_k) = \arg(w_j)$ for all pairs (j, k).

10. If $z_1, z_2, \ldots, z_n$ and $w_1, w_2, \ldots, w_n$ are complex numbers, show that

$$\left[\sum_{i=1}^{n} |z_i + w_i|^2 \right]^{1/2} \leq \left[\sum_{i=1}^{n} |z_i|^2 \right]^{1/2} + \left[\sum_{i=1}^{n} |w_i|^2 \right]^{1/2}.$$

   [*Hint:* For each i, write

$$(z_i + w_i)^2 = z_i(z_i + w_i) + w_i(z_i + w_i).$$

   Sum this equality from 1 to n and apply Exercise 8 to each of the summands on the right. Divide both sides by the resulting common factor of the right side.]

0.10 CONVERGENCE OF SEQUENCES

The material just reviewed [Sections (0.1) through (0.9)] is elementary and is generally covered in precalculus mathematics courses. The concept of convergence, which will occupy our attention for the next several sections, is not so elementary. Series and convergence of sequences are generally covered in calculus. The basic problems here are the same, only the numbers are different.

A function from the positive integers, denoted by $\mathbf{Z}^{++}$, to $\mathbf{C}$ is called a sequence in $\mathbf{C}$ or a sequence of complex numbers. That is, for every $n \in \mathbf{Z}^{++}$, there is specified a unique element $z_n \in \mathbf{C}$. A sequence $(z_n)_{n=1}^{\infty}$ in $\mathbf{C}$ converges to $z \in \mathbf{C}$ if z_n can be made as close as desired to z for all sufficiently large n. For z_n to be close to z means that $|z - z_n|$ is small. A precise definition follows.

(i) *Definition.* A sequence $(z_n)_{n=1}^{\infty}$ in $\mathbf{C}$ *converges* to z (written $\lim_{n \to \infty} z_n = z$) if for each $\varepsilon > 0$ there is an integer $n_0 = n_0(\varepsilon)$ such that $|z - z_n| < \varepsilon$ for all $n \geq n_0$. In geometric terms this means that for each circle centered at z there is an n_0 such that z_n lies in this circle if $n \geq n_0$.

The definition is thus formally the same as for sequences of real numbers.

(ii) *Examples.*

(a) $\lim_{n \to \infty} [z + (i/n)] = z$ for all $z \in \mathbf{C}$.

PROOF. Given $\varepsilon > 0$, there is an integer $n_0 > \varepsilon^{-1}$. We have

$$\left|\left(z + \frac{i}{n}\right) - z\right| = \frac{1}{n} \leq \frac{1}{n_0} < \varepsilon$$

if $n \geq n_0$.

(b) $\lim_{n \to \infty} [(n + i)/n] = 1$.

PROOF. Let $\varepsilon > 0$ be given. Then $|(n + i)/n - 1| = |i/n| < \varepsilon$ if $n \geq n_0$ and $n_0 > \varepsilon^{-1}$.

(c) If $|z| < 1$, then $\lim_{n \to \infty} z^n = 0$.

PROOF. Since the sequence $\{|z|^n\}_{n=1}^{\infty}$ of positive numbers is decreasing, $\lim_{n \to \infty} |z^n| = a \geq 0$ exists. Since the equality

$$\lim_{n \to \infty} |z|^{n+1} = a$$

is also valid and

$$\lim_{n \to \infty} |z|^{n+1} = |z| \lim_{n \to \infty} |z|^n = |z|a$$

we have $|z|a = a$. Therefore, since $|z| \neq 1$, a is zero. Thus $|z|^n$ converges to 0. Since $|z^n - 0| = |z|^n$, z^n also converges to zero.

(d) If $\lim_{n \to \infty} z_n = z$, then $\lim_{n \to \infty} |z_n| = |z|$.

PROOF. Let $\varepsilon > 0$. By definition, there is an n_0 such that $n \geq n_0$ implies $|z - z_n| < \varepsilon$.   Since $||z| - |z_n|| \leq |z - z_n|$, $n \geq n_0$ also implies that $||z| - |z_n|| < \varepsilon$.

(e) The converse (of example (d)) is false. For instance, if $z_n = i^n$ then $(z_n)_{n=1}^{\infty}$ does not converge, but $\lim_{n \to \infty} |z_n| = 1$.

(iii) *Theorem.* Let $(z_n)_{n=1}^{\infty}$ be a sequence in $\mathbf{C}$. Then $\lim_{n \to \infty} z_n = z$ if and only if $\lim_{n \to \infty} \operatorname{Re}(z_n) = \operatorname{Re}(z)$ and $\lim_{n \to \infty} \operatorname{Im}(z_n) = \operatorname{Im}(z)$.

PROOF. The assertions follow from the inequalities

$$\left.\begin{matrix} |\operatorname{Re}(z - z_n)| \\ |\operatorname{Im}(z - z_n)| \end{matrix}\right\} \leq |z - z_n| \leq |\operatorname{Re}(z - z_n)| + |\operatorname{Im}(z - z_n)|.$$

(iv) *Theorem.* Let $\lim_{n \to \infty} z_n = z$ and $\lim_{n \to \infty} w_n = w$. Then:

(a) $\lim_{n \to \infty} (z_n + w_n) = z + w$.

(b) $\lim_{n \to \infty} (z_n w_n) = zw$.

(c) $\lim_{n \to \infty} (z_n/w_n) = z/w$ if $w_n w \neq 0$ for all n.

The proofs are as in the calculus [see (B.6)]. We give some more examples, using algebraic simplification and Theorems (iii) and (iv). Generally speaking, the same techniques used for real sequences apply for complex sequences. See Exercise 16, p. 11.

(v) *More examples.* The following examples use the properties in (iv).

(a) $\displaystyle \lim_{n \to \infty} \frac{1}{n} \cdot \frac{n^2 + 1}{n - i} = \lim_{n \to \infty} \frac{1}{n} \cdot \frac{(n + i)(n - i)}{(n - i)}$

$$= \lim_{n \to \infty} \frac{(n + i)}{n} = \lim_{n \to \infty} \left(1 + \frac{i}{n}\right) = 1.$$

Notice that (iii) or (iv.a) can be used to establish the final limit above.

(b) $\displaystyle \lim_{n \to \infty} e^{i(1/n)} = \lim_{n \to \infty} \left(\cos \frac{1}{n} + i \sin \frac{1}{n}\right) = \cos 0 = 1.$

(c) $\displaystyle \lim_{n \to \infty} \frac{1 + \dfrac{1}{n}}{i + \dfrac{1}{n}} = \frac{\lim_{n \to \infty} \left(1 + \dfrac{1}{n}\right)}{\lim_{n \to \infty} \left(i + \dfrac{1}{n}\right)} = \frac{1}{i} = -i.$

(d) $\lim\limits_{n \to \infty} \dfrac{z^n + 2z^{n-1}}{z^n + 3z^{n-1}} = \lim\limits_{n \to \infty} \dfrac{z^{n-1}(z + 2)}{z^{n-1}(z + 3)} = \lim\limits_{n \to \infty} \dfrac{z + 2}{z + 3} = \dfrac{z + 2}{z + 3}$,

where $z \neq 0$, $z \neq -3$.

(e) $\lim\limits_{n \to \infty} \dfrac{5n^5 + 4n^4 + 2n^3 + n^2 + 5n + 100}{(3 + 2i)n^5 + 3n + 2} = \dfrac{5}{3 + 2i}$.

To see this, write the ratio as

$$\frac{5n^5}{(3 + 2i)n^5} \cdot \frac{\left(1 + \dfrac{4}{5}\dfrac{1}{n} + \dfrac{2}{5}\dfrac{1}{n^2} + \dfrac{1}{5}\dfrac{1}{n^3} + \dfrac{1}{n^4} + \dfrac{20}{n^5}\right)}{\left(1 + \dfrac{3}{3 + 2i}\dfrac{1}{n^4} + \dfrac{2}{3 + 2i}\dfrac{1}{n^5}\right)}.$$

The leading term is $5/(3 + 2i)$ and each of the terms in parentheses converges to 1.

0.11 INFINITE SERIES

A sequence $(a_n)_{n=1}^{\infty}$ in **C** generates a sequence of *partial sums* $(S_n)_{n=1}^{\infty}$ defined by

$$S_n = \sum_{k=1}^{n} a_k.$$

If $S = \lim_{n \to \infty} S_n$ exists, it is denoted $\sum_{n=1}^{\infty} a_n$ and the infinite series $\sum_{n=1}^{\infty} a_n$ is said to converge.

One often refers to the symbol $\sum_{n=1}^{\infty} a_n$ as an infinite series even though $\lim_{n \to \infty} S_n$ does not exist. Of course, in this case $\sum_{n=1}^{\infty} a_n$ does not represent any number, and the series is said to diverge. If $S = \sum_{n=1}^{\infty} a_n$ and $T = \sum_{n=1}^{\infty} b_n$ are convergent infinite series, then $S + T = \sum_{n=1}^{\infty} (a_n + b_n)$, by (0.10.iv). The most important thing for the novice to remember about infinite series is that an expression such as

$$\sum_{n=1}^{\infty} a_n$$

represents the limit of a sequence of finite sums if the limit exists.

The beginning index for an infinite series need not be 1. If a_k is a complex number for each integer $k \geq m$ (m a fixed integer), then the symbol $\sum_{k=m}^{\infty} a_k$ means the number $\lim_{n \to \infty} \sum_{k=m}^{n} a_k$ provided that the limit exists.

(i) *Example (Geometric Series).* $\sum_{n=0}^{\infty} z^n = 1/(1 - z)$ if $|z| < 1$.

PROOF. We have

$$\lim_{n \to \infty} \sum_{k=0}^{n} z^k = \lim_{n \to \infty} \frac{1 - z^{n+1}}{1 - z} = \frac{\lim_{n \to \infty} (1 - z^{n+1})}{1 - z} = \frac{1}{1 - z}.$$

Problem 12, P.11.

Notice that we have used the formula $\sum_{k=0}^{n} z^k = (1 - z^{n+1})/(1 - z)$ and (0.10.ii.c). This limit is easy to evaluate because there is an explicit formula for the partial sums.

0.12 CAUCHY SEQUENCES

The student will doubtless recall that for real-valued sequences and series it is often possible to prove convergence without actually exhibiting a limit. Indeed many important numbers and functions can be defined as the limits of particular sequences or series; for example, e and π. Proofs that such definitions are meaningful are necessarily based upon the completeness property of the real numbers. A first instance is the following theorem.

(i) *Theorem.* Every bounded increasing sequence of real numbers converges and every bounded decreasing sequence of real numbers converges.

There is a proof of (i) in Appendix B. In practice, it is often easier to have completeness expressed in terms of *Cauchy sequences*.

(ii) *Definition.* A sequence $(z_n)_{n=1}^{\infty}$ in $\mathbf{C}$ is a *Cauchy sequence* if for every $\varepsilon > 0$ there is an integer n_0 such that

$$n \geq n_0 \text{ and } m \geq n_0 \quad \text{imply} \quad |z_n - z_m| < \varepsilon.$$

(iii) *Theorem.* If $(z_n)_{n=1}^{\infty}$ converges, then it is a Cauchy sequence.

PROOF. Let $\lim_{n \to \infty} z_n = z$, and suppose that $\varepsilon > 0$. Select n_0 such that $|z - z_n| < \varepsilon/2$ if $n \geq n_0$. If $n, m \geq n_0$, then we have

$$|z_n - z_m| \leq |z_n - z| + |z - z_m| < \varepsilon/2 + \varepsilon/2 = \varepsilon.$$

(iv) *Theorem (Cauchy criterion).* If $(z_n)_{n=1}^{\infty}$ is a Cauchy sequence, then it converges.

PROOF. Let $z_n = x_n + iy_n$, x_n and y_n real numbers. Since

$$|x_n - x_m| \leq |z_n - z_m| \quad \text{and} \quad |y_n - y_m| \leq |z_n - z_m|,$$

the sequences $(x_n)_{n=1}^{\infty}$ and $(y_n)_{n=1}^{\infty}$ are Cauchy sequences of real numbers. Hence they both converge (see Appendix B for a proof), say $x = \lim_{n \to \infty} x_n$ and $y = \lim_{n \to \infty} y_n$. By (0.10.iii), we have

$$\lim_{n \to \infty} z_n = \lim_{n \to \infty} (x_n + iy_n) = x + iy.$$

(v) *Cauchy criterion for series.* A complex infinite series $\sum_{n=1}^{\infty} a_n$ converges if and only if for every $\varepsilon > 0$ there is an n_0 such that $|\sum_{k=m+1}^{n} a_k| < \varepsilon$ for every pair n, m such that $n > m \geq n_0$.

This criterion is valid because $S_n - S_m = \sum_{k=m+1}^{n} a_k$ if $n > m$.

EXERCISES

TYPE I

If the student needs more practice and/or review on convergence of real sequences, he should consult the exercises in Appendix B.

1. Find the limits (they exist) of the following sequences of complex numbers.

   (a) $\dfrac{n^2 + 2n + i}{5(n + 1)^2 + 5in + 3}$

   (b) $\dfrac{(in)^2}{2n^2 + 5in + 1}$

   (c) $\sqrt{n^2 + n} - n$

   (d) $\dfrac{1}{n} + i\left(1 - \dfrac{1}{n}\right)$ (Sketch the sequence of points.)

   (e) $\dfrac{i^n}{n^2}$

   (f) $\dfrac{(n + i)^2}{(1 + i)^n}$.

2. (a) Show that $\lim_{n\to\infty} [-1 + i(1/n)] = \lim_{n\to\infty} [-1 - i(1/n)] = -1$.
   (b) Find $\lim_{n\to\infty} \text{Arg} [-1 + i(1/n)]$ and $\lim_{n\to\infty} \text{Arg} [-1 - i(1/n)]$.

3. Define a function S on $\mathbf{C}$ by $S(z) = \sqrt{|z|} e^{it/2}$ $(z \neq 0)$, $S(0) = 0$ where $t = \text{Arg} (z)$. Find

   $$S(-1), \quad \lim_{n\to\infty} S[-1 + i(1/n)], \quad \text{and} \quad \lim_{n\to\infty} S[-1 - i(1/n)].$$

4. If $S_n = \sum_{m=0}^{n} z^{2m}$ and $T_n = \sum_{m=0}^{n} z^{2m+1}$, find $\lim_{n\to\infty} S_n$ and $\lim_{n\to\infty} T_n$ for $|z| < 1$.

5. Prove that $\sum_{n=1}^{\infty} 1/[n(n + 1)]$ converges and find its limit by finding an explicit formula for the partial sums. [*Hint:* $1/[n(n + 1)] = 1/n - 1/(n + 1)$.]

6. Find $\sum_{k=0}^{\infty} (\tfrac{1}{2})^k$.

7. Find $\lim_{n\to\infty} (1/n^2) \sum_{k=1}^{n} k$.

8. (a) Sketch the sets

$$\{z \mid \lim_{n\to\infty} (1 + z)^n = 0\} \quad \text{and} \quad \{z \mid \lim_{n\to\infty} (z - 1)^n = 0\}.$$

(b) Sketch the set

$$\left\{ z \mid \lim_{n\to\infty} \frac{(z + 1)^n}{z^n} = 0 \right\}.$$

9. Use (0.10.i) to formulate a precise statement of the meaning of the phrase "the sequence $(z_n)_{n=1}^{\infty}$ in **C** does not converge."

0.13 CONVERGENCE TESTS FOR INFINITE SERIES

From earlier courses in analysis, the student is probably familiar with the tests of this section. The section will thus serve both to review and to show that the results extend easily from the real to the complex case.

(i) *Remark.* If $S = \sum_{n=1}^{\infty} a_n$ (convergent), then for each m the equality

$$\sum_{n=m}^{\infty} a_n = S - \sum_{n=1}^{m-1} a_n$$

is valid; that is, the series on the left converges and has the value on the right. To see this, simply observe that the kth partial sum of the left is $\sum_{n=m}^{k} a_n$, which is equal to $S_k - \sum_{n=1}^{m-1} a_n$.

(ii) *Theorem.* If $\sum_{n=1}^{\infty} a_n$ converges, then $\lim_{m\to\infty} \sum_{n=m}^{\infty} a_n = 0$ and $\lim_{n\to\infty} a_n = 0$.

PROOF. The first equality is a result of (i) and the second a result of the equality

$$a_n = [S_n - S_{n-1}]$$

(iii) *Theorem (Comparison test).* If $(x_n)_{n=1}^{\infty}$ is a sequence of nonnegative real numbers such that $\sum_{n=1}^{\infty} x_n$ converges and if $(z_n)_{n=1}^{\infty}$ is a complex sequence such that $|z_n| \leq x_n$ for all n, then $\sum_{n=1}^{\infty} z_n$ converges.

PROOF. The inequality $|\sum_{k=n}^{m} z_k| \leq \sum_{k=n}^{m} x_k$ shows that the partial sums for $\sum_{n=1}^{\infty} z_n$ form a Cauchy sequence.

(iv) *Definitions.* An infinite series $\sum_{n=1}^{\infty} a_n$ *converges absolutely* if $\sum_{n=1}^{\infty} |a_n|$ converges. It is clear from (iii) that an absolutely convergent series is convergent. An infinite series which converges but does not converge absolutely is called *conditionally convergent*. The most famous example is $\sum_{n=1}^{\infty} (-1)^n/n$; we will discuss this series later (2.18) at greater length.

(v) *Examples*.   The comparison test is often used with a geometric series as the majorant series $\sum_{n=1}^{\infty} x_n$. Another common majorant is a series of the form $\sum_{n=1}^{\infty} 1/n^a$; recall from elementary calculus that these series converge if $a > 1$ and diverge if $a \leq 1$.

(a) The series $\sum_{n=1}^{\infty} e^{it_n}/2^n$ converges for any sequence of real numbers $(t_n)_{n=1}^{\infty}$ because $|e^{it_n}/2^n| = 1/2^n$, and the geometric series $\sum_{n=1}^{\infty} 1/2^n$ converges.

(b) The series $\sum_{n=1}^{\infty} (\sqrt{n+1} - \sqrt{n})/n$ converges, for we have

$$\left| \frac{\sqrt{n+1} - \sqrt{n}}{n} \right| = \left| \frac{(\sqrt{n+1} - \sqrt{n})(\sqrt{n+1} + \sqrt{n})}{(\sqrt{n+1} + \sqrt{n})n} \right|$$

$$= \frac{1}{(\sqrt{n+1} + \sqrt{n})n} \leq \frac{1}{n^{3/2}}.$$

Two simple but useful tests for convergence of series are the *root test* and the *ratio test*. To discuss these, it is convenient to introduce the limit superior and limit inferior of a sequence of real numbers.

(vi) *Limit superior and limit inferior*.   Let $(x_n)_{n=1}^{\infty}$ be a sequence of real numbers. Suppose first that there is a constant M such that $x_n < M$ for all n. Using the completeness property of real numbers (B.4), one can show that there is a unique real number x^0 having the property that for any real numbers r and s satisfying $r < x^0 < s$, the set $\{n: r < x_n < s\}$ is infinite and the set $\{x_n: x_n > s\}$ is finite. That is, x^0 is the largest real number such that infinitely many x_n are close to it. Similarly, if there is a constant m such that $x_n > m$ for all n, then there is an x_0 such that for any u and v satisfying $u < x_0 < v$, the set $\{n: u < x_n < v\}$ is infinite and $\{x_n: x_n < u\}$ is finite. The number x^0 is called the *limit superior* of $(x_n)_{n=1}^{\infty}$ and x_0 is called the *limit inferior*; they are denoted

$$x^0 = \overline{\lim_{n \to \infty}} \, x_n, \qquad x_0 = \underline{\lim_{n \to \infty}} \, x_n.$$

Another way to say that $\{x_n: x_n > s\}$ is finite is that there is an n_0 such that $x_n \leq s$ for all $n \geq n_0$. We will use this statement in the proof of the root test.

As an example, let $x_n = (-1)^n + 1/n$. We have $|x_n| < 2$ for all n. The first few values of x_n are

$$x_1 = 0 \qquad\qquad x_5 = -\tfrac{4}{5}$$

$$x_2 = \tfrac{3}{2} \qquad\qquad x_6 = \tfrac{7}{6}$$

$$x_3 = -\tfrac{2}{3} \qquad\qquad x_7 = -\tfrac{6}{7}$$

$$x_4 = \tfrac{5}{4} \qquad\qquad x_8 = \tfrac{9}{8}$$

It is clear that $x_n = -(n - 1)/n$ if n is odd and $x_n = (n + 1)/n$ if n is even. If $r < 1 < s$, then there is an n_0 such that $(n + 1)/n < s$ if $n \geq n_0$. Hence $r < x_n < s$ for all even n larger than n_0. It follows that $\overline{\lim}_{n\to\infty} x_n = 1$. Similarly, $\underline{\lim}_{n\to\infty} x_n = -1$.

We allow $-\infty$ and ∞ as possible values, and then define $\overline{\lim}_{n\to\infty} x_n$ and $\underline{\lim}_{n\to\infty} x_n$ for all sequences $(x_n)_{n=1}^{\infty}$, as follows. If for any M there is an n_0 such that $x_n > M$ if $n \geq n_0$, then we let

$$\lim_{n\to\infty} x_n = \overline{\lim}_{n\to\infty} x_n = \infty$$

and say that $(x_n)_{n=1}^{\infty}$ converges to infinity. If for any M there is an n_0 such that $x_n < M$ for all $n \geq n_0$, then we let

$$\lim_{n\to\infty} x_n = \overline{\lim}_{n\to\infty} x_n = -\infty.$$

If the set $\{x_n: x_n > M\}$ is infinite for all M, then we let $\overline{\lim}_{n\to\infty} x_n = \infty$; if $\{x_n: x_n < M\}$ is infinite for all M, then we let $\underline{\lim}_{n\to\infty} x_n = -\infty$.

As examples we have

$$\lim_{n\to\infty} (n) = \overline{\lim}_{n\to\infty} n = \infty,$$

$$\overline{\lim}_{n\to\infty} (n + (-1)^n n) = \infty; \quad \underline{\lim}_{n\to\infty} (n + (-1)^n n) = 0.$$

(vii) *The root test.* Let $(a_n)_{n=1}^{\infty}$ be a sequence in **C**.

(a) If $\overline{\lim}_{n\to\infty} |a_n|^{1/n} < 1$, then $\sum_{n=1}^{\infty} |a_n|$ converges.

(b) If $\overline{\lim}_{n\to\infty} |a_n|^{1/n} > 1$, then $\sum_{n=1}^{\infty} a_n$ does not converge.

PROOF.   Since $\overline{\lim}_{n\to\infty} |a_n|^{1/n} < 1$, there is a μ such that

$$\overline{\lim}_{n\to\infty} |a_n|^{1/n} < \mu < 1.$$

By the definition of $\overline{\lim}$, there is an n_0 such that $|a_n|^{1/n} \leq \mu$ if $n \geq n_0$. Thus, $|a_n| \leq \mu^n$ holds if $n \geq n_0$. By comparison with $\sum_{n=n_0}^{\infty} \mu^n$, $\sum_{n=n_0}^{\infty} |a_n|$ (and therefore $\sum_{n=1}^{\infty} |a_n|$) converges.

If $\overline{\lim}_{n\to\infty} |a_n|^{1/n} > 1$, then $\lim_{n\to\infty} a_n = 0$ cannot hold, so $\sum_{n=1}^{\infty} a_n$ cannot converge.

(viii) *The ratio test.* Let $(a_n)_{n=1}^{\infty}$ be a sequence of nonzero terms in **C**.

(a) If $\overline{\lim}_{n\to\infty} |a_{n+1}/a_n| < 1$, then $\sum_{n=1}^{\infty} |a_n|$ converges.

(b) If $\underline{\lim}_{n\to\infty} |a_{n+1}/a_n| > 1$, then $\sum_{n=1}^{\infty} a_n$ does not converge.

PROOF. If $\overline{\lim}_{n\to\infty} |a_{n+1}/a_n| < 1$, then there is a $\mu < 1$ and an n_0 such that

$$\left| \frac{a_{n+1}}{a_n} \right| \leq \mu$$

if $n \geq n_0$. We have

$$|a_{n_0+2}| \leq |a_{n_0+1}|\mu \leq |a_{n_0}|\mu^2,$$

and in general, $|a_{n_0+k}| \leq |a_{n_0}| \mu^k$; or, indexing differently,

$$|a_n| \leq |a_{n_0}|\mu^{n-n_0} = |a_{n_0}|\mu^{-n_0}\mu^n$$

if $n > n_0$. Since $\mu < 1$, the geometric series $\sum_{n=1}^{\infty} \mu^n$ converges, and (iii) shows that $\sum_{n=1}^{\infty} a_n$ converges.

For a proof of (b), note that $\underline{\lim}_{n\to\infty} |a_{n+1}/a_n| > 1$ implies that there is an n_0 and a $\mu > 1$ such that $|a_{n+1}/a_n| > \mu$ if $n \geq n_0$. It follows that $|a_n| > |a_{n_0}| \mu^{n-n_0}$ if $n > n_0$, and so $\lim_{n\to\infty} a_n = 0$ cannot hold; hence, by (ii) $\sum_{n=1}^{\infty} a_n$ cannot converge.

(ix) *Comparison of tests.* The root test is stronger than the ratio test. (See Exercise 17, p. 25.) But in many cases the ratio test is easier to apply. Neither test is conclusive, giving no indication of what happens when the limits equal 1. In practice, these are often the interesting cases. Circumstantial evidence that the root test is stronger is given by the following example.

(x) *Example.* $a_n = (2i)^{-n}$ if n is even; $a_n = (3i)^{-n}$ if n is odd.

$$\left| \frac{a_{n+1}}{a_n} \right| = \begin{cases} \dfrac{1}{3}\left(\dfrac{2}{3}\right)^n & \text{if } n \text{ is even} \\[2ex] \dfrac{1}{2}\left(\dfrac{3}{2}\right)^n & \text{if } n \text{ is odd.} \end{cases}$$

Hence $0 = \underline{\lim}_{n\to\infty} |a_{n+1}/a_n| < \overline{\lim}_{n\to\infty} |a_{n+1}/a_n| = \infty$, and the ratio test gives no information. However, it is clear that $\underline{\lim}_{n\to\infty} |a_n|^{1/n} = \frac{1}{3}$ and $\overline{\lim}_{n\to\infty} |a_n|^{1/n} = \frac{1}{2}$, so the root test gives convergence.

*(xi) *Rearrangement of series.* Consider a complex sequence $(a_n)_{n=0}^{\infty}$. Let $\phi : \mathbf{Z}^+ \to \mathbf{Z}^+$ be bijective; that is, one-to-one and onto. Then $(b_n)_{n=0}^{\infty}$ defined by $b_n = a_{\theta(n)}$ is a *rearrangement* of the sequence $(a_n)_{n=0}^{\infty}$ and $\sum_{n=0}^{\infty} b_n$ is a rearrangement of $\sum_{n=0}^{\infty} a_n$.

Theorem. If $\sum_{n=0}^{\infty} a_n$ converges absolutely, then $\sum_{n=0}^{\infty} b_n = \sum_{n=0}^{\infty} a_n = S$ for any rearrangement of $\sum_{n=0}^{\infty} a_n$.

PROOF.   Let $\varepsilon > 0$.   Choose an n_0 such that $\sum_{n=k}^{\infty} |a_n| < \varepsilon/2$ for $k \geq n_0$. In particular, we have

$$\left| \sum_{n=0}^{n_0} a_n - S \right| < \frac{\varepsilon}{2}.$$

Let N be such that $\{b_1, \ldots, b_N\} \supset \{a_1, \ldots, a_{n_0}\}$.   We have

$$\sum_{n=0}^{N} b_n = \sum_{n=0}^{n_0} a_n + \sum_{k \in F} a_k$$

where

$$F = \{\phi^{-1}(1), \ldots, \phi^{-1}(N)\} \backslash \{1, 2, \ldots, n_0\}).$$

Since $k > n_0$ for $k \in F$, the inequalities

$$\left| \sum_{n=0}^{N} b_n - S \right| \leq \left| \sum_{n=0}^{\infty} a_n - S \right| + \sum_{n=n_0+1}^{\infty} |a_n| < \varepsilon$$

hold, proving that $\sum_{n=0}^{\infty} b_n = S$.

EXERCISES

TYPE I

1. Apply the ratio test to prove that $\sum_{n=0}^{\infty} 1/n!$ converges. The series defines e. Show also that $\sum_{n=0}^{\infty} z^n/n!$ converges for all complex z. The series defines e^z. The important function e^z is defined and treated in detail in Section (2.11).

2. Establish convergence or divergence for the following series.
   (a) $\sum_{n=1}^{\infty} n/2^n$.
   (b) $\sum_{n=1}^{\infty} 1/z^n$, $z \neq 0$.
   (c) $\sum_{n=1}^{\infty} 2^n/n!$.

3. Examine $\sum_{n=1}^{\infty} a_n$ for the following sequences.
   (a) $a_n = \sqrt{n+1} - \sqrt{n}$.
   (b) $a_n = (\sqrt[n]{n} - 1)^n$.
   (c) $a_n = 1/(1 + z^n)$ $z = x + iy$.

4. Show that $\sum_{n=1}^{\infty} 1/[n(n+1)] = 1$.
   [*Hint:* Show that $S_n = 1 - 1/(n+1)$.]

5. Find $\sum_{n=1}^{\infty} 1/[n^2(n^2 + 1)]$.

6. Show that $\sum_{n=2}^{\infty} 1/(n^2 - 1) = 3/4$.
   [*Hint:* Obtain an explicit expression for the partial sums S_n.]

7. Find $\overline{\lim}_{n \to \infty} s_n$ and $\underline{\lim}_{n \to \infty} s_n$ if $s_1 = 0$, $s_{2m} = (2m - 1)/2^m$, $s_{2m+1} = 1/2 + s_{2m}$.

8. The root test fails for $\sum_{n=1}^{\infty} n^{-a}$, $a > 0$. Show by integral comparisons (or otherwise) that these series converge if $a > 1$, and fail to converge if $a \leq 1$. ($\sum_{n=1}^{\infty} 1/n$ is called the harmonic series.)

9. Establish convergence or divergence of

$$\sum_{n=2}^{\infty} \frac{1}{(\log n)^{\alpha}}, \qquad \alpha > 0.$$

   [*Hint:* Compare $(\log n)^{\alpha}$ with n.]

10. Show that $\sum_{n=2}^{\infty} 1/(\log n)^{\log n}$ converges.
    [*Hint:* Compare the summands with $1/n^2$.]

11. Establish convergence or divergence of the following series.
    (a) $\sum_{n=1}^{\infty} 1/(\sqrt{n^2 + 1})$.
    (b) $\sum_{n=1}^{\infty} 1/(\sqrt{n^3 + 1})$.
    (c) $\sum_{n=2}^{\infty} 1/[\sqrt{n(n^2 - 1)}]$.

TYPE II

1. Prove that $\lim_{n \to \infty} (1 + 1/n)^n = \sum_{n=0}^{\infty} 1/n! (= e)$.
   [*Hint:* In the equality

$$\left(1 + \frac{1}{n}\right)^n = \sum_{k=0}^{n} \binom{n}{k}\left(\frac{1}{n}\right)^k,$$

   the equality

$$\binom{n}{k}\left(\frac{1}{n}\right)^k = \frac{1}{k!}\left(1 - \frac{1}{n}\right)\left(1 - \frac{2}{n}\right)\cdots\left(1 - \frac{k-1}{n}\right)$$

   holds.]

2. If $a_n \geq 0$ and if $\sum_{n=1}^{\infty} a_n$ converges, then show that $\sum_{n=1}^{\infty} \sqrt{a_n}/n$ converges.

3. Let $\{s_n\}_{n=1}^{\infty}$ be such that $\lim_{n \to \infty} s_n = s$. If $t_n = (s_1 + s_2 + \cdots + s_n)/n$, show that $\lim_{n \to \infty} t_n = s$.

4. In Exercise 3, is the converse true?

5. Let $(a_n)_{n=1}^{\infty}$ be a positive sequence for which $a_{n+1} \leq a_n$, all n, and $\sum_{n=1}^{\infty} a_n$ converges. Prove that $\lim_{n \to \infty} na_n = 0$.
   [*Hint:* For any n and m, $na_{n+m} \leq a_{m+1} + a_{m+2} + \cdots + a_{m+n}$ is valid.]

6. Use Exercise 5 to prove that the harmonic series $\sum_{n=1}^{\infty} 1/n$ diverges.

7. Using the same hypothesis as in Exercise 5, prove that $\sum_{n=1}^{\infty} a_n = \sum_{n=1}^{\infty} (a_n - a_{n+1})n$.

8. If $(a_n)_{n=1}^{\infty}$ is a complex sequence for which $\sum_{n=1}^{\infty} (a_n - a_{n+1})n$ converges and for which $\lim_{n \to \infty} na_n = 0$, prove that $\sum_{n=1}^{\infty} a_n$ converges.

9. Give an example of a sequence $(a_n)_{n=1}^\infty$ for which $\sum_{n=1}^\infty a_n$ converges but $(na_n)_{n=1}^\infty$ does not converge to 0.

10. *Condensation Test.* Let $a_n > 0$ for all n and suppose that $(a_n)_{n=1}^\infty$ is decreasing. Let k_n be a sequence of integers such that

$$k_{n+1} - k_n \le M(k_n - k_{n-1})$$

for a constant M. The series $\sum_{n=1}^\infty a_n$ converges if and only if $\sum_{n=1}^\infty (k_{n+1} - k_n)a_{k_n}$ converges.

OUTLINE OF PROOF.   Let $S_n = \sum_{j=1}^n a_j$, $T_n = \sum_{j=1}^n (k_{j+1} - k_j)a_{k_j}$. Show that

$$S_n \le \sum_{j=1}^{k_0 - 1} a_j + T_m \qquad \text{if } n < k_m,$$

and that

$$T_m - T_0 \le MS_n \qquad \text{if } n > k_m.$$

11. Let $\alpha > 0$. Use Exercise 10 and the choice $k_n = 2^n$ to determine the behavior of $\sum_{n=1}^\infty 1/n^\alpha$.

12. (a) Apply the condensation test with $a_n = 1/(n \log n)$ and $k_n = 2^n$ to prove that $\sum_{n=2}^\infty 1/(n \log n)$ converges.
    (b) Apply the condensation test with $a_n = 1/n(\log n)^\alpha$ $(\alpha > 1)$ and $k_n = 3^n$ to prove that $\sum_{n=1}^\infty 1/n(\log n)^\alpha$ converges.

13. *Kummer's test.* Let $(u_i)_{i=1}^\infty$ and $(a_i)_{i=1}^\infty$ be sequences of positive numbers. Show that if for some C and N we have

$$a_n \frac{u_n}{u_{n+1}} - a_{n+1} \ge C > 0$$

for all $n > N$, then $\sum_{i=1}^\infty u_i$ converges. Also show that if $\sum_{i=1}^\infty 1/a_i$ diverges and if $a_n(u_n/u_{n+1}) - a_{n+1} \le 0$ for all $n > N$, then $\sum_{i=1}^\infty u_i$ diverges.
[*Hint:* Establish the inequality

$$\sum_{j=1}^k u_{N+j} \le \frac{a_N u_N - a_{N+k}u_{N+k}}{C} \le \frac{a_N u_N}{C} .]$$

14. Deduce the ratio test from the result in Exercise 13.

15. Use Kummer's test to study the convergence properties of

$$\sum_{n=1}^{\infty} \frac{1}{n^p}, \qquad p > 0.$$

16. Show that, for every sequence of real numbers $(x_n)_{n=1}^{\infty}$, we have $\underline{\lim}_{n\to\infty} x_n = \lim_{k\to\infty} (\text{glb } \{x_n \mid n \geq k\})$. (Refer to Appendix B for the definition of glb.)

17. *An inequality.* If $(a_n)_{n=1}^{\infty}$ is a sequence in $\mathbf{C}\backslash\{0\}$, then we have

$$\varliminf_{n\to\infty} \left| \frac{a_{n+1}}{a_n} \right| \leq \varliminf_{n\to\infty} \sqrt[n]{|a_n|} \leq \varlimsup_{n\to\infty} \sqrt[n]{|a_n|} \leq \varlimsup_{n\to\infty} \left| \frac{a_{n+1}}{a_n} \right|.$$

OUTLINE. Begin with the inequality on the right. If $\varlimsup_{n\to\infty} a_{n+1}/a_n = \infty$, there is nothing to prove. Let μ satisfy $\varlimsup_{n\to\infty} |a_{n+1}/a_n| < \mu < \infty$ instead. There is an n_0 such that $|a_n/a_{n-1}| < \mu$ for $n \geq n_0$. We have

$$|a_n| < \mu^{n-n_0+1} |a_{n_0-1}| \qquad \text{if } n \geq n_0.$$

Since $\lim_{n\to\infty} \mu^{(n-n_0+1)/n} |a_{n_0} - 1|^{1/n} = \mu$, we have $\varlimsup_{n\to\infty} |a_n|^{1/n} \leq \mu$. But μ is arbitrary, so the right inequality follows.

The middle inequality is a triviality; the left inequality is proved much the same as the right one.

Remark. The inequalities show that the root test will work any time that the ratio test works. Explain why.

0.14 SEQUENCE OF FUNCTIONS

We have just reviewed the notion of a sequence of numbers. Formally, the notion of a sequence of functions is similar. A mapping which assigns to each $n \in \mathbf{Z}^{++}$ a complex-valued function f_n is called a sequence of functions. If all the functions f_n are defined on a particular set S, then for each $z \in S$ they form a sequence of complex numbers; that is, $(f_n(z))_{n=1}^{\infty}$. It makes sense to ask whether the sequence of numbers $(f_n(z))_{n=1}^{\infty}$ converges.

If it converges for every $z \in S$, then a limit function f is defined by

$$f(z) = \lim_{n\to\infty} f_n(z) \qquad (z \in S),$$

and the sequence $(f_n)_{n=1}^{\infty}$ is said to converge pointwise to f. We have already encountered sequences of functions and their limits. For example, if $f_n(z) = z^n$ on $D(1, 0)$, then we have seen that $\lim_{n\to\infty} f_n = 0$, the zero function. If we let $P_n(x) = \sum_{k=0}^{n} (1/k!) x^k$, then $(P_n)_{n=1}^{\infty}$ is a sequence of functions on $\mathbf{R}$ and $\lim_{n\to\infty} P_n(x) = \exp(x)$ for all $x \in \mathbf{R}$.

If S is a subset of $\mathbf{C}$ and $(f_n)_{n=1}^\infty$ is a sequence of functions on S and $f = \lim_{n \to \infty} f_n$, it very often happens that we want to know whether f has the same general properties as the f_n. If the f_n are continuous, is f continuous? If f'_n exists on S for each n, does f' exist? Does $\lim_{n \to \infty} f'_n = f'$ hold? Does $\lim_{n \to \infty} \int_S f_n = \int_S f$ hold? Examples in the exercises show that all the answers are "no, not in general." An important property that frequently leads to affirmative answers follows.

Definition. A sequence of functions $(f_n)_{n=1}^\infty$ on a set $S \subset \mathbf{C}$ *converges uniformly on S to f* if for every $\varepsilon > 0$ there is an n_0 such that $|f_n(z) - f(z)| < \varepsilon$ for all $n \geq n_0$ and for all $z \in S$.

Notice that the difference between pointwise convergence of f_n to f and uniform convergence is that in the definition of uniform convergence the index n_0 depends only on ε and not upon the particular $z \in S$ that is under consideration.

(i) *Example.* Let $f_n(z) = z^n$, $z \in D(1, 0)$. We have seen in (0.10.ii.c) that $\lim_{n \to \infty} f_n(z) = 0$. However, the convergence is not uniform. To see this, we will assume that the convergence is uniform and derive a contradiction. Take $\varepsilon = \frac{1}{2}$. Then there is an n_0 such that

$$|z^n| < \tfrac{1}{2}$$

holds for all $z \in D(1, 0)$ and all $n \geq n_0$. Hence

$$|z| < (\tfrac{1}{2})^{1/n} \qquad (n \geq n_0; z \in D(1, 0)) \tag{*}$$

would hold, but (*) clearly fails if $|z| \geq 1/2^{1/n_0}$.

(ii) *Example.* Let $P_n(z) = \sum_{k=0}^n (1/k!)z^k$ ($z \in \mathbf{C}$), so that $\lim_{n \to \infty} P_n(x) = e^x$ on $\mathbf{R}$. Since $|e^x - P_n(x)| = \sum_{k=n+1}^\infty (1/k!)x^k$, for fixed n the equality $\lim_{x \to \infty} \sum_{k=n+1}^\infty (1/k!)x^k = \infty$ is valid. Thus P_n does not converge uniformly on $\mathbf{R}$.

Notation. We will let $[a, b]$ denote the line segment joining a to b including the endpoints a and b. Also we will let $]a, b[$ denote the line segment joining a to b but excluding the endpoints. There are two other possible combinations $]a, b]$ and $[a, b[$, the first is the line segment excluding a and including b and the second includes a and excludes b.

(iii) *Example.* Define a sequence of functions f_n with domain $[0, 1]$ and range in $\mathbf{C}$ by

$$f_n(t) = \begin{cases} t + i & \text{if } t \in \left[0, \dfrac{1}{4n}\right] \\[2ex] \dfrac{1}{4n} + i\left(\dfrac{1}{3n} - t\right)12n & \text{if } t \in \left[\dfrac{1}{4n}, \dfrac{1}{3n}\right] \\[2ex] t & \text{if } t \in \left[\dfrac{1}{3n}, \dfrac{1}{2n}\right] \\[2ex] \dfrac{1}{2n} + i\left(t - \dfrac{1}{2n}\right)2n & \text{if } t \in \left[\dfrac{1}{2n}, \dfrac{1}{n}\right] \\[2ex] t + i & \text{if } t \in \left[\dfrac{1}{n}, 1\right] \end{cases}$$

The image of $[0, 1]$ under f_n is shown in Figure 0.5. For each f_n it is clear that there are points t_n such that $|f_n(t_n) - (t_n + i)| = 1$, so that lub $\{|f_n(t) - (t + i)| : t \in [0, 1]\} = 1$ for all n. Nevertheless, we have $\lim_{n \to \infty} f_n(t) = t + i$ for all t. Hence $(f_n)_{n=1}^{\infty}$ converges to $t + i$, but not uniformly.

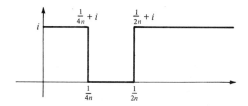

Figure 0.5

(iv) *Example.* Define f_n on **R** by $f_n(x) = (1/n) \sin nx$. We have

$$|f_n(x) - 0| = \frac{1}{n}|\sin nx| < \frac{1}{n}.$$

For given $\varepsilon > 0$, $|f_n(x) - 0| < \varepsilon$ holds for all n satisfying $n > 1/\varepsilon$. Hence the sequence $(f_n)_{n=1}^{\infty}$ converges uniformly to 0 on **R**.

(v) *Example.* Define f_n on $[0, 1]$ as 1 on $[1/n, 1]$ and linear from 0 to 1 on $[0, 1/n]$. Then,

$$\lim_{n \to \infty} f_n(x) = \begin{cases} 1 & \text{if } 0 < x \leq 1 \\ 0 & \text{at } 0. \end{cases}$$

(vi) *Example.* Define f_n on $[0, 1]$ by $f_n(x) = x^n$. Then

$$\lim_{n \to \infty} f_n(x) = \begin{cases} 0 & \text{if } 0 \le x < 1 \\ 1 & \text{if } x = 1. \end{cases}$$

In the last two examples, the functions f_n are continuous but the limit function is not continuous.

EXERCISES

TYPE I

1. On **R**, let $f_n(x) = (1/\sqrt{n}) \sin (nx)$. Show that $\lim_{n \to \infty} f_n(x) = f(x) = 0$ uniformly on **R**, that $f_n'(x)$ exists for all x for each n, but that $\lim_{n \to \infty} f_n'(x) \ne f'(x)$.

2. Construct a sequence of continuous functions $(f_n)_{n=1}^\infty$ on $[0, 1]$ such that $\int_0^1 f_n(x) \, dx = 1$ for every n, $\lim_{n \to \infty} f_n(x) = f(x)$, and $\int_0^1 f(x) \, dx \ne 1$. (Hence $\lim_{n \to \infty} \int_0^1 f_n(x) \, dx = \int_0^1 \lim_{n \to \infty} f_n(x) \, dx$ does not always hold.) [*Hint:* Let the graph of f_n be as shown.]

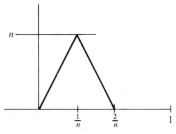

3. Show that $\lim_{n \to \infty} z^n = 0$ uniformly on $D(\eta, 0)$ for every $\eta < 1$.

4. Let $S = \{x \mid 0 \le x \le 1\}$ and $f_n(x) = x^n$. Show that $f(x) = \lim_{n \to \infty} f_n(x)$ is not continuous on S.

5. Let $f_n(x) = \int_0^x g_n(t) \, dt$, $0 \le x \le 2$ where

$$g_n(t) = \begin{cases} t^n & 0 \le t \le 1 \\ 1 - (t - 1)^n & 1 \le t \le 2. \end{cases}$$

Find $\lim_{n \to \infty} f_n(x) = f(x)$. Show that $f_n'(x)$ exists for all x between 0 and 2 but that $f(x)$ is not continuous at $x = 1$.

TYPE II

1. *Double sequences.* A mapping from all pairs of positive integers (n, m) to the complex numbers is called a *double sequence* of complex numbers.

Denote the value of the double sequence at (n, m) by a_{mn}. There are three limits to consider: the two iterated limits

$$\lim_{n\to\infty} \lim_{m\to\infty} a_{mn}, \qquad \lim_{m\to\infty} \lim_{n\to\infty} a_{mn}$$

and the double limit

$$\lim_{(n,m)\to\infty} a_{mn}.$$

The latter is equal to A if for every $\varepsilon > 0$ there is an m_0 such that $|a_{mn} - A| < \varepsilon$ if $n, m \geq m_0$.

(a) If $a_{nm} = 0$ for $n < m$ and $a_{nm} = 1$ for $n \geq m$, show that

$$\lim_{n\to\infty} \lim_{m\to\infty} a_{nm} \neq \lim_{m\to\infty} \lim_{n\to\infty} a_{nm}.$$

(b) Show that both iterated limits exist and are equal if the double limit exists.

(c) Suppose that $\lim_{n\to\infty} a_{nm} = b_m$ and that the convergence is uniform in m. Prove that the double limit exists.

CHAPTER 1

Introduction to Functions
of a Complex Variable

1.0

We assume that the reader has some familiarity with the elementary properties of complex numbers that are reviewed in Sections (0.0) through (0.9). From time to time we shall also make use of certain fundamental algebraic and analytic properties of the real numbers. This latter information may be found in Appendix B. We will use standard notation from set theory as appropriate. We denote the set of integers by $\mathbf{Z}$, the positive integers by $\mathbf{Z}^{++}$, and the nonnegative integers by $\mathbf{Z}^{+}$. That is,

$$\mathbf{Z} = \{0, \pm 1, \pm 2, \dots\}; \mathbf{Z}^{+} = \{0, 1, 2, 3, \dots\}; \mathbf{Z}^{++} = \{1, 2, 3, \dots\}.$$

The symbol $\mathbf{Q}$ denotes the rational numbers; $\mathbf{R}$ the real numbers; the definitions of $\mathbf{Q}^{+}, \mathbf{Q}^{++}, \mathbf{R}^{+}, \mathbf{R}^{++}$ are parallel to those of $\mathbf{Z}^{+}$ and $\mathbf{Z}^{++}$, respectively. The symbol $\mathbf{C}$ denotes the complex numbers described in sections (0.0) through (0.9). All numbers are presumed to be complex; that is, they lie in $\mathbf{C}$.

The *open disk* in the x-y plane of radius r centered at $(0, 0)$ is defined as the set $\{(x, y) \mid \sqrt{x^2 + y^2} < r\}$. In our notation, with $z = x + iy$, it is $D(r, 0) = \{z \mid |z| < r\}$. Recall that $|z| = (x^2 + y^2)^{1/2}$. Similarly, the open disk of radius r centered at a is

$$D(r, a) = \{z \mid |z - a| < r\}.$$

The *circle* of radius r centered at a is

$$C(r, a) = \{z \mid |z - a| = r\}.$$

Other notation is explained when it is first introduced. Appendix A is devoted to notation.

1.1 FUNCTIONS ON C

The graph of a function from **R** to **R** is the set $\{(x, f(x)) \mid x \in \text{domain } f\}$; it is easily visualized mentally and in many cases easy to sketch. Even for functions from $\mathbf{R}^2$, the plane, to **R**, there is a helpful geometric picture. The graph of the function is a two-dimensional surface lying in 3-space. The graph of $f: \mathbf{C} \to \mathbf{C}$ is a subset of $\mathbf{C} \times \mathbf{C}$, the set of all pairs of complex numbers, and thus requires two complex planes or four real dimensions for a sketch. This is difficult to manage in 3-space. We get an idea of the geometric properties of f by using two complex planes and describing the image of certain sets in the domain of f. Some examples follow.

Example 1. $f(z) = z^2$, where $z = x + iy$.

To understand something of the behavior of this function we examine the image of some familiar curves.

(i) The circle $C(\rho, 0)$ is mapped onto a circle centered at the origin with radius ρ^2. To see this, let $z = x + iy$ where $|z| = \rho$ and compute

$$|z^2| = |z|^2 = \rho^2.$$

If we use formula (i) of Section (0.8) we can write a point on $C(\rho^2, 0)$ as $w = \rho^2 e^{i\psi}$. There are two distinct points of $C(\rho, 0)$ mapped onto w; namely, $\rho e^{i\psi/2}$ and $\rho e^{i(\psi/2 + \pi)}$.

(ii) If ϕ is fixed and r varies from 0 to ∞ ($0 \le r < \infty$), then $\{z \mid z = re^{i\phi}\}$ is a ray. Evidently this ray is mapped onto the ray $r^2 e^{2i\phi}$.

(iii) To see what happens to the lines $x = $ constant and $y = $ constant, let $z^2 = u + iv = w$. Since $z = x + iy$, $z^2 = x^2 - y^2 + 2ixy$ and, by Section (0.0), $x^2 - y^2 = u$ and $2xy = v$. For fixed $x \ne 0$ we have $u = (-v^2/4x^2) + x^2$ and for fixed $y \ne 0$ we have $u = (v^2/4y^2) - y^2$. See Figure 1.1.

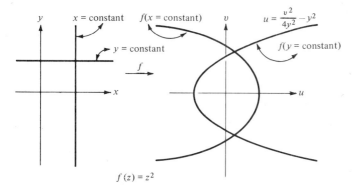

Figure 1.1

(iv) Finally, since

$$x^2 - y^2 = u \quad \text{and} \quad 2xy = v,$$

we see that the hyperbolas $x^2 - y^2 = $ constant and $xy = $ constant are mapped onto the curves $u = $ constant and $v = $ constant, respectively. See Figure 1.2.

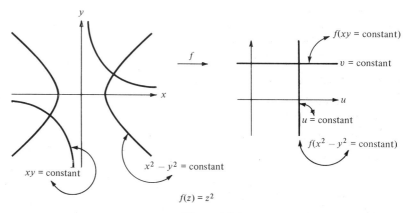

Figure 1.2

Example 2. $f(z) = z^n$, $n \in \mathbf{Z}^{++}$.

For now we will content ourselves with some simple observations about this function. First compute $|z^n| = |z|^n$. This is most efficiently done using the polar form for z [see Section (0.8)], to wit:

$$z = re^{it}$$

so

$$z^n = r^n e^{int},$$

and

$$|z^n| = r^n = |z|^n.$$

The fact that $|z^n| = |z|^n$ shows that f maps the disk $D(R, 0)$ to $D(R^n, 0)$. If k is an integer, then the region

$$P_k = \{z \mid z = re^{i\theta}, 0 \le r \le R, \pi(2k - 1)/n < \theta \le \pi(2k + 1)/n\}$$

is pie-shaped (the reader should make a sketch). We will show that f maps P_k onto $D(R^n, 0)$ for each k satisfying $0 \le k < n$. If $w \in D(R^n, 0)$ write $w = \rho e^{i\psi}$ where

$$\psi = \text{Arg}(w) + 2k\pi.$$

Then, since $-\pi < \text{Arg}(w) \leq \pi$, we have

$$\frac{(2k-1)\pi}{n} < \frac{\psi}{n} = \frac{\text{Arg}(w)}{n} + \frac{2k\pi}{n} \leq \frac{(2k+1)\pi}{n}.$$

Hence $z = \rho^{1/n}e^{i\psi/n}$ lies in P_k. Evidently $f(z) = w$. Recall that a z which satisfies $z^n = w$ is called an nth root of w. We have shown that each $w \in D(R^n, 0)$ has an nth root in each P_k. Since the n sets P_k do not overlap and each w has precisely n distinct roots [Section (0.9)], each P_k contains exactly one n^{th} root of w. Thus the function f is one-to-one on each P_k.

Example 3. $f(z) = 1/z$.

If $|z| > R$, then $|1/z| < 1/R$, so f maps the exterior of the disk of radius R to the interior of the disk $D(R^{-1}, 0)$, with 0 deleted. In order to learn more about f, consider the expression

(i) $az\bar{z} - \beta\bar{z} - \bar{\beta}z + c = 0$,

where a is 1 or 0, c is a real number, and β is a complex number. If $a = 0$ and if we write $\beta = \beta_1 + i\beta_2$ and $z = x + iy$, then (i) can be written

(ii) $\beta_1 x \overset{+}{} \beta_2 y = \tfrac{1}{2}c$,

which is the equation of a line provided that $\beta \neq 0$. If $a = 1$, then (i) can be written

(iii) $|z - \beta|^2 = \beta\bar{\beta} - c$,

the equation of a circle of radius $\sqrt{|\beta|^2 - c}$ centered at β provided that $|\beta|^2 - c \geq 0$. Thus the locus of points satisfying (i) is a line, a circle, or the empty set. Conversely, it is easy to see that any line or circle satisfies an expression of the form (i).

If z satisfies (i), by dividing both sides of (i) by $|z|^2 = z\bar{z}$ we obtain

(iv) $a - \beta(1/z) - \bar{\beta}(1/\bar{z}) + (c/z\bar{z}) = 0$.

If we let $w = 1/z$, then we have

(v) $a - \beta w - \bar{\beta}\bar{w} + cw\bar{w} = 0$,

which is an expression of the same form as (i) (divide by c if $c \neq 0$). Hence the function f maps the collection of lines and circles into the collection of lines and circles. Observe that circles through the origin ($c = 0$) are mapped into lines and that lines that fail to pass through the origin ($c \neq 0$, $a = 0$) are mapped into circles which pass through the origin. If β is real, the reader should show as an exercise that the circle of radius $|\beta|$ centered at β goes to the line parallel to y-axis with x-intercept $1/(2\beta)$. What happens to this circle if β is not real? Properties of f are illustrated in Figure 1.3.

Other Examples. Some information about four more mappings is given in Figures 1.4 through 1.7.

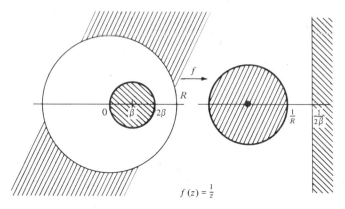

$$f(z) = \frac{1}{z}$$

Figure 1.3

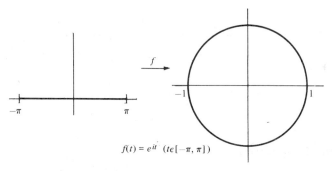

$$f(t) = e^{it} \quad (t \in [-\pi, \pi])$$

Figure 1.4   Starting at $-\pi$, f traces out the circle counterclockwise from -1 as t moves from $-\pi$ to π. The function is 1 to 1 on $]-\pi, \pi]$ and on $[-\pi, \pi[$. On $[-\pi, \pi]$, $f(\pi) = f(-\pi) = -1$.

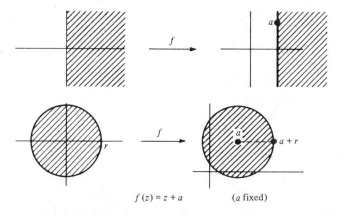

$$f(z) = z + a \qquad (a \text{ fixed})$$

Figure 1.5   The function in translation by a fixed a. The image of a set under f has the same geometric relation to a as the set has to 0.

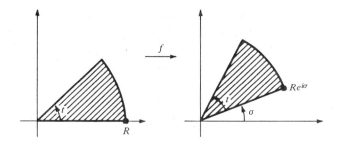

$$f(z) = e^{i\sigma} z$$
f is rotation about 0 through the angle σ

Figure 1.6

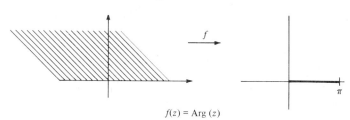

$$f(z) = \text{Arg } (z)$$

Figure 1.7

1.2 DEFINITION OF CONTINUITY

We will use the following definition of continuity. A function f mapping a set $S \subset \mathbf{C}$ into $\mathbf{C}$ is continuous on S if the equality

(i) $\lim_{n \to \infty} f(s_n) = f(s)$

holds for every $s \in S$ and every sequence $(s_n)_{n=1}^{\infty}$ in S for which $\lim_{n \to \infty} s_n = s$.

This is not the only possible definition. The following theorem shows that the definition we are using is the same as the "$\delta - \varepsilon$" definition used by many authors.

Theorem.   Let $S \subset \mathbf{C}, f: S \to \mathbf{C}, z \in S$, and $A \in \mathbf{C}$. The conditions (i) and (ii) are equivalent.

(i) $\lim_{n \to \infty} f(z_n) = A$ for every sequence $(z_n)_{n=1}^{\infty}$ in S such that $\lim_{n \to \infty} z_n = z$.

(ii) For every $\varepsilon > 0$ there is $\delta > 0$ such that

$$|s - z| < \delta \quad \text{and} \quad s \in S \text{ imply that } |f(s) - A| < \varepsilon.$$

PROOF.   Suppose $\varepsilon > 0$, (ii) holds for ε and δ, and $\lim_{n \to \infty} z_n = z$ $(z_n \in S)$. There is an n_0 such that $|z_n - z| < \delta$ if $n \geq n_0$; hence $|f(z_n) - A| < \varepsilon$ if $n \geq n_0$.

If (ii) fails then there is an $\varepsilon > 0$ and a sequence $(z_n)_{n=1}^{\infty}$ in S such that $|z_n - z| < 1/n$ and $|f(z_n) - A| \geq \varepsilon$. For this sequence $\lim_{n \to \infty} z_n = z$ holds and $\lim_{n \to \infty} f(z_n) = A$ fails.

Notation. The notation $\lim_{s \to z, \, s \in S} f(s) = A$ means that (i) and (ii) hold.

1.3 ELEMENTARY PROPERTIES AND EXAMPLES OF CONTINUITY

The definition of continuity just given is the same as that given in calculus so it is not surprising that continuous functions on **C** have many familiar properties. For example,

(i) If f and g are continuous on a set $S \subset \mathbf{C}$, then $f + g$ and fg are continuous on S.

(ii) If f and g are continuous on S and $g(s) \neq 0$ for all $s \in S$, then f/g is continuous on S.

Of course, $(f + g)(z) = f(z) + g(z)$, $(fg)(z) = f(z)g(z)$ and $(f/g)(z) = f(z)/g(z)$. Also $(g \circ f)(z) = g(f(z))$.

(iii) If f is continuous on S and if g is continuous on $f(S) = \{w \mid w = f(z)$ for $z \in S\}$, then $g \circ f$ is continuous on S.

As a sample, we shall prove item (iii). Let $s \in S$ and let $\lim_{n \to \infty} s_n = s$, $s_n \in S$. Since $(g \circ f)(z) = g(f(z))$ we must show that $\lim_{n \to \infty} g(f(s_n)) = g(f(s))$. However, this is easy to do. If $w_n = f(s_n)$ and $w = f(s)$, then the sequence $(w_n)_{n=1}^{\infty}$ is in $f(S)$ and $\lim_{n \to \infty} w_n = w$ because f is continuous. Since g is continuous, $\lim_{n \to \infty} g(w_n) = g(w)$; that is, $\lim_{n \to \infty} g(f(s_n)) = g(f(s))$.

(iv) The function $f(z) = z^m$ is continuous on **C** if $m \in \mathbf{Z}^+$.

An inductive proof of (iv) is as easily constructed as in elementary calculus.

Later we will have use for the following notion of continuity. Suppose that for every sequence $\{s_n\}$ in S with the property that $\lim_{n \to \infty} |s_n| = \infty$, we have $\lim_{n \to \infty} f(s_n) = L$, where $L \in \mathbf{C}$ and does not depend upon the particular sequence $\{s_n\}$. In this case we write $f(\infty) = L$ and say that f is continuous at ∞.

(v) *Polynomials.* A polynomial in z is a finite linear combination of non-negative integral powers of z with complex coefficients. If P and Q are polynomial functions of z and $S = \{z \in \mathbf{C} \mid Q(z) \neq 0\}$, then (i), (ii), and (iv) show that $f = P/Q$ is continuous on S. The reader can easily verify that the functions with values Re(z) and Im(z) are continuous. It follows that $g = R/T$ is continuous on $S = \{z \in \mathbf{C} \mid T(z) \neq 0\}$, if R and T are polynomials in Re(z) and Im(z); that is, polynomials in the two real variables

x and y. It is important to distinguish between these types of functions: A polynomial P in z has the form

$$P(z) = a_n z^n + \cdots + a_0 \qquad (\text{complex } a\text{'s}, \ n \in \mathbf{Z}^+);$$

a polynomial R in x and y has the form

$$R(x, y) = \sum_{m,\, k \in F} c_{m,\,k} x^m y^k \qquad (c\text{'s complex}, \ F \text{ a finite subset of } \mathbf{Z}^+).$$

Each polynomial in z is a polynomial in x and y; but, for example, $R(x, y) = x$ is not a polynomial in z. Neither is $R(z) = \bar{z}$.

Any function $f(z) = f(x, y) = G(g(x), h(y))$, where G is continuous on $\mathbf{C}$ (or some suitable subset) and g and h are continuous real-valued functions on $\mathbf{R}$ (or suitable subsets), is continuous by (iii).

(vi) If f is continuous on S, then $|f|$ is also continuous on S. This follows easily from the inequality

$$||f(z)| - |f(w)|| \leq |f(z) - f(w)|,$$

z and w in S.

(vii) (*Discontinuity of Arg*). The function Arg, as defined in Section (0.8), is continuous on $\mathbf{C} \backslash R^-$, R^- the nonpositive x-axis, but it is discontinuous at every point on R^-. For example,

$$\lim_{n \to \infty} \operatorname{Arg}\left(-1 + i\,\frac{1}{n}\right) = \pi; \ \lim_{n \to \infty} \operatorname{Arg}\left(-1 - i\,\frac{1}{n}\right) = -\pi,$$

although

$$\lim_{n \to \infty}\left(-1 + i\,\frac{1}{n}\right) = \lim_{n \to \infty}\left(-1 - i\,\frac{1}{n}\right) = -1.$$

This discontinuity of Arg can be shifted by selecting a different interval of length 2π for the image, but it cannot be removed. For any interval $]a, a + 2\pi]$ of length 2π, every nonzero z has a unique representation

$$z = |z|e^{i\psi}, \ \psi \in \,]a, a + 2\pi].$$

The corresponding function $\alpha(z) = \psi$ is discontinuous on $\{z \mid \psi = a\}$.

EXERCISES

TYPE I

1. If S is a subset of the domain of a function f, then $f(S)$ is by definition $\{f(s) \mid s \in S\}$. Find $f(S)$ for the following f and S.

(a) $f(z) = \bar{z}$, $S = \{z \mid \text{Im}(z) > 0\}$.
(b) $f(z) = z^n$ and
    (i) $S = \{z \mid |z| \leq R\}$;
    (ii) $S = \{z \mid 0 < \text{Arg}(z) < \pi/4\}$.
(c) $f(z) = \log |z|$, $S = D(r, 0)\backslash\{0\}$.
(d) $f(z) = \log |z| + i \, \text{Arg}(z)$ and
    (i) $S = D(r; 0)\backslash\{0\}$;
    (ii) $S = \{x + iy \mid 0 < \max[|x|, |y|] < 1\}$.
(e) $f(z) = 1/(1 - z)$ and
    (i) $S = \{z \mid |z| \neq 1\}$;
    (ii) $S = \{z \mid |z| < 1\}$.
(f) $f(z) = \cos|z| + i\sin|z|$ and
    (i) $S = \{z \mid |z| = r\}$;
    (ii) $S = \{z \mid |z| \leq r\}$.

2. Let $f(z) = 1/[(1 - z)^2]$. Show that f maps $\{z \mid z \neq 1\}$ onto **C**.

TYPE II

1. Define f on **C** by

$$f(z) = \begin{cases} \dfrac{xy}{x^2 + y^2} & \text{if } z = x + iy \neq 0 \\[2mm] 0 & \text{if } z = 0. \end{cases}$$

(a) Show that f is continuous on $\mathbf{C}\backslash\{0\}$.
(b) Show that f is not continuous at 0.
(c) Show that $x \rightarrow f(x, y)$ and $y \rightarrow f(x, y)$ are everywhere continuous functions in x and y (respectively) if y and x are fixed (respectively).
(d) Show that it is impossible to define $f(0)$ in a way that will make the resulting function continuous at 0.

2. With respect to (1.3.iii) discuss the validity of the assertions

(i) $g \circ f$ and g continuous $\Rightarrow f$ continuous;

(ii) $g \circ f$ and f continuous $\Rightarrow g$ continuous.

1.4 UNIFORM CONVERGENCE AND CONTINUITY

The only practical way that numerical values of many important functions (for example, the exponential, logarithm, sine, and cosine functions) can be estimated is by using the fact that there exists a sequence of simpler functions (for example, polynomials) whose limit is the desired function. The notions of convergence of a sequence of functions and uniform convergence of a sequence of functions are generally studied briefly in a course in calculus. These notions, which are central to the development of the material in the rest of this book, are reviewed in Section (0.14).

To say that a sequence of functions $(f_n)_{n=1}^\infty$ converges on S to f means that for each $z \in S$ (S some fixed subset of $\mathbf{C}$) $\lim_{n \to \infty} f_n(z) = f(z)$. However, if we are estimating $f(z)$ by $f_n(z)$, the choice of n required to obtain the desired precision in general depends upon the variable z. But saying that $(f_n)_{n=1}^\infty$ converges uniformly on S to f means that $\lim_{n \to \infty} f_n(z) = f(z)$, and that the choice of n required to obtain the desired precision can be made independently of the variable z (if $z \in S$).

Precisely, a sequence of functions $(f_n)_{n=1}^\infty$ on S converges *uniformly on S to f* if for every $\varepsilon > 0$ there is an n_0 such that $|f_n(z) - f(z)| < \varepsilon$ for all $n \geq n_0$ and all $z \in S$.

Examples (v) and (vi) of section (0.14) showed that even though the functions f_n are continuous and $\lim f_n(z) = f(z)$, the function f need not be continuous. The following theorem states that this phenomenon cannot occur if the convergence is uniform.

Theorem. Let $(f_n)_{n=1}^\infty$ be a sequence of continuous functions on S which converges uniformly on S to f. Then f is continuous on S.

PROOF. Fix $z \in S$, and let $\varepsilon > 0$. By hypothesis there is a k such that

$$|f_k(s) - f(s)| \leq \frac{\varepsilon}{3}$$

holds for all $s \in S$. Since f_k is continuous, there is a $\delta > 0$ such that the condition $|w - z| < \delta$ implies that $|f_k(w) - f_k(z)| < \varepsilon/3$. Hence the inequality

$$|f(w) - f(z)| \leq |f(w) - f_k(w)| + |f_k(w) - f_k(z)| + |f_k(z) - f(z)|$$

proves that

$$|f(w) - f(z)| < \frac{\varepsilon}{3} + \frac{\varepsilon}{3} + \frac{\varepsilon}{3} = \varepsilon \qquad \text{if } |w - z| < \delta.$$

Thus f is continuous at z.

This last theorem involves some fundamental notions that are often difficult to grasp immediately. However, the importance of this theorem can hardly be overemphasized. Very often it is essentially the only way to prove that a particular function is continuous. We will see examples in Chapter 2.

1.5 A DOMINATED CONVERGENCE THEOREM (WEIERSTRASS M-TEST)

Let $(f_n)_{n=1}^\infty$ be a sequence of functions on a set $S \subset \mathbf{C}$, and suppose that there is a sequence $(M_n)_{n=1}^\infty$ in $\mathbf{R}^+$ such that

(i) $|f_n(z)| \leq M_n$ for all $z \in S$;
(ii) $\sum_{n=1}^\infty M_n$ converges.

Then the sequence of partial sums $\sum_{k=1}^{n} f_k$ converges uniformly on S. In particular, if all the f_n are continuous on S, then $f(z) = \sum_{n=1}^{\infty} f_n(z)$ is continuous on S.

PROOF. If $f(z) = \sum_{n=1}^{\infty} f_n(z)$, then the inequality

$$\left| f(z) - \sum_{k=1}^{n} f_k(z) \right| \leq \sum_{k=n+1}^{\infty} M_n$$

shows that $\lim_{n \to \infty} \left[\sum_{k=1}^{n} f_k \right] = f$ uniformly on S.

Remark. If all the f_n are continuous, the continuity of f means that

(iii) $$\lim_{\substack{z \to z_0 \\ z \in S}} \sum_{n=1}^{\infty} f_n(z) = \sum_{n=1}^{\infty} f_n(z_0) = \sum_{n=1}^{\infty} \lim_{\substack{z \to z_0 \\ z \in S}} f_n(z)$$

for all $z_0 \in S$.

1.6 REMARKS

A function f is continuous on a set if the function values $f(s)$ are close to $f(z)$ when s is close to z; that is, for given $\varepsilon > 0$, there is a $\delta > 0$ such that $|f(s) - f(z)| < \varepsilon$ if $|s - z| < \delta$. Just how close s must be made to z to make $f(s)$ within ε of $f(z)$ sometimes depends on z; that is, δ depends on both ε and z. It is crucial in many parts of analysis that it be possible to make $f(s)$ within ε of $f(z)$ whenever s and z are sufficiently close; that is, that it be possible to make δ depend only on ε, and *not* on z. The possibility of doing this often depends upon the nature of the set S. The properties in question are called "topological." Therefore, we now give a sketch of the topology of the complex plane. The principal theorem we are aiming for is (1.13).

On the first reading the student may be content just to learn the definitions and statements of the theorems in Sections (1.7) through (1.14). The proof of the Fundamental Theorem of Algebra (one of the most famous theorems of mathematics) given in Section (1.15) shows how some of the ideas just learned may be used. Another option is to skip (for now) this material altogether and turn to Section (1.16) with the idea of referring back as needed.

1.7 DEFINITIONS

A set $S \subset \mathbf{C}$ is *open* if for every $a \in S$ there is an $\varepsilon > 0$ such that $D(\varepsilon, a) \subset S$. Recall that $D(\varepsilon, a)$ denotes the open disk about a of radius ε: $D(\varepsilon, a) = \{z \mid |z - a| < \varepsilon\}$. The condition is satisfied if $S = \varnothing$, so $\varnothing$ is open ($\varnothing$ denotes the empty set).

A set T is *closed* if its complement $T' = \mathbf{C}\backslash T$ is open.

A point z is a *cluster point* (or *limit point*) of a set T if $D(\varepsilon,z) \cap (T\backslash\{z\}) \neq \emptyset$ for all $\varepsilon > 0$. That is, every disk about z contains points of T other than z. Note that z does not have to be in T to be a cluster point of T. Examples are given in Type I exercises at the end of this section. The reader unfamiliar with closed sets and cluster points would do well to do those problems before continuing.

1.8 TWO EQUIVALENCES

(i) *A set T is closed if and only if it contains all its cluster points.*

PROOF.   If t is a cluster point of T, then no disk $D(\varepsilon,t)$ can lie in T'; thus if T' is open (T closed), t cannot lie in T'.

Conversely, if T contains all its cluster points, then for every $t \in T'$ there is some $\varepsilon > 0$ such that $D(\varepsilon,t)$ does not intersect T; that is, $D(\varepsilon,t) \subset T'$. Thus T' is open.

(ii) *A point t is a cluster point of T if and only if there is a sequence $(t_n)_{n=1}^{\infty}$ in T such that $t_n \neq t$ for all n and $\lim_{n \to \infty} t_n = t$.*

The proof is left to the reader.

1.9 THE CLOSURE OF A SET

If T is any subset of $\mathbf{C}$, the *closure* $\bar{T}$ of T is the set $T \cup S$, where S denotes the set of cluster points of T. The reader should verify that the set $\bar{T}$ contains all its cluster points, and so is closed. As an example, the reader should verify that $\overline{D(1, 0)} = \{z \mid |z| \leq 1\}$. A set A is *dense* in a set B if $\bar{A} \cap B = B$.

1.10 MORE DEFINITIONS

A *family of sets* F is simply a set whose elements are sets. If F is a family of sets, then $\cup\{F \mid F \in \mathsf{F}\}$ is all elements x such that $x \in F$ for at least one F in F. We will sometimes denote this union by $\cup_{\mathsf{F}}F$. Let A be a set. A family of sets F is a *cover* for A if $A \subset \cup_{\mathsf{F}}F$. A *subcover* of F for A is a subfamily E of F such that $A \subset \cup_{\mathsf{E}}F$. We will need these ideas in the next section. We will also need the definition of a subsequence. Let $(z_n)_{n=1}^{\infty}$ be a sequence in $\mathbf{C}$. Let $k \to n(k)$ be a mapping from $\mathbf{Z}^{++}$ to itself which satisfies $n(k) < n(j)$ if $k < j$. Then the sequence $(z_{n(k)})_{k=1}^{\infty}$ is said to be a *subsequence* of the sequence $(z_n)_{n=1}^{\infty}$. For example, if we let $n(k) = 2k$, then the subsequence $(z_{2k})_{k=1}^{\infty}$ is obtained from the original sequence by removing the terms with odd indices. In general, to obtain a subsequence of a sequence, one removes terms from the original sequence in such a way that infinitely many terms are left.

1.11 COMPACTNESS

The following conditions on a set $S \subset \mathbf{C}$ are equivalent. A set having them is said to be *compact*.

(i) S is closed and bounded.

(ii) Every sequence $(s_n)_{n=1}^{\infty}$ in S has a subsequence convergent to a point of S.

(iii) Every infinite set in S has a cluster point which lies in S.

(iv) Every family F of open sets which covers S has a finite subcover. That is, if $S \subset \cup_F U$, then there are sets $U_1, U_2, \ldots, U_N$ in F such that $S \subset \cup_{n=1}^{N} U_n$.

(v) Every family E of closed sets with the property that $\cap_{n=1}^{N} (S \cap E_n) \neq \varnothing$ for every finite subfamily $\{E_1, E_2, \ldots, E_N\}$ of E satisfies $S \cap [\cap_E E] \neq \varnothing$.

PROOF OF EQUIVALENCE

(i) $\Rightarrow$ (ii). Let $s_n = x_n + iy_n$ and let $x = \overline{\lim}_{n \to \infty} x_n$. For each $k \in \mathbf{Z}^{++}$ there is an integer $n(k)$ such that $x_{n(k)} \in D(1/k, x)$. The $n(k)$ can also be chosen so that $n(k) > n(k')$ if $k > k'$. The sequence $\{x_{n(k)}\}_{k=1}^{\infty}$ converges to x. Let $y = \overline{\lim}_{k \to \infty} y_{n(k)}$ and select integers $k(m)$ such that $k(m) > k(m')$ if $m > m'$ and $\lim_{m \to \infty} y_{n(k(m))} = y$. The subsequence $(s_{n(k(m))})_{m=1}^{\infty}$ converges to $x + iy$.

(ii) $\Rightarrow$ (iii). If S is infinite, it contains a sequence $(s_n)_{n=1}^{\infty}$. If a subsequence converges to s, then s is a cluster point of S.

(iii) $\Rightarrow$ (iv). If condition (iii) holds for S then S is bounded. For if S is not bounded there is a sequence in S such that each term of the sequence is separated from every other term by a fixed amount. Evidently such a sequence has no cluster points.

Since S is bounded, there is a closed square, σ, containing S in its interior. Partition σ into four closed squares with equal sides. If S *cannot* be covered by a finite number of sets from F, then for at least one of these smaller squares, say σ_1, $S \cap \sigma_1$ cannot be covered by a finite number of sets from F. Partition σ_1 into four closed equal squares. As before, for at least one of these new smaller squares, say σ_2, $S \cap \sigma_2$ cannot be covered by a finite number of sets from F. Continuing this process, we obtain a sequence of squares $(\sigma_i)_{i=1}^{\infty}$ such that none of the sets $(S \cap \sigma_i)_{i=1}^{\infty}$, can be covered by a finite number of sets from F. There are points $s_i \in S \cap \sigma_i$ such that for each i the point s_i is not equal to any s_k for $k < i$. Since the diagonal of σ_{i+1} is one-half the length of the diagonal of σ_i, it is easy to see that the sequence $(s_i)_{i=1}^{\infty}$ is a Cauchy sequence in S. By (0.12.iv) it converges, say $\lim_{i \to \infty} s_i = s_0$. Since s_0 is the only cluster point of the infinite set $\{s_i\}_{i=1}^{\infty}$, it is in S by (iii). Hence we have $s_0 \in U$ for some U in F. Since U contains a disk which contains s_0 and the diameter of σ_i goes to zero as i goes to ∞, we have $\sigma_i \cap S \subset U$ for all sufficiently large *i*. This contradicts the fact that $\sigma_i \cap S$ is not covered by any finite number of sets from F.

(iv) $\Rightarrow$ (v). Let E be as in (v). Let F = $\{C \backslash E \mid E \in$ E$\}$. Hence F is a family of open sets. If $S \cap [\cap_E E] = \varnothing$ holds, then

$$S \subset C \backslash \left[\bigcap_E E \right] = \bigcup_E (C \backslash E) = \bigcup_F F$$

holds; that is, F is a cover of S by open sets. By (iv), there are sets $(E_n)_{n=1}^N$ in E such that $S \subset \cup_{n=1}^N (C \backslash E_n)$. Hence

$$\bigcap_{n=1}^N (S \cap E_n) = S \cap \left[\bigcap_{n=1}^N E_n \right] = \varnothing$$

holds, contradicting the assumption in (v).

(v) $\Rightarrow$ (i). If S is not bounded, there is a sequence $(s_n)_{n=1}^\infty$ in S such that $|s_n| > n$ and $|s_{n+1} - s_n| > 1$ for all n. The sets $E_k = \{s_n \mid n \geq k\}$ are closed and have void intersection $\cap_{k=1}^\infty E_k$, but $\cap_{k=1}^m E_k \neq \phi$ for every m. This proves that S is bounded if (v) holds. Next we prove that S is closed. If s is a cluster point of S, there is a sequence (s_n) in S with $\lim_{n \to \infty} s_n = s$. If $E_k = \overline{\{s_n \mid n \geq k\}}$, then each E_k is closed and $E_k \cap S$ is nonvoid. Hence the set $\cap_{k=1}^\infty E_k \cap S$ is nonvoid and it can only contain s.

All of the equivalent definitions of compactness are useful; (i) is probably the easiest to visualize but sometimes difficult to apply.

The following definition and theorem play important roles in mathematics.

1.12 DEFINITION

Let $S \subset$ C. A function $f: S \to$ C is *uniformly continuous on* S if for every $\varepsilon > 0$ there is a $\delta > 0$ such that

$$z, w \in S \quad \text{and} \quad |z - w| < \delta \text{ imply } |f(z) - f(w)| < \varepsilon.$$

[Compare with (1.2).]

1.13 CONTINUITY AND COMPACTNESS

A continuous function on a compact set $S \subset$ C is uniformly continuous on S.

PROOF. Let $\varepsilon > 0$. For every $x \in S$ there is a δ_x such that

$$z \in D(\delta_x, x) \text{ implies } |f(z) - f(x)| < \varepsilon/2.$$

The family $\{D(\tfrac{1}{2}\delta_x, x) \mid x \in S\}$ covers S, and so has a finite subcover $\{D(\tfrac{1}{2}\delta_y, y) \mid y \in F\}$, F a finite subset of S. Let $\delta = \min \{\tfrac{1}{2}\delta_y \mid y \in F\}$. Let

z, w in S satisfy $|z - w| < \delta$. There is a $y \in F$ such that $z \in D(\frac{1}{2}\delta_y, y)$. Hence we have $|w - y| \leq |w - z| + |z - y| < \delta_y$ and

$$|f(z) - f(w)| \leq |f(z) - f(y)| + |f(y) - f(w)| < \varepsilon.$$

1.14 THEOREM

A real-valued continuous function f on a compact set attains its maximum and minimum, that is, there are points $s, t \in S$ such that $f(s) = \text{lub}$ $\{f(x) \mid x \in S\}$ and $f(t) = \text{glb} \{f(x) \mid x \in S\}$. (lub and glb are defined in Appendix A.)

PROOF.   There is a sequence $(s_n)_{n=1}^{\infty}$ in S such that $\lim_{n \to \infty} f(s_n) = \text{lub}$ $\{f(x) \mid x \in S\}$, and this sequence contains a subsequence which converges to a point $s \in S$ (1.11.ii). By the continuity of f, we have

$$f(s) = \text{lub}\{f(x) \mid x \in S\}.$$

Similarly, there is a $t \in S$ with $f(t) = \text{glb}\{f(x) \mid x \in S\}$.

We will make frequent use of (1.13) and (1.14), most prominently in integration. To illustrate the use of these ideas we will use (1.14) to prove the Fundamental Theorem of Algebra.

Recall that there are n distinct solutions of the polynomial equation $z^n - a = 0$, $a \neq 0$; the n n^{th} roots of a. It is a fact that every polynomial equation

$$z^n + a_{n-1}z^{n-1} + \cdots + a_0 = 0$$

with complex coefficients has a complex solution, and counting multiple roots it has exactly n solutions. This means that $\mathbf{C}$ is *algebraically complete*; every polynomial equation with coefficients in $\mathbf{C}$ has a root in $\mathbf{C}$. The same assertion is not true with $\mathbf{R}$ replacing $\mathbf{C}$, as we have seen. We give an elementary proof of this fundamental theorem below, based on (1.14), the binomial expansion, and roots of complex numbers. There are easier proofs, which we will give in Chapter 4, based on Cauchy's theorem.

*1.15 FUNDAMENTAL THEOREM OF ALGEBRA

Let $P(z) = a_0 + a_1 z + \cdots + z^n$ be a polynomial with complex coefficients, $n \geq 1$. The equation $P(z) = 0$ has a solution in the complex numbers.

PROOF.   Write $P(z) = z^n(1 + a_{n-1}/z + \cdots + a_0/z^n)$ $(z \neq 0)$. Since $\lim_{|z| \to \infty}$ $|z^{-1}| = \lim_{|z| \to \infty} |z^{-2}| = \cdots = \lim_{|z| \to \infty} |z^{-n}| = 0$, the equality $\lim_{|z| \to \infty}$ $|P(z)| = \infty$ holds.   Let $\alpha = \text{glb}\{|P(z)| \mid z \in \mathbf{C}\}$ and let r be such that $|P(z)| > \alpha + 1$ if $|z| > r$; hence, $\alpha = \text{glb}\{|P(z)| \mid z \in \overline{D(r, 0)}\}$. By (1.14), there is a $z_0 \in \overline{D(r, 0)}$ such that $|P(z_0)| = \alpha$. Thus $|P(z_0)|$ is the absolute minimum of P. We claim that $P(z_0) = 0$; to prove it, assume that $|P(z_0)| > 0$.

[The idea of the following proof is that we can move in any direction we like from $P(z_0)$ by moving in an intelligently chosen direction from z_0.] For complex ξ,

$$P(z_0 + \xi) = P(z_0) + \sum_{k=1}^{n} a_k \sum_{j=1}^{k} \binom{k}{j} z_0^{k-j}\xi^j$$

$$= P(z_0) + b_1\xi + b_2\xi^2 + \cdots + b_n\xi^n$$

holds for constants b_i. If $b_1 = b_2 = \cdots = b_n = 0$, then P would be constant, which it is not. Let m be the smallest integer such that $b_m \neq 0$; let $b = b_m$. Hence we have

$$P(z_0 + \xi) = P(z_0) + b\xi^m + Q(\xi)\xi^{m+1}$$

where $Q(\xi) = b_{m+1} + b_{m+2}\xi + \cdots + b_n\xi^{n-m-1}$ if $m < n$ and $Q(\xi) = 0$ if $m = n$. Select ζ so that $b\zeta^m = -P(z_0)$; this is possible, for $-P(z_0)/b$ is a fixed nonzero complex number and so has m m^{th} roots (0.9). For any positive ρ, we have

$$P(z_0 + \zeta\rho) = P(z_0) - P(z_0)\rho^m + (\zeta\rho)^{m+1}Q(\zeta\rho)$$

$$= P(z_0)(1 - \rho^m) + (\zeta\rho)^{m+1}Q(\zeta\rho).$$

The function of ρ given by $\zeta^{m+1}Q(\zeta\rho)$ is bounded for the fixed ζ and $0 < \rho < 1$, say by A. For $0 < \rho < 1$, the inequality

$$|P(z_0 + \zeta\rho)| \leq |P(z_0)|(1 - \rho^m) + \rho^{m+1}A = |P(z_0)| + \rho^m\left[\rho - \frac{P(z_0)}{A}\right]A$$

is valid. It follows that $|P(z_0 + \zeta\rho)| < |P(z_0)|$ if $0 < \rho < |P(z_0)|/A$, contradicting the minimality of $|P(z_0)|$.

EXERCISES

TYPE 1

1. Determine which of the following subsets of $\mathbf{C}$ are open, which are closed, and which are neither open nor closed.
   (a) $S = \{z \mid \operatorname{Im} z > 0\}$
   (b) $S = \{z \mid \operatorname{Im} z \geq 0\}$
   (c) $T = \{z \mid |z| = 1\}$
   (d) $S = \{z \mid \operatorname{Im} z > 0\} \cup \{0\}$
   (e) $S = \{z \mid |z| > 1\}$
   (f) $\mathbf{R}$
   (g) $\{x + iP(x) \mid x \in R\}$, P a fixed polynomial with real coefficients
   (h) $\{x + iy \mid |x| < 1 \text{ and } |y| \leq 1\}$
   (i) $D(1, 0)$.

   Find the cluster points of each set and describe the closure of each. Which sets are compact? Which have compact closures?

2. Prove (1.8.ii).

3. Exhibit an open cover of each of the following sets which has no finite open subcover.
   (a) $D(1, 0)$
   (b) **C**
   (c) $\{z = x + iy \mid 0 \le xy < 1 \text{ and } \max [|x|, |y|] < 1\}$
   (d) $\{z = x + iy \mid \max [|x|, |y|] < 1\}$.

4. Show that $f(z) = 1/z$ is not uniformly continuous on $D(1, 0)\backslash\{0\}$ and that $f(z) = z^2$ is not uniformly continuous on **C**.

5. Let $(f_n)_{n=1}^{\infty}$ be a sequence of complex-valued continuous functions on a compact set $S \subset \mathbf{C}$, and suppose that $(f_n)_{n=1}^{\infty}$ converges uniformly on S to f. Prove that f is uniformly continuous.

6. Let U be the family of all open subsets of **C**. Show that U satisfies the conditions (i)–(iii).

   (i) $\emptyset \in$ U.
   (ii) If $U \in$ U and $V \in$ U, then $U \cap V \in$ U.
   (iii) If $\{U_\alpha \mid \alpha \in I\}$ is a family in U, then $\cup_{\alpha \in I} U_\alpha \in$ U.

   Give an example of a sequence $\{U_n\}_{n=1}^{\infty}$ of sets in U such that $\cap_{n=1}^{\infty} U_n \notin$ U.

7. Let C be the family of closed subsets of **C**. Translate the results of the above problem into properties of C.

8. Let T be a subset of **C**. Prove that $\overline{T}$ is the smallest closed set containing T.

9. Let f be a continuous complex-valued function on a compact set $S \subset \mathbf{C}$. Use (1.14) to prove that f is bounded.

TYPE II

1. Open and closed sets on the line **R** are defined as in **C**; disks become intervals. Prove that every open set in **R** has a unique expression as the union of a pairwise disjoint countable family of intervals.

2. Let $(x_n)_{n=1}^{\infty}$ be a bounded sequence of real numbers such that $x_n \ne x_m$ if $n \ne m$. Prove that $l = \overline{\lim}_{n \to \infty} x_n$ is the largest cluster point of $\{x_n \mid n \ge 1\}$ and that $s = \underline{\lim}_{n \to \infty} x_n$ is the smallest.

3. Let U be an open set in **C**. Prove that a complex-valued function f on U is continuous on U if and only if the inverse image set $f^{-1}(V)$ is open for every open set $V \in \mathbf{C}$. [The definition of $f^{-1}(V)$ is $\{u \in U \mid f(u) \in V\}$.]

4. If a sequence of functions $(f_n)_{n=1}^{\infty}$ on a compact set S converges monotonically $(f_{n+1}(z) \ge f_n(z)$ for all $z \in S$ and $n)$ to f, the convergence is then uniform.

5. Let Y be dense in the set S. Suppose that $f: Y \to \mathbf{C}$ is uniformly continuous.

   (a) Show that $\lim_{y \to s, \, y \in Y} f(y) = F(s)$ exists for all $s \in S$.

   (b) Show that F is uniformly continuous on S.

6. Let S be a noncompact nonvoid subset of $\mathbf{C}$. Construct a function f which is continuous on S but not uniformly continuous on S. Construct a real-valued continuous function on S for which the conclusion of (1.14) fails.

1.16 COMPLEX DIFFERENTIATION

Most of the concepts discussed up to this point are not particularly intrinsic to complex analysis. They deal with fundamental topological and convergence features of Euclidean space $\mathbf{R}^2$. Essentially the same results hold for arbitrary Euclidean spaces $\mathbf{R}^n$, and even for more general spaces. The sharp distinction between "real analysis" and "complex analysis" results from the properties of the complex derivative.

(i) *Definition.* Let U be an open set in $\mathbf{C}$ and let $z_0 \in U$. If

$$\lim_{\substack{z \to z_0 \\ z \in U \setminus \{z_0\}}} \frac{f(z) - f(z_0)}{z - z_0} = f'(z_0)$$

exists, then f is said to have the *derivative* $f'(z_0)$ at z_0. If f has a derivative at every point of U, then f is called *analytic* (or *holomorphic*) on U.

   The definition of f' is formally the same as in the real case. Notice that we have to be able to divide in order to form the difference quotients in the definition of f'. In $\mathbf{R}^2$, the complex numbers provide a method of dividing elements of $\mathbf{R}^2$ by elements of $\mathbf{R}^2$. Such division is not possible in higher dimensional spaces (except in $\mathbf{R}^4$, which can be equipped with a multiplication which makes it a noncommutative field, the quaternions).

(ii) *Analyticity implies continuity.* If f has a derivative at $z_0 \in U$, then there is a $\delta > 0$ such that

$$\left| \frac{f(z) - f(z_0)}{z - z_0} - f'(z_0) \right| < 1$$

if $|z - z_0| < \delta$, $z \neq z_0$. Thus

$$|f(z) - f(z_0)| < |z - z_0|[1 + |f'(z_0)|] \quad \text{if} \quad |z - z_0| < \delta;$$

from this inequality, continuity of f at z_0 follows easily.

1.17 PROPERTIES OF THE DERIVATIVE

Let U be an open subset of $\mathbf{C}$. The following implications hold.

(i) f, g analytic on $U \Rightarrow f + g$ is analytic on U and $(f + g)' = f' + g'$.

(ii) f, g analytic on $U \Rightarrow fg$ is analytic on U and $(fg)' = gf' + fg'$.

(iii) f, g analytic on U and $g(z) \neq 0$ for $z \in U \Rightarrow f/g$ is analytic on U.

(iv) f analytic on U, $f(U)$ open, and g analytic on $f(U)$ implies $g \circ f$ is analytic on U and $(g \circ f)' = [g' \circ f] \cdot f'$.

(v) Suppose that f is analytic in a disk $D(\varepsilon, a) \subset U$. If $f'(z) = 0$ for all z in D, then f is a constant on $D(\varepsilon, a)$.

Properties (i)–(iii) are consequences of (0.10.iv) and (1.2). The proofs are as in the real case. We give a proof of the chain rule (iv). (In Chapter 5 we will see that the analyticity of f actually implies that $f(U)$ is open, provided that f is not constant on any disk contained in U.) Fix $z_0 \in U$, and let $\lim_{n \to \infty} z_n = z_0$, $z_n \in U$ and $z_n \neq z_0$. Write

(a)
$$\frac{g \circ f(z_n) - g \circ f(z_0)}{z_n - z_0} = \begin{cases} \dfrac{g(f(z_n)) - g(f(z_0))}{f(z_n) - f(z_0)} \cdot \dfrac{f(z_n) - f(z_0)}{z_n - z_0}, & f(z_n) \neq f(z_0); \\ 0, & \text{if } f(z_n) = f(z_0). \end{cases}$$

If there is an n_0 such that $f(z_n) \neq f(z_0)$ for $n \geq n_0$, then the top expression in (a) is valid for $n \geq n_0$, the limit of the left side exists by the continuity and analyticity of f at z_0, the analyticity of g at $f(z_0)$, and (0.10.iv); and the limit is $g'(f(z_0)) \cdot f'(z_0)$. If there is a subsequence of (z_n) for which $f(z_n) = f(z_0)$, then $f'(z_0) = 0$; (a) shows that the limit of the left side of (a) is also 0. Hence, $(g \circ f)'(z_0)$ exists and equals $g'(f(z_0)) \cdot f'(z_0)$ in all cases.

PROOF OF (v).   Fix $z \in D$. Define a function F on $[0, 1]$ by

$$F(t) = f(a + t(z - a)).$$

Observe that F is a differentiable function of the real variable t on $]0, 1[$, and hence we have $F'(t) = (z - a)f'(a + t(z - a))$ for all $t \in]0, 1[$. Hence $F'(t) = 0$, and F is constant on $[0, 1]$. In particular, $F(0) = F(1)$, so that we have $f(a) = f(z)$. Since z is arbitrary, f is identically equal to $f(a)$.

1.18 REMARKS

A complex-valued function f defined on an open subset U of $\mathbf{C}$ can be written uniquely as

$$f(z) = u(z) + iv(z),$$

where u and v are real-valued functions defined on U; u and v are the real and imaginary parts of f. Writing $z = x + iy$, we see that f, u, and v are all functions of the two real variables x and y, for $x + iy = z \in U$. Accordingly, we can ask if they have partial derivatives with respect to x and y, second partials, if they are differentiable, and so forth. The partial derivative of f with respect to x at $x_0 + iy_0 = z_0$ is

$$\frac{\partial f}{\partial x}(z_0) = \lim_{x \to x_0} \frac{f(x + iy_0) - f(x_0 + iy_0)}{x - x_0},$$

if the limit exists. By (0.10.iii), $\partial f/\partial x\,(z_0)$ exists if and only if $\partial u/\partial x\,(z_0)$ and $\partial v/\partial x\,(z_0)$ exist, and then

$$\frac{\partial f}{\partial x}(z_0) = \frac{\partial u}{\partial x}(z_0) + i\frac{\partial v}{\partial x}(z_0).$$

In the next three sections, we will study the relation between the analyticity of f and its partials with respect to x and y.

1.19 CAUCHY-RIEMANN EQUATIONS

If $f = u + iv$ (u and v real) is analytic on an open set U, then for $z_0 \in U$

$$f'(z_0) = \lim_{z \to z_0} \frac{f(z) - f(z_0)}{z - z_0}$$

can be computed by letting z approach z_0 from any direction. In particular going parallel to the x-axis we have

$$f'(z_0) = \lim_{x \to x_0} \frac{f(x + iy_0) - f(x_0 + iy_0)}{x - x_0}$$

and going parallel to the y-axis we have

$$f'(z_0) = \lim_{y \to y_0} \frac{f(x_0 + iy) - f(x_0 + iy_0)}{i(y - y_0)}.$$

The limit as $x \to x_0$ is also the definition of $(\partial f/dx)(z_0)$. The limit as $y \to y_0$ differs from the definition of $(\partial f/\partial y)(z_0)$ by a factor $1/i$, that is, $-i$. Thus the equality

$$f'(z_0) = \frac{\partial f}{\partial x}(z_0) = -i\frac{\partial f}{\partial y}(z_0)$$

holds at every point z_0 in U. Equating real and imaginary parts and expressing the equality in terms of u and v, we obtain

(i) $\quad \dfrac{\partial u}{\partial x} = \dfrac{\partial v}{\partial y}\,; \dfrac{\partial v}{\partial x} = -\dfrac{\partial u}{\partial y} \qquad$ (Cauchy-Riemann equations).

The real and imaginary parts of analytic functions must satisfy the Cauchy-Riemann equations. Thus we see that analyticity is much more restrictive than simply the existence of partial derivatives; an "arbitrary" pair of real-valued functions of x and y is unlikely to satisfy (i).

1.20 DIFFERENTIABLE FUNCTIONS

Let U be an open set in $\mathbf{R}^2$ and let $w : U \to \mathbf{R}$. Suppose that $\partial w/\partial x$ and $\partial w/\partial y$ exist on U. For $z = x + iy$ and $z_0 = x_0 + iy_0$ in U, write

$$w(z) - w(z_0) = \left[\frac{\partial w}{\partial x}(z_0) \right] \Delta x + \left[\frac{\partial w}{\partial y}(z_0) \right] \Delta y + R(\Delta z, z_0)$$

where $\Delta z = \Delta x + i\Delta y = z - z_0$. If the remainder R satisfies

$$\lim_{\Delta z \to 0} \left| \frac{R(\Delta z, z_0)}{\Delta z} \right| = 0,$$

then w is said to be *differentiable at* z_0, with *differential dw* defined by

(i) $dw(\Delta x, \Delta y) = \dfrac{\partial w}{\partial x}(z_0) \, \Delta x + \dfrac{\partial w}{\partial y}(z_0) \, \Delta y.$

The definition of differentiability extends to $f = u + iv$ by simply demanding that u and v be differentiable. The differential df is then $du + idv$, and (i) holds for f.

Notice that a function f which is differentiable at z_0 need not have a (complex) derivative at z_0. For example, let $f(z) = x - iy$ ($z = x + iy$). We have $\partial f/\partial x = 1$ and $\partial f/\partial y = -i$. Hence we have

$$(x - x_0) - i(y - y_0) = f(z) - f(z_0) = \frac{\partial f}{\partial x} \Delta x + \frac{\partial f}{\partial y} \Delta y$$

for any z_0. Thus we have $R(\Delta x, \Delta y, z_0) = 0$, and f is differentiable at every point. However, to compute $f'(z_0)$ we must examine

$$\frac{\bar{z} - \bar{z}_0}{z - z_0} = e^{-2i \, \text{Arg} \, (z - z_0)}$$

which clearly has no limit as z goes to z_0. The next theorem tells exactly which differentiable functions are analytic.

1.21 THEOREM

If $f = u + iv$ is differentiable on U (open in $\mathbf{C}$) and u and v satisfy the Cauchy-Riemann equations, then f is analytic on U.

PROOF. Write

$$\frac{f(z) - f(z_0)}{z - z_0} = \frac{\partial f}{\partial x}(z_0)\frac{\Delta x}{\Delta z} + \frac{\partial f}{\partial y}(z_0)\frac{\Delta y}{\Delta z} + \frac{R(\Delta z, z_0)}{\Delta z}$$

$$= \frac{\partial f}{\partial x}(z_0)\left[\frac{\Delta z + \overline{\Delta z}}{2\,\Delta z}\right] + \frac{\partial f}{\partial y}(z_0)\left[\frac{\Delta z - \overline{\Delta z}}{2i\,\Delta z}\right] + \frac{R(\Delta z, z_0)}{\Delta z}$$

$$= \frac{1}{2}\left[\frac{\partial f}{\partial x}(z_0) - i\frac{\partial f}{\partial y}(z_0)\right] + \frac{1}{2}\left[\frac{\partial f}{\partial x}(z_0) + i\frac{\partial f}{\partial y}(z_0)\right]\frac{\overline{\Delta z}}{\Delta z} + \frac{R(\Delta z)}{\Delta z}.$$

The second term of the last expression is 0 because the Cauchy-Riemann equations are satisfied; the third term goes to zero as $\Delta z \to 0$, by differentiability.

A function u on an open set U is said to be *harmonic* on U if u has first and second partial derivatives and $\partial^2 u/\partial x^2 + \partial^2 u/\partial y^2 = 0$.

1.22 COROLLARY

Let $f = u + iv$ be analytic on an open set U, and suppose that u and v are twice continuously differentiable. Then u and v are harmonic functions.

PROOF. By the Cauchy-Riemann equations and continuity of the second partials, we have

$$\frac{\partial^2 u}{\partial x^2} = \frac{\partial^2 v}{\partial x\,\partial y} = \frac{\partial^2 v}{\partial y\,\partial x} = -\frac{\partial^2 u}{\partial y^2}.$$

Hence we have $\partial^2 u/\partial x^2 + \partial^2 u/\partial y^2 = 0$, and this is the defining equality for harmonicity of u

The subject of Chapter 6 is harmonic functions. (See also Type II Exercises on p. 103.)

EXERCISES

TYPE I

1. Use the definition of the derivative to compute $f'(z)$ for the following functions f; indicate the values of z for which $f'(z)$ exists.
   (a) $f(z) = z^2$
   (b) $f(z) = 1/z$
   (c) $f(z) = z^2/(1 - z)$
   (d) $f(z) = \sum_{k=0}^{N} a_k z^k$, $a_N \neq 0$ \qquad (complex polynomial)
   (e) $f(z) = (\sum_{k=0}^{N} a_k z^k)/(\sum_{k=0}^{M} b_k z^k)$, $a_N b_M \neq 0$ \qquad (complex rational function)
   (f) $f(z) = z\bar{z}$.

2. Each complex number $z \neq 0$ has exactly two square roots [see (0.9)]. Write $z = |z|e^{it}$ ($t = \text{Arg}(z)$), and let $S(z) = \sqrt{|z|}e^{it/2}$. Find the set $U = \{z \mid S \text{ is continuous}\}$. Use the definition of S' to show that S' exists on U and find $S'(z)$.

3. Which of the following functions of $z = x + iy$ are analytic? Which are differentiable?

   (a) $x^2 + iy$ (f) $x^3 - 3xy^2 + i(3x^2y - y^3)$
   (b) $\sqrt{x^2 + y^2}$ (g) $x - iy$ ($= \bar{z}$)
   (c) $(x^2 - y^2) + 2ixy$ (h) $\sin x + i\cos y$
   (d) $(x^2 - y^2) + 4ixy$ (i) $\sin x - i\cos y$.
   (e) $(x^2 - y^2) - 2ixy$

4. Express the Cauchy-Riemann equations in polar coordinates.

5. If $f = u + iv$ is analytic on an open set U, and $0 \notin U$, use Exercise 4 to express the harmonicity of u in polar coordinates.

6. Define the function Log by $\text{Log}(z) = \log|z| + i\text{Arg}(z)$ ($z \neq 0$). Show that Log is analytic on $\{z \in \mathbf{C} \mid |z| \neq 0 \text{ and } z \text{ is not on the negative real axis}\}$.

7. Verify directly the Cauchy-Riemann equations for a polynomial function.

8. *The operators $\partial f/\partial z$ and $\partial f/\partial \bar{z}$.* The Cauchy-Riemann equations are succinctly expressed by $[\partial/\partial x + i(\partial/\partial y)](f) = 0$; and then, f' exists and $f' = \frac{1}{2}[\partial/\partial x - i(\partial/\partial y)](f)$. Let $\partial/\partial \bar{z} \equiv \partial/\partial x + i(\partial/\partial y)$ and $\partial/\partial z \equiv \partial/\partial x - i(\partial/\partial y)$.

   These are *formal* definitions of the operators $\partial/\partial \bar{z}$ and $\partial/\partial z$. Make the substitutions $x = (z + \bar{z})/2$ and $y = (z - \bar{z})/(2i)$, treat z and $\bar{z}$ as independent variables, take partials with respect to z and $\bar{z}$, and see that your result is consistent with the definition. An even more succinct expression of the Cauchy-Riemann expressions is $\partial f/\partial \bar{z} = 0$.

9. Use the Cauchy-Riemann equations for an analytic function to find some other expressions for f'; we already have $f' = \frac{1}{2}(\partial f/\partial x - i\,\partial f/\partial y)$.

10. Let $u(x, y) = e^x\cos y$ and $v(x, y) = e^x\sin y$. Show that $u + iv$ is analytic on $\mathbf{C}$ by verifying the hypothesis of (1.21).

11. Let $f(z) = (az + b)/(cz + d)$ where a, b, c, and d are constants. Find a necessary and sufficient condition on a, b, c, and d that assures that $f'(z) \neq 0$.

12. Let f be a nonconstant real-valued function defined on an open (complex) set U. Prove that f is not analytic on U.

13. Let f be analytic on an open set U. Let $\tilde{f}$ be defined by $\tilde{f}(z) = \overline{f(\bar{z})}$, $z \in U$. For each point $a \in U$, prove that either

(i) f is constant on a disk containing a,

or

(ii) $\bar{f}$ is not analytic on any open set containing a.

Also prove that either (i) or

(iii) $|f|$ is not analytic on any open set containing a.

14. Recall that a real-valued function u with continuous first and second partial derivatives on an open set U of $\mathbf{R}^2$ has a critical point at $(x_1, y_1) \in U$ if

$$\frac{\partial u}{\partial x}(x_1, y_1) = \frac{\partial u}{\partial y}(x_1, y_1) = 0.$$

Use this fact to show that the real and imaginary parts of an analytic function have the same critical points.

15. Let $f(z) = (ax + by) + i(cx + dy)$, $z = x + iy$ and a, b, c, and d real constants. For fixed a and b, find c and d such that f is analytic on $\mathbf{C}$.

16. Let $u(z) = x^3 - 3xy^2$, $z = x + iy$. Find a real-valued function v on $\mathbf{C}$ such that $u + iv$ is analytic on $\mathbf{C}$.

17. Let f be analytic and not constant on an open disk U. Let $\tilde{U} = \{\bar{z} \mid z \in U\}$. Define F on $\tilde{U}$ by $F(z) = f(\bar{z})$. Prove that F is not analytic on $\tilde{U}$.

18. Let R be a polynomial in x, y. Hence

$$R(x, y) = \sum_{m, n \in F} c_{mn} x^m y^n$$

for a finite set F and complex coefficients c_{mn}. Show that R satisfies the Cauchy-Riemann equations if and only if there is a polynomial P of z such that $P(z) = R(x, y)$ $(z = x + iy)$ for all z.

19. Let $f \in \mathbb{C}^1(U)$, U an open set in $\mathbf{C}$, f complex-valued. (The symbol $\mathbb{C}^1(U)$ denotes the set of all complex-valued functions on U which have continuous first partial derivatives throughout U.) Let $z = x + iy \in U$ and let $\Delta z = \Delta x + i\Delta y$ be small enough so that $z + \eta \in U$ if $|\eta| \leq |\Delta z|$. Show that

$$\frac{\partial f}{\partial x}\Delta x + \frac{\partial f}{\partial y}\Delta y = \frac{1}{2}\left[\frac{\partial f}{\partial z}\Delta z + \frac{\partial f}{\partial \bar{z}}\overline{\Delta z}\right].$$

TYPE II

1. (a) Let $f = u + iv$ be analytic on an open subset U of $\mathbf{C}$. Show that if a curve determined by the condition $u =$ constant intersects a curve determined by the condition $v =$ constant, then their tangents are perpendicular at the point of intersection.

(b) Suppose that f has an inverse function g $(g(f(z)) = z)$ which is analytic on $f(U)$. Assume that $f(U)$ is open. What can you say about the curves in $f(U)$ determined by the two families of lines $x = $ constant and $y = $ constant?

2. Let v be a real-valued function on a disk $D(r, a) = D$ such that $(1/v) + iv$ is analytic on D. Prove that v is constant on D.

CHAPTER 2

Power Series and the Exponential Function

2.1 INTRODUCTION

Perhaps the conceptually simplest analytic functions are the polynomial functions:

$$P(z) = a_0 + a_1 z + \cdots + a_n z^n.$$

If we let the sums for polynomials become infinite, we get the analytic functions defined by power series; those of the form

$$f(z) = \sum_{n=0}^{\infty} a_n z^n \quad \text{or} \quad g(z) = \sum_{n=0}^{\infty} a_n (z - z_0)^n.$$

The first series is a power series with center 0, and the second with center z_0. Results about f are clearly transferable to results about g, so we will concentrate on power series centered at the origin. Power series are special cases of limits of infinite sequences of functions, the terms of which are $S_n(z) = \sum_{k=0}^{n} a_k z^k$ [see (0.14) and (1.4)]. Power series play a more important role in complex analysis than in real analysis because every analytic function has a local power series representation. That is, if f is analytic on an open set U and $z_0 \in U$, then there is an $r > 0$ and a sequence $(a_n)_{n=0}^{\infty}$ such that $D(r, z_0) \subset U$ and $f(z) = \sum_{n=0}^{\infty} a_n (z - z_0)^n$ whenever $z \in D(r, z_0)$. This fact is one of the important consequences of *Cauchy's theorem*, the subject of Chapter 4. In this chapter, we concentrate on power series themselves. Suppose that for some z_0 we know that $\sum_{n=0}^{\infty} a_n z_0^n$ converges. The terms of the series are then bounded, say $|a_n z_0^n| < M$ for all n. If $|z| < |z_0|$, we have $|z| = r|z_0|$ for $0 \leq r < 1$ and $|a_n z^n| = |a_n z_0^n| r^n < M r^n$. Thus by comparison with the geometric series $M \sum_{n=1}^{\infty} r^n$ we decide that $\sum_{n=0}^{\infty} a_n z^n$ converges absolutely for all z in the disk $D(|z_0|, 0)$. It is natural to ask what is the largest disk in which $\sum_{n=0}^{\infty} a_n z^n$ converges. The next theorem answers this question.

2.2 THEOREM

Let $(a_n)_{n=0}^{\infty}$ be a sequence in $\mathbf{C}$. If $\mu = \overline{\lim}_{n \to \infty} \sqrt[n]{|a_n|}$ satisfies $0 < \mu < \infty$, then let

$$\rho = \frac{1}{\overline{\lim_{n \to \infty}} \sqrt[n]{|a_n|}}.$$

If $\mu = 0$, let $\rho = \infty$; if $\mu = \infty$, let $\rho = 0$. Let $0 \le r < \rho$. Then

(i) the series $\sum_{n=0}^{\infty} a_n z^n$ converges uniformly and absolutely on $D(r, 0)$,

(ii) the series $\sum_{n=0}^{\infty} a_n z^n$ does not converge if $|z| > \rho$.

PROOF.   By the root test (0.13.vii), $\sum_{n=0}^{\infty} a_n z^n$ converges absolutely if $\overline{\lim}_{n \to \infty}$ $\sqrt[n]{|a_n z^n|} < 1$; that is, if $|z| < (\overline{\lim}_{n \to \infty} \sqrt[n]{|a_n|})^{-1}$. In particular, $\sum_{n=0}^{\infty} |a_n r^n|$ converges, so, by the Weierstrass M-test (1.5) the series $\sum_{n=0}^{\infty} a_n z^n$ converges uniformly on $D(r, 0)$. (Take $M_n = |a_n| r^n$.)

If $|z| > \rho$, then we have $\overline{\lim}_{n \to \infty} \sqrt[n]{|a_n z^n|} > 1$ and so the series cannot converge, as the general term does not go to zero.

The number ρ is the *radius of convergence* of the power series $\sum_{n=0}^{\infty} a_n z^n$; $D(\rho, 0)$ is its *disk of convergence*. The case $\rho = \infty$ is possible. If $\rho = \infty$, then the series converges for all z and the function $f(z) = \sum_{n=0}^{\infty} a_n z^n$ is said to be an entire function. We use the theory developed in sections (1.4) and (1.5) to show that $f(z) = \sum_{n=0}^{\infty} a_n z^n$ is continuous on $D(\rho, 0)$. First, recall that $f(z) = \lim_{k \to \infty} \sum_{n=0}^{k} a_n z^n$ and for each k the polynomial $\sum_{n=0}^{k} a_n z^n$ is continuous. These polynomials converge uniformly to f on each $D(r, 0)$ for $r < \rho$, and so by (1.4) f is continuous on each $D(r, 0)$. Since each z in $D(\rho, 0)$ lies in some $D(r, 0)$ for $r < \rho$, f is continuous at each z in $D(\rho, 0)$. We show in (2.4) that $\sum a_n z^n$ is also analytic and can be differentiated term-by-term.

If $\lim_{n \to \infty} \sqrt[n]{|(a_n)|}$ exists, then, of course, it is equal to $\overline{\lim}_{n \to \infty} \sqrt[n]{|a_n|}$ and in this case the formula

$$\rho = \frac{1}{\lim_{n \to \infty} \sqrt[n]{|a_n|}}$$

holds for the radius of convergence. We now give another expression for ρ, valid in some cases.

2.3 THEOREM

If $(a_n)_{n=1}^{\infty} \subset \mathbf{C} \backslash \{0\}$ and $\rho = \lim_{n \to \infty} (|a_n|/|a_{n+1}|)$ exists (∞ allowed), then ρ is the radius of convergence of $\sum_{n=0}^{\infty} a_n z^n$.

PROOF. If $|z| < \rho$, then $|z|/\rho < 1$ and hence $\overline{\lim}_{n \to \infty} |(a_{n+1}z^{n+1})/(a_n z^n)| < 1$. Convergence follows from the ratio test. Similarly, if $|z| > \rho$, then $\sum a_n z^n$ diverges. Thus ρ must be the radius of convergence.

For example, to compute the radius of convergence of $\sum_{n=0}^{\infty} z^n/n!$ we simply write

$$\lim_{n \to \infty} \left| \frac{1}{n!} \middle/ \frac{1}{(n+1)!} \right| = \lim_{n \to \infty} (n+1) = \infty.$$

Thus the series $\sum_{n=0}^{\infty} z^n/n!$ converges for all z. This series defines one of the most important functions in both applied and pure mathematics and we will study it extensively in subsequent sections.

2.4 THEOREM

Let $f(z) = \sum_{n=0}^{\infty} a_n z^n$ have disk of convergence $D(\rho, 0)$. Then f' exists on $D(\rho, 0)$ and

(i) $\quad f'(z) = \sum_{n=1}^{\infty} n a_n z^{n-1} = \sum_{n=0}^{\infty} (n+1) a_{n+1} z^n.$

The new series also has radius of convergence ρ. Consequently, all derivatives of f exist, and

(ii) $\qquad\qquad f^{(k)}(z) = \sum_{n=0}^{\infty} \frac{(n+k)!}{n!} a_{n+k} z^n.$

PROOF. Fix $z_0 \in D(\rho, 0)$, let $z \in D(\rho, 0)$, and write

$$\frac{f(z) - f(z_0)}{z - z_0} = \sum_{n=0}^{\infty} a_n \frac{z^n - z_0^n}{z - z_0}$$

$$= \sum_{n=0}^{\infty} a_n (z^{n-1} + \cdots + z^{n-k} z_0^{k-1} + \cdots + z_0^{n-1}).$$

If $|z_0| < r < \rho$ and $|z - z_0| < r - |z_0|$ hold, then $|z| < r$ also holds and

$$|(z^{n-1} + \cdots + z^{n-k} z_0^{k-1} + \cdots + z_0^{n-1})| < n r^{n-1}.$$

The series $\sum_{n=0}^{\infty} a_n n r^{n-1}$ converges by the root test; so, by (1.5) we have

$$\lim_{z \to z_0} \frac{f(z) - f(z_0)}{z - z_0} = \lim_{z \to z_0} \sum_{n=0}^{\infty} a_n \frac{z^n - z_0^n}{z - z_0}$$

$$= \sum_{n=0}^{\infty} \lim_{z \to z_0} a_n \frac{z^n - z_0^n}{z - z_0} = \sum_{n=0}^{\infty} n a_n z^{n-1}.$$

2.5 COROLLARY (TAYLOR FORM OF COEFFICIENTS)

If $f(z) = \sum_{n=0}^{\infty} a_n z^n$ has radius of convergence $\rho > 0$, then $a_n = f^{(n)}(0)/n!$ for $n \geq 0$.

To prove (2.5), use the equality (ii) just proved to find $f^{(n)}(0)$.

2.6 EXAMPLES

(i) To find the radius of convergence of $\sum_{n=1}^{\infty} n^\alpha z^n$ for any $\alpha \geq 0$, consider $|a_n/a_{n+1}|$. We have

$$\left| \frac{a_n}{a_{n+1}} \right| = \left(\frac{n}{n+1} \right)^\alpha.$$

The terms on the right converge to 1 and so the radius of convergence of each of the series $\sum_{n=1}^{\infty} n^\alpha z^n$ is 1, by (2.3). By (2.4), we have

$$\frac{d}{dz} \left(\sum_{n=1}^{\infty} n^\alpha z^n \right) = \sum_{n=1}^{\infty} n^{\alpha-1} z^{n-1}.$$

(ii) We have already observed that $\sum_{n=0}^{\infty} (1/n!)z^n$ has radius of convergence ∞. Using (2.4), we have

$$\frac{d}{dz} \left(\sum_{n=0}^{\infty} \frac{1}{n!} z^n \right) = \sum_{n=1}^{\infty} \frac{n}{n!} z^{n-1} = \sum_{n=1}^{\infty} \frac{1}{(n-1)!} z^{n-1} = \sum_{n=0}^{\infty} \frac{1}{n!} z^n.$$

Thus the derivative of the power series $\sum (1/n!)z^n$ is the original power series.

(iii) The power series $\sum_{n=0}^{\infty} n! \, z^n$ has radius of convergence zero, for we have $a_n/a_{n+1} = 1/n$, which, of course, converges to 0. Thus $\sum_{n=0}^{\infty} n! \, z^n$ converges only in the trivial case $z = 0$.

(iv) To find the radius of convergence ρ of $\sum_{n=1}^{\infty} (1/n^n)z^n$, it is easiest to use (2.2). We have

$$\rho = \lim_{n \to \infty} \frac{1}{\sqrt[n]{1/n^n}} = \lim_{n \to \infty} n = \infty.$$

EXERCISES

TYPE I

1. Find the radius of convergence of $\sum_{n=0}^{\infty} a_n z^n$ for the $(a_n)_{n=0}^{\infty}$ given below.
   (a) $a_n = n^n$, $n \geq 1$; $a_0 = 0$
   (b) $a_{2n} = [(-1)^n/(2n)!]$; $a_k = 0$ if k is odd
   (c) $a_{2n+1} = (-1)^n/[(2n+1)!]$; $a_k = 0$ if k is even
   (d) $a_{2n} = (\frac{2}{3})^{2n}$, $a_{2n+1} = (\frac{1}{3})^{2n+1}$
   (e) $a_n = (\log n)/n$, $n \geq 1$; $a_0 = 0$

(f) $a_n = 1/\log n$, $n \geq 2$, $a_0 = a_1 = 0$

(g) $a_n = 1/[\log (\log (n))]$, $n \geq 4$, $a_i = 0$, $i < 4$

(h) $a_n = \binom{n + k}{n}$, k fixed for all n

(i) $a_n = n^{\log n}$, $n \geq 2$; $a_0 = a_1 = 0$

(j) $a_n = n!/n^n$, $n \geq 1$; $a_0 = 0$

(k) $a_n = \begin{cases} 1/n & \text{if } n \text{ is even} \\ 1/n! & \text{if } n \text{ is odd} \end{cases}$

(l) $a_n = \begin{cases} 0 & \text{if } n \text{ is not a prime} \\ n & \text{if } n \text{ is a prime} \end{cases}$

(m) $a_n = $ coefficient of 2^{-n} in the 2-adic expansion of e.

2. Suppose that $\sum_{n=0}^{\infty} a_n z^n$ has radius of convergence $\rho > 0$. What is the radius of convergence of $\sum_{n=0}^{\infty} n^{\alpha} a_n z^n$, $\alpha > 0$? Of $\sum_{n=0}^{\infty} \sqrt[n]{n}\, a_n z^n$?

3. Suppose that $(a_n)_{n=0}^{\infty}$ is bounded. Prove that the radius of convergence ρ of $\sum_{n=0}^{\infty} a_n z^n$ satisfies $\rho \geq 1$.

4. Suppose that $f(z) = \sum_{n=0}^{\infty} a_n z^n$ has radius of convergence $\rho > 0$ and that f is an even function ($f(z) = f(-z)$ for all $z \in D(\rho, 0)$). Prove that $a_n = 0$ for odd n.

5. (a) Show that the radius of convergence of $\sum_{n=1}^{\infty} [(-1)^{n+1}/n]z^n$ is 1.

   (b) Hence the disk of convergence of the series about 1 is $D(1, 1)$; that is $\sum_{n=1}^{\infty} [(-1)^{n+1}/n](z - 1)^n$ converges if $|z - 1| < 1$. Show that the series diverges at 0.

   (c) Let $L(z) = \sum_{n=1}^{\infty} [(-1)^{n+1}/n](z - 1)^n$ on $D(1, 1)$. Show that $L'(z) = 1/z$, $z \in D(1, 1)$. [Later we will see that $L(z) = \text{Log}(z)$ on $D(1, 1)$.]

6. The power series $\sum_{n=0}^{\infty} z^n$ has radius of convergence 1; that is, it fails to converge if $|z| > 1$. Nevertheless, show that there is a function f analytic on $\mathbf{C}\backslash\{1\}$ such that $f(z) = \sum_{n=0}^{\infty} z^n$ on $D(1, 0)$.

TYPE II

1. On the real interval $]-1, 1[$, define $f(x) = \frac{1}{2}sg(x)x^2$. The definition of sg is $sg(x) = x/|x|$ if $x \neq 0$; $sg(0) = 0$.

   (a) Show that f' exists and is continuous on $]-1, 1[$, but that $f''(0)$ does not exist.

   Hence f is differentiable, but it has no derivatives of higher order at 0.

   (b) Show that f has no real power series about 0; that is, $f(x) = \sum_{n=0}^{\infty} a_n x^n$ for all x in some $]-\varepsilon, \varepsilon[$ is impossible for all sequences $(a_n)_{n=1}^{\infty}$ and all $\varepsilon > 0$.

2. Let $\sum a_n z^n$ have radius of convergence ρ and then prove that $\rho = \text{lub} \{|z| \mid \sum_{n=0}^{\infty} a_n z^n \text{ converges}\}$.

3. Let $f(z) = \sum_{n=1}^{\infty} [1/n(n + 1)]z^n$ on $\overline{D(1, 0)}$. Show that $[f(z) - f(1)]/$ $(z - 1)$ has no limit as z goes to $1(z \in D(1, 0))$, even though the series converges uniformly on $\overline{D(1, 0)}$.

2.7 IDENTITY THEOREM FOR POWER SERIES

By (2.5) two power series defining the same analytic function in some $D(\rho, 0)$ must have identical coefficients. The following theorem provides more information.

Theorem. Suppose that $f(z) = \sum_{n=0}^{\infty} a_n z^n$ and $g(z) = \sum_{n=0}^{\infty} b_n z^n$ have radii of convergence ρ and μ, $0 < \rho \leq \mu$. If there is a sequence $(w_k)_{k=1}^{\infty}$ such that $w_k \neq 0$ for all k, $\lim_{k \to \infty} w_k = 0$ and $f(w_k) = g(w_k)$ (all k), then $a_n = b_n$ holds, for $n = 0, 1, \ldots$.

PROOF.   If $\{n \mid a_n \neq b_n\}$ is nonvoid, let m be its smallest element. For each k, we have

$$w_k^m \left[(a_m - b_m) + \sum_{n=m+1}^{\infty} (a_n - b_n)w_k^{n-m} \right] = 0.$$

Since $w_k \neq 0$, the equality

$$(a_m - b_m) + \sum_{n=m+1}^{\infty} (a_n - b_n)w_k^{n-m} = 0$$

follows. Since $\lim_{k \to \infty} \sum_{n=m+1}^{\infty} (a_n - b_n)w_k^{n-m} = 0$, it follows that $a_m = b_m$. Hence $\{n \mid a_n \neq b_n\}$ must be empty.

2.8 COROLLARY

Let $f(z) = \sum_{n=0}^{\infty} a_n z^n$ have positive radius of convergence. If 0 is a cluster point of $\{z \mid f(z) = 0\}$, then f is identically zero.

2.9 MULTIPLICATION OF POWER SERIES

Let $f(z) = \sum_{n=0}^{\infty} a_n z^n$ and $g(z) = \sum_{n=0}^{\infty} b_n z^n$ have radii of convergence α and β, respectively. Since f and g are functions, the formula $h(z) = f(z)g(z)$ defines a function on $D(\rho, 0)$ where ρ satisfies $\rho \leq \alpha$ and $\rho \leq \beta$. It is not obvious how to prove that h is a power series, but if we think of f and g as being approximately equal to polynomials, it is easy to guess what the power series for h ought to be. If we multiply two polynomials together, the coefficient of z^n is $[\sum_{k=0}^{n} a_k b_{n-k}]$. The guess, then, is $h(z) = \sum_{n=0}^{\infty} (\sum_{k=0}^{n} a_k b_{n-k})z^n$. We now prove that this guess is correct.

Theorem. Let $\sum_{n=0}^{\infty} A_n$ and $\sum_{n=0}^{\infty} B_n$ be convergent series and suppose in addition that $\sum_{n=0}^{\infty} A_n$ converges absolutely. Define

(i) $C_n = \sum_{k=0}^{n} A_k B_{n-k}.$

Then $\sum_{n=0}^{\infty} C_n$ converges and

(ii) $\left[\sum_{n=0}^{\infty} A_n\right]\left[\sum_{n=0}^{\infty} B_n\right] = \sum_{n=0}^{\infty} C_n.$

Remarks. The series $\sum_{n=0}^{\infty} C_n$ is called the *Cauchy product* of $\sum_{n=0}^{\infty} A_n$ and $\sum_{n=0}^{\infty} B_n$.

For power series $\sum a_n z^n$ and $\sum b_n z^n$, let $A_n = a_n z^n$ and $B_n = b_n z^n$. Then $C_n = \sum_{k=0}^{n} a_k z^k b_{n-k} z^{n-k} = \left(\sum_{k=0}^{n} a_k b_{n-k}\right)z^n$. Equality (ii) thus becomes

(iii) $\left[\sum_{n=0}^{\infty} a_n z^n\right]\left[\sum_{n=0}^{\infty} b_n z^n\right] = \sum c_n z^n$ where $c_n = \sum_{k=0}^{n} a_k b_{n-k}.$

PROOF OF THE THEOREM. Let $S_0, S_1, \cdots, S_n$ denote the partial sums of $S = \sum_{n=0}^{\infty} B_n$, and let $R_n = S - S_n$. We have

$$\sum_{k=0}^{n} C_k = A_0 B_0 + [A_1 B_0 + A_0 B_1] + \cdots + \sum_{i+j=n} A_i B_j$$
$$= A_0 S_n + A_1 S_{n-1} + \cdots + A_n S_0$$
$$= A_0(S - R_n) + A_1(S - R_{n-1}) + \cdots + A_n(S - R_0)$$
$$= (A_0 + A_1 + \cdots + A_n)S - [A_0 R_n + A_1 R_{n-1} + \cdots + A_n R_0].$$

The proof will be completed by showing that $\lim_{n \to \infty} \sum_{k=0}^{n} A_k R_{n-k} = 0$. [This equality is valid (roughly) because for small k, R_{n-k} is small if n is large; if k is large A_k is small.] For a proof, let $\varepsilon > 0$. Since the sequence $(R_k)_{k=0}^{\infty}$ converges, there is an M such that $|R_k| < M$ for all $k \geq 0$. Fix m such that

$$n \geq m \text{ implies } \sum_{k=m}^{n} |A_k| < \frac{\varepsilon}{2M}.$$

Let $A > \sum_{k=0}^{\infty} |A_k|$. Since $\lim_{k \to \infty} R_k = 0$, there is a k_0 such that

$$k \geq k_0 \text{ implies } |R_k| < \frac{\varepsilon}{2A}.$$

If $n \geq m + k_0 - 1$, then $n - m + 1 \geq k_0$ and hence

$$\left|\sum_{k=0}^{n} A_k R_{n-k}\right| = |(A_0 R_n + \cdots + A_{m-1} R_{n-m+1})$$

$$+ (A_m R_{n-m} + \cdots + A_n R_0)| < \frac{\varepsilon}{2} + \frac{\varepsilon}{2} = \varepsilon.$$

The proof of the theorem is thus complete.

Equality (iii) holds whenever the series on the left converge. Hence the radius of convergence f of $\sum c_n z^n$ is at least the minimum of the radius of convergence α of $\sum_{n=0}^{\infty} a_n z^n$ and the radius of convergence β of $\sum_{n=0}^{\infty} b_n z^n$. Examples in the exercises show that ρ can be larger than min (α, β). In fact $\rho = \infty$ is possible, even if α and β are both finite.

*2.10 CHANGE OF CENTER FOR POWER SERIES

Consider $f(z) = \sum_{n=0}^{\infty} a_n z^n$ with radius of convergence $\rho > 0$ and suppose that b is interior to $D(\rho, 0)$ ($|b| < \rho$). We will find a power series expansion for $f(z)$ with center b. To this end write, for z in $D(\rho, 0)$ and a positive integer N,

$$f(z) = \sum_{n=0}^{\infty} a_n z^n$$

$$= \sum_{n=0}^{\infty} a_n (z - b + b)^n$$

$$= \sum_{n=0}^{\infty} a_n \left(\sum_{k=0}^{n} \binom{n}{k} (z - b)^k b^{n-k} \right)$$

$$= \sum_{n=0}^{N} a_n \left(\sum_{k=0}^{n} \binom{n}{k} (z - b)^k b^{n-k} \right) + \sum_{n=N+1}^{\infty} a_n \left(\sum_{k=0}^{n} \binom{n}{k} (z - b)^k b^{n-k} \right)$$

$$= \sum_{k=0}^{N} \left(\sum_{n=k}^{N} \binom{n}{k} a_n b^{n-k} \right) (z - b)^k + \sum_{n=N+1}^{\infty} a_n \left(\sum_{k=0}^{n} \binom{n}{k} (z - b)^k b^{n-k} \right)$$

$$= \sum_{k=0}^{N} \left(\sum_{n=k}^{\infty} \binom{n}{k} a_n b^{n-k} \right) (z - b)^k + \sum_{n=N+1}^{\infty} a_n \left(\sum_{k=0}^{n} \binom{n}{k} (z - b)^k b^{n-k} \right)$$

$$\quad - \sum_{k=0}^{N} \left(\sum_{n=N+1}^{\infty} \binom{n}{k} a_n b^{n-k} \right) (z - b)^k$$

$$= \sum_{k=0}^{N} \left(\sum_{n=k}^{\infty} \binom{n}{k} a_n b^{n-k} \right) (z - b)^k + \sum_{n=N+1}^{\infty} a_n \left(\sum_{k=N+1}^{n} \binom{n}{k} b^{n-k} (z - b)^k \right).$$

The crucial equality in the calculation (the next to last one) is valid because

$$\sum_{n=k}^{\infty} \binom{n}{k} a_n b^{n-k}$$

converges for each k. To see this observe that

$$\frac{n!}{(n - k)!} > \binom{n}{k}$$

and that

$$f^{(k)}(b) = \sum_{n=0}^{\infty} \frac{(n + k)!}{n!} a_{n+k} b^n = \sum_{n=k}^{\infty} \frac{n!}{(n - k)!} a_n b^{n-k}$$

holds whenever $|b| < \rho$. The second term in the final expression is dominated by

$$\sum_{n=N+1}^{\infty} |a_n| \left(\sum_{k=0}^{n} \binom{n}{k} |z - b|^k |b|^{n-k} \right) = \sum_{n=N+1}^{\infty} |a_n| (|z - b| + |b|)^n.$$

For z such that $|z - b| < \rho - |b|$, the last series is the remainder of a convergent series. Thus given $\varepsilon > 0$, there is an N_0 such that this remainder is less than ε for $N \ge N_0$; that is, we have

$$\left| \sum_{n=0}^{\infty} a_n z^n - \sum_{k=0}^{N} \left(\sum_{n=k}^{\infty} \binom{n}{k} a_n b^{n-k} \right) (z - b)^k \right| < \varepsilon.$$

The equality

$$\sum_{k=0}^{\infty} b_k (z - b)^k = \sum_{n=0}^{\infty} a_n z^n \qquad \left(b_k = \sum_{n=k}^{\infty} \binom{n}{k} a_n b^{n-k} \right)$$

is thus valid if $|z - b| < \rho - |b|$; that is, the power series on the left has radius of convergence not less than $\rho - |b|$.

It is of interest to note that often

$$(\overline{\lim_{k \to \infty}} \sqrt[k]{|b_k|})^{-1} > \rho - |b|;$$

that is, the disk of convergence of $\sum_{k=0}^{\infty} b_k (z - b)^k$ may extend beyond $D(\rho, 0)$. Whenever this happens we have a representation (or continuation) of f outside the initial disk of convergence $D(\rho, 0)$.

Example. The series $f(z) = \sum_{n=0}^{\infty} z^n$ converges if and only if $|z| < 1$. Let $b = -\frac{1}{2}$. We want to find b_n so that $f(z) = \sum_{n=0}^{\infty} b_n (z + \frac{1}{2})^n$. From the previous calculation, we know that

$$b_n = \sum_{m=n}^{\infty} \left(-\frac{1}{2} \right)^{m-n} \frac{m!}{n! \, (m - n)!}.$$

Instead of summing the series directly, however, note that the coefficients b_n are unique (2.5) and argue as follows. For $|z| < 1$,

$$f(z) = \sum_{n=0}^{\infty} z^n = \frac{1}{1 - z} = \frac{1}{1 - (z + \frac{1}{2} - \frac{1}{2})} = \frac{1}{\frac{3}{2} - (z + \frac{1}{2})}$$

$$= \frac{2}{3} \frac{1}{1 - \frac{2}{3}(z + \frac{1}{2})} = \frac{2}{3} \sum_{n=0}^{\infty} \left(\frac{2}{3} \right)^n (z + \frac{1}{2})^n$$

$$= \sum_{n=0}^{\infty} \left(\frac{2}{3} \right)^{n+1} (z + \frac{1}{2})^n.$$

The calculation is valid if $|z + \frac{1}{2}| < \frac{3}{2}$. The final power series has radius of convergence $\frac{3}{2}$, by (2.3). Hence $b_n = (\frac{2}{3})^{n+1}$, and we have also proved the equality

$$\left(\frac{2}{3}\right)^{n+1} = \sum_{m=n}^{\infty} \left(-\frac{1}{2}\right)^{m-n} \frac{m!}{n!\,(m-n)!}.$$

The disk of convergence $D(\frac{3}{2}, -\frac{1}{2})$ of $\sum_{n=0}^{\infty} b_n(z + \frac{1}{2})^n$ includes that of $\sum_{n=0}^{\infty} z^n$, and represents the same function in the latter disk. See Exercise 7, p. 65, for further development.

EXERCISES

TYPE I

1. Let $A_n = B_n = (-1)^n/\sqrt{n+1}$. Show that $\sum_{n=1}^{\infty} A_n$ converges but that the Cauchy product of $\sum_{n=1}^{\infty} A_n$ and $\sum_{n=1}^{\infty} B_n$ does not converge.

2. Use (2.9.ii) to show that $\sum_{n=0}^{\infty} a^n = 1/(1-a)$ if $|a| < 1$.
   [*Hint:* Consider the Cauchy product of $\sum_{n=0}^{\infty} a^n$ and $(1-a)$.]

TYPE II

1. Let $f(z) = \sum_{n=0}^{\infty} a_n z^n$ and $g(z) = \sum_{n=0}^{\infty} b_n z^n$ have positive radii of convergence. If 0 is a cluster point of $\{z \mid f(z) \cdot g(z) = 0\}$, then f or g is identically 0.

2. (a) With reference to (2.9), show that $\alpha = 1$, $\beta = \infty$, $\rho = 1$ is possible.
   (b) Let $a_0 = 2$; $a_n = 1$, $n > 0$; $b_0 = \frac{1}{2}$; $b_n = -2^{-(n+1)}$, $n > 0$. Find the radii of convergence of $\sum a_n z^n$, $\sum b_n z^n$, and $\sum_{n=1}^{\infty} c_n z^n$ (the Cauchy product).

3. With reference to Exercise (2.b), can you find some more examples?

4. Let $0 < \alpha \leq \beta < \infty$, $\alpha \leq \rho \leq \infty$. Can you find power series $\sum a_n z^n$ and $\sum b_n z^n$ with radii of convergence α and β such that the Cauchy product $\sum c_n z^n$ has radius of convergence ρ?

5. Suppose that $g(z) = \sum_{n=1}^{\infty} a_n z^n$ has infinite radius of convergence. Suppose that there are constants K, N, and R such that $\text{Re}[g(z)] \leq K|z|^N$ for all z such that $|z| \geq R$. Show that g is a polynomial of degree $\leq N$.

6. Suppose that $f(z) = \sum_{n=0}^{\infty} a_n z^n$ has radius of convergence $\rho > 0$. If $(w_k)_{k=1}^{\infty}$ is a sequence in $D(\rho, 0)$ which converges to a point $w \in D(\rho, 0)$ such that $f(w_k) = 0$ for all k, then $a_n = 0$ for all n.
   [*Hint:* Use (2.10).]

Remark.   This result is a slight improvement over (2.7).

7. Let b_n be as in the example of (2.10). The series $f(z) = \sum_{n=0}^{\infty} b_n(z + \tfrac{1}{2})^n$ includes in its disk of convergence the points -1 and i. Find the power series representations $\sum_{n=0}^{\infty} c_n(z + 1)^n$ and $\sum_{n=0}^{\infty} d_n(z - i)^n$ of f centered at -1 and i, respectively. Determine the disks of convergence of these series and make a sketch of the four disks of convergence discussed in (2.10) and in this exercise. What would you guess about further power series representation centered at other points?

8. Show that

$$\sum_{k=0}^{\infty} \left(\sum_{n=k}^{\infty} \binom{n}{k} a_n b^{n-k} \right)(z - b)^k$$

is not a rearrangement of

$$\sum_{n=0}^{\infty} a_n \left(\sum_{k=0}^{n} \binom{n}{k}(z - b)^k b^{n-k} \right).$$

2.11 THE EXPONENTIAL AND TRIGONOMETRIC FUNCTIONS

In Sections (2.11)–(2.17) we study several important functions whose definitions and properties depend in part on power series. We begin with the exponential and trigonometric functions.

As we have remarked, the function

$$\exp(z) = \sum_{n=0}^{\infty} \frac{1}{n!} z^n$$

is a very important function. In this section, we will study this function and some of its close relatives. The series defining $\exp(z)$ converges absolutely for all z by (2.3). For real numbers a and b, the formula

$$e^{a+b} = e^a e^b$$

distinguishes the exponential function from other functions. We first show that $\exp(z)$ has this property for complex numbers.

(i) *Theorem.* The formula $\exp(z + w) = \exp(z)\exp(w)$ holds for all complex z and w.

PROOF. We use the theorem in Section (2.9) to compute

$$\exp(z)\exp(w) = \left(\sum_{n=0}^{\infty} \frac{z^n}{n!} \right)\left(\sum_{n=0}^{\infty} \frac{w^n}{n!} \right)$$

$$= \sum_{n=0}^{\infty} \left(\sum_{k=0}^{n} \frac{z^k}{k!} \frac{w^{n-k}}{(n-k)!} \right) = \sum_{n=0}^{\infty} \left(\frac{1}{n!} \sum_{k=0}^{n} \frac{n!}{k!(n-k)!} z^k w^{n-k} \right).$$

From the binomial expansion, we have

$$\sum_{k=0}^{n} \frac{n!}{k!\,(n-k)!}\, z^k w^{n-k} = (z+w)^n.$$

Thus,

$$\sum_{n=0}^{\infty} \frac{1}{n!} \sum_{k=0}^{n} \left(\frac{n!}{k!\,(n-k)!}\, z^k w^{n-k} \right) = \sum_{n=0}^{\infty} \frac{(z+w)^n}{n!} = \exp(z+w).$$

We use the notation e^z for complex z also; that is, we let $e^z = \exp(z)$. We will continue to use both notations. Using the formula just proved, we have

$$e^{x+iy} = e^x e^{iy} = e^x \left(\sum_{k=0}^{\infty} \frac{(iy)^k}{k!} \right)$$

$$= e^x \left(\sum_{n=0}^{\infty} \frac{i^{2n} y^{2n}}{(2n)!} + \sum_{n=0}^{\infty} \frac{i^{2n+1} y^{2n+1}}{(2n+1)!} \right)$$

$$= e^x \left(\sum_{n=0}^{\infty} \frac{(-1)^n y^{2n}}{(2n)!} + i \sum_{n=0}^{\infty} \frac{(-1)^n y^{2n+1}}{(2n+1)!} \right).$$

From calculus, we recognize that

$$\cos(y) = \sum_{n=0}^{\infty} \frac{(-1)^n y^{2n}}{(2n)!} \quad \text{and} \quad \sin(y) = \sum_{n=0}^{\infty} \frac{(-1)^n y^{2n+1}}{(2n+1)!}.$$

Thus the formula

$$e^z = e^{x+iy} = e^x (\cos y + i \sin y).$$

holds. In particular, the definition of e^{iy} given here is consistent with that in (0.8).

The above series for cos and sin are used to *define* these functions for complex variables.

(ii) *Definitions.*

$$\cos(z) = \sum_{n=0}^{\infty} \frac{(-1)^n}{(2n)!}\, z^{2n},$$

$$\sin(z) = \sum_{n=0}^{\infty} \frac{(-1)^n}{(2n+1)!}\, z^{2n+1}$$

and

$$\exp(z) = \sum_{n=0}^{\infty} \frac{z^n}{n!}.$$

These definitions extend the domain of the sine, cosine, and exponential functions from the real numbers to the complex numbers. The ratio test shows that the extended series converge for all z. The identity theorem for power series (2.7) shows that these extensions are unique.

Repeated applications of Theorem (2.4) show that these functions have derivatives of all orders for all z. We now systematically develop properties of exp, sin, and cos. We include some "real variable" properties for the sake of logical completeness.

(iii) *Elementary properties of* exp, cos, *and* sin. The following equalities are easy to verify using the definitions just given. The reader is urged to verify them.

(a) $\exp' (z) = \exp (z)$; $\cos' (z) = -\sin (z)$; $\sin' (z) = \cos (z)$.
(b) $\exp (iz) = \cos (z) + i \sin (z)$.
(c) $\cos (z) = (e^{iz} + e^{-iz})/2$; $\sin (z) = (e^{iz} - e^{-iz})/2i$.
(d) $\overline{\exp (z)} = \exp (\bar{z})$.
(e) $\cos (z) = \cos (-z)$; $\sin (z) = -\sin (-z)$.

(iv) *Further identities.* It is now easy to obtain the addition formulas

(a) $\cos (z + w) = \cos (z) \cos (w) - \sin (z) \sin (w)$;
 $\sin (z + w) = \sin (z) \cos (w) + \sin (w) \cos (z)$.

Since $1 = \cos (0) = \cos (z - z) = \cos^2 (z) + \sin^2 (z)$, we have

(b) $\cos^2 (z) + \sin^2 (z) = 1$,

from which it follows that

(c) $|\exp (it)| = 1$ if t is real,

because $\cos (t)$ and $\sin (t)$ are the real and imaginary parts of $\exp (it)$.

(v) *Theorem.* The function exp is strictly increasing on **R** and

$$\lim_{t \to \infty} (\exp (t)) = \infty; \quad \lim_{t \to -\infty} (\exp (t)) = 0.$$

PROOF. By the definition, $\exp (0) = 1$ and $\exp (t) > 1$ if $t > 0$. Thus $\exp (y - x) > 1$ if $y > x$, proving that exp is strictly increasing. The equality $\lim_{t \to \infty} \exp (t) = \infty$ follows from the inequality $\exp (t) > 1 + t$. We have

$$\lim_{t \to -\infty} \exp (t) = \lim_{t \to \infty} \exp (-t) = \lim_{t \to \infty} [\exp (t)]^{-1} = 0.$$

(vi) *The definition of* π. (a) First we prove that $\cos (2) < 0$.

PROOF.

$$\cos{(2)} = \left[1 - 2 + \frac{2^4}{4!}\right] - \left[\frac{2^6}{6!} - \frac{2^8}{8!} + \cdots + (-1)^{n+1}\frac{2^{2n}}{(2n)!} + \cdots\right].$$

The first bracketed expression is negative. To see that the second is positive, group the terms in the following pattern: $(2^6/6! - 2^8/8!) + (2^{10}/10! - 2^{12}/12!) + \cdots$.

(b) *Definition.* The number $\pi/2$ is defined as the smallest positive number x which satisfies $\cos{(x)} = 0$. This is a meaningful definition because $\cos{(0)} > 0$ and $\cos{(2)} < 0$ so that the intermediate value theorem tells us that there are numbers x between 0 and 2 such that $\cos{(x)} = 0$. The completeness property of the real numbers then tells us that there is a smallest positive x. In Section (3.13) we will show that π as defined above equals the ratio of the circumference of a circle to its diameter.

(vii) *Properties of π.* From the definition of π and the addition formulas (iv.a), the graphs of sin and cos on **R** can be obtained. The pertinent features are listed below, without detailed proofs. The reader should convince himself of their validity, using only (i–vi), and the fact from elementary calculus that a function with a positive derivative is strictly increasing.

(a) $\cos{(x)} > 0$ if $x \in [0, \pi/2[$.

(b) $\sin{(\pi/2)} = 1$.
   [By (a), sin is monotonically increasing on $[0, \pi/2[$. Since $\sin^2 + \cos^2 = 1$, $\sin{(\pi/2)} = 1$.]

(c) $\cos{(\pi)} = -1$; $\cos{(n\pi)} = (-1)^n$; $\cos{(n\pi/2)} = 0$ if n is odd.

(d) $\sin{(n\pi/2)}$ is 0 for even n, 1 for $n \equiv 1 \pmod 4$, -1 for $n \equiv 3 \pmod 4$.

(e) sin is strictly decreasing on $[-\pi, -\pi/2]$ and $[\pi/2, \pi]$; strictly increasing on $[-\pi/2, \pi/2]$; cos is strictly increasing on $[-\pi, 0]$, strictly decreasing on $[0, \pi]$.

(f) $\exp{(z + 2\pi i n)} = \exp{(z)}$ for all $z \in \mathbf{C}$ and all $n \in \mathbf{Z}$.

(g) If $\exp{(z)} = 1$, then $z = 2\pi i n$ for some $n \in \mathbf{Z}$.
   To prove (g), write $z = x + iy$. Since $\exp{(x)} = |\exp{(z)}|$, we have $x = 0$ and $z = iy$. By (iii.b), $\cos{(y)} = 1$ and $\sin{(y)} = 0$ must hold. By (c), (d), and (e), $y = 2\pi n$ for some $n \in \mathbf{Z}$.

(viii) *Polar form.*

(a) If $|w| = 1$, then there is a unique real t such that $-\pi < t \le \pi$ and $w = e^{it}$.

PROOF. Let $w = u + iv$. Since $|u| \leq 1$, there are two t's such that $u = \cos t$ and $t \in \,]-\pi, \pi]$. Each of the two satisfies $\cos^2 t + \sin^2 t = 1$, and $\sin (t)$ is positive for one and negative for the other. Thus $v = \sin (t)$ for exactly one of the two.

(b) If $w \in \mathbf{C}$ and $w \neq 0$, then $w = |w| \exp (it)$ for a unique $t \in \,]-\pi, \pi]$. In fact,

$$w = |w|e^{it} = e^{\log|w|}e^{it} = e^{\log|w| + it},$$

t as in (a) for $w/|w|$. The unique number t is defined as Arg (w).

We have, of course, already used the above result several times, relying on the reader's knowledge of trigonometry to obtain $t = $ Arg (z). The above proof relies only on results in this chapter.

2.12 TANGENT AND COTANGENT

The sets of zeros of cos and sin are $\{\pi n/2 \mid n \text{ is an odd integer}\}$ and $\{\pi n \mid n \in \mathbf{Z}\}$, respectively. Hence the functions

$$\tan (z) = \frac{\sin (z)}{\cos (z)} \quad \text{and} \quad \cot (z) = \frac{\cos (z)}{\sin (z)}$$

are analytic on $\{z \neq \pi n/2 \mid n \text{ is odd}\}$ and $\{z \neq n\pi \mid n \in Z)$, respectively. The singularities $\pi n/2$ and $n\pi$ of these functions will be discussed in more detail in Chapter 5. Elementary trigonometric identities are proved as in the real case.

It is of interest to obtain a power series expansion for tan (z). To do this we suppose first that there are complex numbers $(a_n)_{n=0}^{\infty}$ for which

(i) $\quad \tan (z) = \displaystyle\sum_{n=0}^{\infty} a_n z^n$

holds for $|z| < \pi/2$. Since $\tan (0) = 0$, we must have $a_0 = 0$. If we replace $\sin (z)$ and $\cos (z)$ in the equality $\sin (z) = \tan (z) \cos (z)$ by their power series centered at 0, then we obtain

(ii) $\quad \displaystyle\sum_{n=0}^{\infty} \frac{(-1)^n z^{2n+1}}{(2n + 1)!} = \left(\sum_{n=0}^{\infty} a_n z^n\right)\left(\sum_{n=0}^{\infty} \frac{(-1)^n z^{2n}}{(2n)!}\right).$

If $\sum_{n=0}^{\infty} c_n z^n$ is the Cauchy product of the series on the right in (ii), then by (2.7) applied to (ii) we see that c_n is zero for even n and

(iii) $\quad c_{2n+1} = \dfrac{(-1)^n}{(2n + 1)!}$

holds for all n. On the other hand, by (2.9) we have

(iv) $c_{2n} = 0 = a_{2n} \cdot 0 + \dfrac{(-1)}{2!} a_{2(n-1)} + \dfrac{(-1)^2}{4!} a_{2(n-2)}$

$$+ \cdots + a_2 \frac{(-1)^{n-1}}{[2(n-1)]!} + a_0 \frac{(-1)^n}{(2n)!}.$$

(v) $c_{2n+1} = a_{2n+1} \cdot 1 + a_{2n-1} \dfrac{(-1)}{2!} + \cdots + a_3 \dfrac{(-1)^{n-1}}{[2(n-1)]!} + a_1 \dfrac{(-1)^n}{(2n)!}$

It follows from (iv) that $a_{2n} = 0$ holds for all n. From (iii) and (v), we obtain

(vi) $a_{2n+1} = \dfrac{1}{2!} a_{2n-1} - \dfrac{1}{4!} a_{2n-3} + \cdots + \dfrac{(-1)^{n+1}}{(2n)!} a_1 + \dfrac{(-1)^n}{(2n+1)!}.$

The recursive relation (vi) uniquely determines the a's. Taking $n = 0$, we have $a_1 = 1$; $n = 1$ and $n = 2$ give

$$a_3 = \left(\frac{1}{2!} - \frac{1}{3!} \right) = \frac{1}{3} \quad \text{and} \quad a_5 = \left(\frac{1}{3!} - \frac{1}{4!} + \frac{1}{5!} \right) = \frac{16}{5!}.$$

It follows easily from (vi) that the sequence $(a_n)_{n=0}^{\infty}$ is bounded by 1. The power series $\sum_{n=0}^{\infty} a_n z^n$ thus has radius of convergence not less than 1. It follows that (ii), [and therefore (i)] holds. The power series expansion of $\tan(z)$ thus begins

$$\tan(z) = z + \frac{1}{3} z^3 + \frac{2}{15} z^5 + \cdots.$$

Actually, (i) holds if $|z| < \pi/2$. This will follow immediately from (4.8) and hence we will not prove it at this point.

The technique of equating coefficients and using (2.7) to obtain a recursive relation for the unknown coefficients (a_n) is known as the *method of undetermined coefficients*. Other applications of the method are suggested in the exercises.

*2.13 HYPERBOLIC FUNCTIONS

As in the real case, define

$$\cosh(z) = \frac{e^z + e^{-z}}{2}; \qquad \sinh(z) = \frac{e^z - e^{-z}}{2};$$

cosh and sinh are clearly analytic in **C**. These functions play a role in the solution of many differential equations, but are unimportant for the main development of complex analysis. Proofs of the properties in the following list may be treated as elementary exercises; the list may be useful for reference.

(i) $(\sinh)' = \cosh$; $(\cosh)' = \sinh$.

(ii) $\cosh^2 - \sinh^2 = 1$.

(iii) $\sinh (z + w) = \sinh (z) \cosh (w) + \cosh (z) \sinh (w)$.

(iv) $\cosh (z + w) = \cosh (z) \cosh (w) + \sinh (z) \sinh (w)$.

(v) $\sinh$ is odd; $\cosh$ is even.

(vi) $\sinh (x + iy) = \sinh (x) \cos (y) + i \cosh (x) \sin (y)$, all $x, y \in \mathbf{C}$.

(vii) $\cosh (x + iy) = \cosh (x) \cos (y) + i \sinh (x) \sin (y)$, all $x, y \in \mathbf{C}$.

(viii) $|\sinh (z)|^2 = [\sinh (x)]^2 + [\sin (y)]^2$ $x + iy$, x and y real.

(ix) $|\cosh (z)|^2 = [\sinh (x)]^2 + [\cos (y)]^2$ $x + iy$, x and y real.

(x) Zeros of $\sinh = \{n\pi i \mid n \in \mathbf{Z}\}$.

(xi) Zeros of $\cosh = \{n\pi i/2 \mid n$ is an odd integer$\}$.

(xii) $\cosh$ and $\sinh$ have period $2\pi i$.

2.14 THE LOGARITHM

The correspondence $\exp(x) = u$ maps the real line, $\{-\infty < x < \infty\}$ one-to-one and onto the ray $\{0 < u < \infty\}$. Hence for each positive u there is exactly one x such that $u = \exp (x)$. We write $\log (u) = x$ and call x the logarithm of u. In the complex case the situation is more complicated.

The function $\exp$ maps $\mathbf{C}$ onto $\mathbf{C}\backslash\{0\}$. To see this, consider a complex number $w \neq 0$. We seek a z for which $\exp (z) = w$. Let $z = x + iy$ and let $\psi = \mathrm{Arg}(w)$. From $e^x e^{iy} = |w|e^{i\psi}$ we see that $x = \log|w|$ must hold. Canceling e^x and $|w|$ we obtain $e^{iy} = e^{i\psi}$ and so $y = \psi + 2\pi in$ holds for some integer n. Thus we must have

$$x + iy = \log|w| + i\mathrm{Arg}(w) + 2\pi in.$$

Conversely, each $z_n = \log|w| + i\mathrm{Arg}(w) + 2\pi in$ satisfies $\exp(z_n) = w$. Thus the mapping taking w to z is not one-to-one and as a function with domain $\mathbf{C}$, $\exp$ does not have an inverse. However, if we are careful we can still discuss $\log (z)$ for complex numbers z.

We define

$$\log (z) = \{w \mid e^w = z\}.$$

For each $z \neq 0$, $\log (z)$ is an infinite set of numbers. Hence $\log$ is *not* a function. Evidently one way to express $\log (z)$ is

$$\log (z) = \{\log |z| + i\mathrm{Arg} (z) + 2\pi in \mid n \in \mathbf{Z}\}.$$

It is often convenient to single out a particular number in this set. One possibility is the number Log (z) = log $|z|$ + iArg (z). As we shall presently see, this is not the only possibility. The number Log (z) is often called the principal value of the logarithm of z. [This is analogous to calling Arcsin(θ) the principal value of arcsin(θ), as is often done in elementary trigonometry.] Evidently Log is a function whose domain is $\mathbf{C}\backslash\{0\}$. Since $-\pi <$ Arg $(z) \leq \pi$, the functions Arg and Log are discontinuous on $\{x \mid x \leq 0\} = -\mathbf{R}^+$.

Other functions inverse to exp on particular sets are also of interest. Pick a real number θ and define Arg$_\theta$ $(z) = \psi$ $(z \neq 0)$ where ψ satisfies $|z|e^{i\psi} = z$ and $\theta - \pi < \psi \leq \theta + \pi$. Using Arg$_\theta$ (z) to write log(z), we have

$$\log (z) = \{\log |z| + i\text{Arg}_\theta (z) + 2\pi in \mid n \in \mathbf{Z}\}.$$

We define

$$\text{Log}_\theta (z) = \log |z| + i\text{Arg}_\theta (z).$$

This function is continuous except at points on the ray $\{re^{i(\theta+\pi)} \mid 0 \leq r < \infty\}$. Evidently exp $\{\text{Log}_\theta z\} = z$. The definition of Log$_\theta$ z is motivated by the observation that exp maps sets of the type

$$\{z \mid \theta - \pi < \text{Im}(z) \leq \pi + \theta\} \qquad (\theta \text{ a fixed real number})$$

one-to-one and onto $\mathbf{C}\backslash\{0\}$. The inverse of exp on this set is Log$_\theta$.

Suppose that we fix a $z_0 \neq 0$ and let $\theta =$ Arg (z_0). Observe that Log$_\theta$ is continuous except on the ray

$$\{re^{i(\text{Arg } z_0+\pi)} \mid 0 \leq r < \infty\}.$$

In particular Log$_\theta$ z is continuous near z_0. For this choice of θ we say that Log$_\theta$ is a branch of log which contains z_0.

Evidently Log$_\theta$ cannot be analytic on the ray $\{re^{i(\theta+\pi)} \mid 0 \leq r < \infty\}$ since, as a function of z, it is not even continuous there. If we knew that $d(\text{Log}_\theta z)/dz$ existed for z not on this ray, then we could write $z = \exp (\text{Log}_\theta z)$ and use the chain rule to compute

$$1 = \exp (\text{Log}_\theta z) \frac{d}{dz} (\text{Log}_\theta z),$$

or,

$$\frac{1}{z} = \frac{d}{dz} (\text{Log}_\theta z).$$

The difficulty, of course, is that we do not know that $d(\text{Log}_\theta z)/dz$ exists. One way to show that Log$_\theta$ is analytic off of $\{re^{i(\theta+\pi)} \mid 0 \leq r < \infty\}$ is as follows.

For $z \notin \{re^{i(\theta+\pi)} \mid 0 \le r < \infty\}$ and Δz of small magnitude, consider the quotient

$$\frac{\text{Log}_\theta \, (z + \Delta z) - \text{Log}_\theta \, z}{\Delta z}.$$

Let $w = \text{Log}_\theta \, z$, so that $e^w = z$. We have $z + \Delta z = e^{w + \Delta w}$, where $\Delta w = \text{Log}_\theta \, (z + \Delta z) - \text{Log}_\theta \, (z)$. We write

$$\frac{\text{Log}_\theta \, (z + \Delta z) - \text{Log}_\theta \, z}{\Delta z} = \frac{\text{Log}_\theta \, (z + \Delta z) - \text{Log}_\theta \, (z)}{e^{w + \Delta w} - e^w} = \frac{\Delta w}{e^{w + \Delta w} - e^w}.$$

Since Log_θ is continuous for z satisfying $|\text{Arg}_\theta \, (z) - \theta| < \pi$, we have $\lim_{\Delta z \to 0} \Delta w = 0$ whenever z satisfies $|\text{Arg}_\theta \, (z) - \theta| < \pi$. Hence for these z we have

$$\lim_{\Delta z \to 0} \frac{\text{Log}_\theta \, (z + \Delta z) - \text{Log}_\theta \, z}{\Delta z} = \lim_{\Delta z \to 0} \frac{\Delta w}{e^{w + \Delta w} - e^w} = \frac{1}{e^w} = \frac{1}{z}.$$

Thus the formula

$$\frac{d}{dz} (\text{Log}_\theta \, z) = \frac{1}{z}$$

holds on $\mathbf{C} \backslash \{re^{i(\theta+\pi)} \mid 0 \le r < \infty\}$.

The function Log_θ on its domain of analyticity is called a branch of the complete logarithm function. For a way to construct other types of branches, see Exercise 7, p. 173.

For positive real numbers a and b the equality $\text{Log} \, (ab) = \text{Log} \, (a) + \text{Log} \, (b)$ holds. However for arbitrary complex numbers z and w the most one can say is that $\text{Arg} \, (zw) = \text{Arg} \, (z) + \text{Arg} \, (w) + \delta 2\pi$, where δ is -1, 0, or 1; hence $\text{Log} \, (zw) = \text{Log} \, (z) + \text{Log} \, w + \delta 2\pi i$, where δ is -1, 0, or 1. For example, we have

$$0 = \text{Log} \, (1) = \text{Log} \, ((-1)(-1)) = \text{Log} \, (-1) + \text{Log} \, (-1) - 2\pi i$$

since $\text{Log} \, (-1) = \pi i$.

2.15 DEFINITION (COMPLEX EXPONENTIATION)

If α and $z \ne 0$ are complex numbers, define

$$z^\alpha = \{e^{\alpha w} \mid w \in \text{log} \, (z)\}.$$

Thus z^α is a set of numbers. Another way to express z^α is

$$z^\alpha = \{e^{\alpha \text{Log} \, (z) + 2\pi i \alpha n} \mid n \, \varepsilon \, \mathbf{Z}\}.$$

From this we see that z^α is an infinite set if α is real and irrational. Also we can readily see that if $\alpha = 1/k$ for $k \in \mathbf{Z}^{++}$, then this definition of z^α agrees with the discussion given in Section (0.9); that is, $z^{1/k}$ is the finite set of kth roots of z. If α is an integer k, then z^α is the single number z^k.

The equality $z^\alpha z^\beta = z^{\alpha+\beta}$ holds in the sense that for any $w \in \log(z)$ we have

$$e^{\alpha w} e^{\beta w} = e^{(\alpha+\beta)w},$$

and if $e^{\alpha w} \in z^\alpha$, and $e^{\beta w} \in z^\beta$ then $e^{(\alpha+\beta)w} \in z^{\alpha+\beta}$. However, if we interpret $z^\alpha z^\beta$ as the set of all products of an element in z^α by an element of z^β, then $z^\alpha z^\beta$ is the set

$$\{e^{\alpha w} e^{\beta v} \mid w, v \in \log(z)\} = \{e^{(\alpha+\beta)\operatorname{Log}(z) + 2\pi i(\alpha m + \beta n)} \mid m, n \in \mathbf{Z}\}$$

while $z^{\alpha+\beta}$ is the set $\{e^{(\alpha+\beta)w} \mid w \in \log(z)\}$. Clearly the inclusion

$$z^{\alpha+\beta} \subset z^\alpha z^\beta$$

holds, but equality may fail. For example, we have

$$\{-1\} = (-1)^1 \subsetneqq (-1)^{1/2}(-1)^{1/2} = \{-1, 1\}.$$

We can define what is called a branch of z^α by restricting our attention to a certain region of the plane and choosing numbers from the set z^α so as to make an analytic function of z on this region. For example, on $\mathbf{C}\backslash\mathbf{R}^-$ we can set

$$F_\alpha(z) = e^{\alpha[\log|z| + i\operatorname{Arg}(z)]} = e^{\alpha\operatorname{Log}(z)}.$$

Since F_α is a composition of analytic functions, F_α is analytic on $\mathbf{C}\backslash\mathbf{R}^-$. Different branches of log yield different branches of z^α.

From our study of real numbers we have some notions of how z^α should behave. For example, if x, y, r, and s are real numbers, the equalities

$$x^r x^s = x^{r+s}, \qquad x^r y^r = (xy)^r, \qquad (x^r)^s = x^{rs}$$

are valid if $x > 0$ and $y > 0$.

The corresponding rules for $F_\alpha(z)$ would be

(i) $F_\alpha(z)F_\beta(z) = F_{\alpha+\beta}(z),$
(ii) $F_\alpha(z)F_\alpha(w) = F_\alpha(zw),$
(iii) $F_\beta(F_\alpha(z)) = F_{\alpha\beta}(z).$

It is evident from the definition of $F_\alpha(z)$ that equality (i) holds. That equality (ii) fails is easily seen from the fact that

$$\operatorname{Log}(zw) = \operatorname{Log}(z) + \operatorname{Log}(w) + \delta 2\pi i, \qquad \delta \in \{-1, 0, 1\}.$$

To see that (iii) fails, note that $F_{1/2}(F_4(i)) = F_{1/2}(i^4) = 1$ and that $F_{(1/2)4}(i) = F_2(i) = i^2 = -1$.

2.16 PRINCIPAL SQUARE ROOT

In the special case $\alpha = \frac{1}{2}$, $F_{1/2}(z)$ is the *principal square root* of z ($z \neq 0$); hence, we have

$$F_{1/2}(z) = |z|^{1/2} e^{i(t/2)}, \qquad t \in \,]-\pi, \pi].$$

Notice that the square root function is discontinuous on the nonpositive x-axis and analytic off of it. (By selecting a different single-valued definition of Arg, the discontinuity can be changed to a different half-line.) The function $F_{1/2}$ has derivative $1/(2F_{1/2})$. This can be determined either by direct calculation as in elementary calculus or by implicitly differentiating the equality $[F_{1/2}(z)]^2 = z$.

Let $E(z) = |z|^{1/2} e^{i(\theta/2)}$, where $z = |z|e^{i\theta}$ and $0 \leq \theta < 2\pi$. The function E also satisfies $E(z)^2 = z$ for all z. It is discontinuous on the nonnegative x-axis and analytic elsewhere. The functions E and $F_{1/2}$ together with their domains of analyticity are called branches of the complete square root function. Additional branches are discussed in (7.7). Notice that $E(z) = F_{1/2}(z)$ for z in the upper half-plane and that $E(z) = -F_{1/2}(z)$ for z in the lower half-plane.

*2.17 BINOMIAL SERIES

For $\alpha \in \,]0, 1[$, define

(i) $\quad \dbinom{\alpha}{n} = \dfrac{\alpha(\alpha - 1) \cdots (\alpha - n + 1)}{n!}$, $n \geq 1$; $\dbinom{\alpha}{0} = 1$.

Fix $a \in \mathbf{C}\backslash\{0\}$. Since

$$\lim_{n \to \infty} \left| \binom{\alpha}{n} F_\alpha(a) a^{-n} \right| \bigg/ \left| \binom{\alpha}{n+1} F_\alpha(a) a^{-(n+1)} \right| = |a|,$$

the *binomial series* $f(z) = \sum_{n=0}^{\infty} \binom{\alpha}{n} F_\alpha(a) a^{-n} z^n$ has radius of convergence $|a|$.
The equality $F_\alpha(z + a) = f(z)$ will be established in solving Exercise 6, p. 139.

EXERCISES

TYPE I

1. If $z = x + iy$, define $\exp(z) = e^x \cos(y) + ie^x \sin(y)$. Use this definition and the Cauchy-Riemann equations to verify that exp is analytic in $\mathbf{C}$. (The equality provides an alternate definition for exp, based on properties of e^x for real x and cos and sin for real variables.)

2. Find the images of the following sets under exp.
   (a) $\{z \mid -\pi < \text{Im}(z) \le \pi\}$
   (b) $\{z \mid 0 < \text{Im}(z) \le 2\pi\}$
   (c) $\{z \mid \text{Re}(z) < 0\}$
   (d) $\{z \mid \text{Re}(z) > 0\}$
   (e) $\{z \mid 0 < \text{Im}(z) < \pi/2\}$.

3. Find the images of the following sets under Log.
   (a) $\{z \mid \text{Im}(z) > 0\}$
   (b) $\{z \mid \text{Im}(z) > 0\} \cap D(0, 1)$
   (c) $\{z \mid \text{Re}(z) > 0\} \cap D(0, 1)$.

4. Prove that
   (a) $\{z \in \mathbf{C} \mid \cos(z) = 0\} = \{(n\pi)/2 \mid n \text{ is an odd integer}\}$.
   (b) $\{z \in \mathbf{C} \mid \sin(z) = 0\} = \{n\pi \mid n \in \mathbf{Z}\}$.
   [This result is used in (2.12).]

5. Describe and sketch the image of the following sets under $F_{1/2}$; see (2.16).
   (a) $\mathbf{C}$
   (b) $\{z = x + iy \mid x > 0\}$
   (c) $\{x + iy \mid y < 0\}$
   (d) $\{z \mid -\pi/4 < \text{Arg}(z) < \pi/4\}$
   (e) $\{x + iy \mid x = 2\} \cup \{x + iy \mid y = 2\}$
   (f) $D(1, 1)$
   (g) $D(1, 0)$.

6. Find the images of the lines $x = $ constant and $y = $ constant $(0 < y < \pi)$ under $w(z) = \cosh(z)$.

7. Show that $u(z) = \log|z|$ and $v(z) = \text{Arg}(z)$ are harmonic on $\mathbf{C}\backslash(-\mathbf{R}^+)$.

8. Use the method illustrated in (2.12) to compute the first few terms of power series expansions for the following functions.

   (a) $\cot(z)$ (c) $\dfrac{\cos(z)}{1 - \sin(z)}$

   (b) $\dfrac{\sin(z)}{1 - \cos(z)}$ (d) $\dfrac{1}{\sin(z)}$

   Indicate where the expansions are valid.

9. Verify that $\text{Log}_{2\pi}(-i) = i(7\pi/2)$ and that $\text{Log}(-i) = -i(\pi/2)$.

10. Show that exp maps $\{z \mid \theta - \pi < \text{Im}(z) \le \pi + \theta\}$ onto $\mathbf{C}\backslash\{0\}$. What does exp do to the set $\{z \mid \theta - \pi \le \text{Im}(z) < \pi + \theta\}$?

TYPE II

1. Let z be a fixed nonzero complex number, and let α be a nonzero real number. Let
$$R_\alpha(z) = \{w \in \mathbf{C} \mid F_\alpha(w) = z\}.$$
Give some distinguishing features of $R_\alpha(z)$ for the following cases.
   (a) α a positive integer
   (b) α a negative integer
   (c) $\alpha = 1/n$, n a positive integer
   (d) $\alpha = p/q$, p and q relatively prime positive integers
   (e) α irrational.

2. Let $R_\alpha(z)$ be as in 1. Let $\cong$ denote group isomorphism. Show that $R_{p/q}(1) \cong \mathbf{Z}_q$ (the integers mod q) if p and q are relatively prime positive integers. Show that $R_\alpha(1) \cong \mathbf{Z}$ if α is irrational.

3. If f is analytic on an open set Ω and $z_0 \in \Omega$, then there is an $r > 0$ such that $D(r, z_0) \subset \Omega$ and $f(z) = \sum_{n=0}^\infty (f^{(n)}(z_0)/n!)(z - z_0)^n$ for $z \in D(r, z_0)$. We will prove this fact in Chapter 4, using Cauchy's theorem.
   (a) Show that $\mathrm{Log}^{(n)}$ exists in $\mathbf{C}\backslash(-\mathbf{R}^+)$ for all n by using (2.14).
   (b) Assuming the result stated above, find the series expansion of Log about 1. What is the radius of convergence of the series? What happens to the series at 0?

4. (a) Prove that $\zeta(z) = \sum_{n=1}^\infty n^{-z}$ converges on $\{z \mid \mathrm{Re}\,(z) > 1\} = \Omega$.
   (b) Prove that ζ is analytic on Ω.
   (c) Find $\zeta^{(k)}(z)$, $k \geq 1$.
   (d) Find the Taylor coefficients of ζ at 2.
   (e) Notice what happens to ζ at $z = 1$.
   (The function ζ is the famous ζ-function of Riemann, where $n^{-z} = e^{-z\mathrm{Log}(n)}$.)

5. Fix $\alpha \neq 0$. Find the Taylor coefficients $(a_n)_{n=0}^\infty$ of $F_\alpha(z)$ at 1. What is the radius of convergence of the series $\sum_{n=0}^\infty a_n(z - 1)^n$?

6. Use the formulas
$$\frac{1 - z^N}{1 - z} = \sum_{n=0}^{N-1} z^n \qquad \text{and} \qquad z = re^{i\theta}\ (z \neq 0)$$
to evaluate
   (a) $\displaystyle\sum_{n=0}^{N-1} \cos(nx)$
   (b) $\displaystyle\sum_{n=0}^{N-1} \sin(nx)$
   for real x.

7. For $-1 < r < 1$ and real x, show that

(a) $\displaystyle\sum_{n=0}^{\infty} r^n \cos(nx) = \frac{1 - r\cos(x)}{1 - 2r\cos(x) + r^2}$

(b) $\displaystyle\sum_{n=0}^{\infty} r^n \sin(nx) = \frac{r\sin(x)}{1 - 2r\cos(x) + r^2}.$

8. (a) Show that

$$\sin\left[-i \operatorname{Log}(iz + F_{1/2}(1 - z^2))\right] = z, \text{ all } z. \qquad (*)$$

By (a), $\operatorname{Sin}^{-1}(z) = \left[-i \operatorname{Log}(iz + F_{1/2}(1 - z^2))\right]$ defines a (single-valued) inverse to sin on **C**.

(b) Find the discontinuities of Sin^{-1}. Notice that $(*)$ holds if other branches of the logarithm are used. The discontinuities of Sin^{-1} are, of course, different for different branches.

9. Find functions $\operatorname{Cos}^{-1}(z)$, $\operatorname{Tan}^{-1}(z)$, $\operatorname{Cot}^{-1}(z)$, $\operatorname{Sinh}^{-1}(z)$, and $\operatorname{Cosh}^{-1}(z)$ which are inverse to their namesakes in terms of Log, as in 8 for $\operatorname{Sin}^{-1}(z)$.

10. Let a and b be real numbers. Prove the identity

$$F_{ib}\left(\frac{ia - 1}{ia + 1}\right) = e^{-b} \operatorname{Cot}^{-1}\left(\frac{a^2 - 1}{2a}\right).$$

*2.18 CONVERGENCE ON THE BOUNDARY

If $\sum_{n=0}^{\infty} a_n z^n$ has radius of convergence $\rho > 0$, what happens for $|z| = \rho$?

Example 1. $\sum_{n=0}^{\infty} z^n$ does not converge if $|z| = 1$ because the general term does not go to zero.

Example 2. $\sum_{n=0}^{\infty} n^{-2} z^n$ converges uniformly on $\{|z| \leq 1\}$ because $|n^{-2} z^n| \leq n^{-2}$, and the Weierstrass M-test (1.5) applies.

Consider the power series $\sum_{n=1}^{\infty} n^{-1} z^n$, having radius of convergence 1. It diverges at $z = 1$ because

$$\lim_{N \to \infty} \sum_{n=1}^{N} \frac{1}{n} > \lim_{N \to \infty} \int_{[1, N]} \frac{1}{x} \, dx = \infty.$$

The reader is probably aware that the series converges at $z = -1$, conditionally or nonabsolutely. What happens on the rest of $\mathbf{T} = \{z \mid |z| = 1\}$? We will show that the series converges on $\mathbf{T} \backslash \{1\}$. The result is a corollary of a more general technique, which we now describe. First we need a technical device called partial summation.

Let $(a_n)_{n=0}^{\infty}$ and $(b_n)_{n=0}^{\infty}$ be sequences in $\mathbf{C}$. The partial sums

$$S_n = \sum_{k=0}^{n} a_k b_k; \qquad T_n = \sum_{k=0}^{n} a_k$$

satisfy the *partial summation formula*

(i) $\quad S_n = \sum_{k=0}^{n-1} T_k(b_k - b_{k+1}) + T_n b_n.$

To prove (i), write

$$\sum_{k=0}^{n-1} T_k(b_k - b_{k+1}) + T_n b_n = \sum_{k=0}^{n-1} T_k b_k - \sum_{k=0}^{n-1} T_k b_{k+1} + T_n b_n$$

$$= \sum_{k=0}^{n-1} T_k b_k - \sum_{k=1}^{n} T_{k-1} b_k + T_n b_n$$

$$= T_0 b_0 + \sum_{k=1}^{n-1} T_k b_k - \sum_{k=1}^{n-1} T_{k-1} b_k - T_{n-1} b_n + T_n b_n$$

$$= a_0 b_0 + \sum_{k=1}^{n-1} a_k b_k + a_n b_n.$$

The equality (i) is useful when $\lim_{n \to \infty} T_n b_n$ and $\sum_{k=0}^{\infty} T_k(b_k - b_{k+1})$ can be proved to exist. (Compare the partial summation formula with the formula from calculus for integration by parts.) As an example, consider the alternating harmonic series $\sum_{n=1}^{\infty} [(-1)^{n+1}]/n$. Taking $a_n = (-1)^{n+1}$ and $b_n = 1/n$, we obtain $T_n = 1$ for odd n and $T_n = 0$ for even n. Hence we have

$$S_n = \sum_{k=1}^{n-1} T_k \left(\frac{1}{k} - \frac{1}{k+1} \right) + \frac{T_n}{n},$$

from which it follows that

$$\sum_{n=1}^{\infty} \frac{(-1)^{n+1}}{n} = \lim_{n \to \infty} S_n = \sum_{p=1}^{\infty} \left(\frac{1}{2p-1} - \frac{1}{2p} \right) = \frac{1}{2} \sum_{p=1}^{\infty} \frac{1}{p(2p-1)}.$$

We now return to the problem of convergence of a power series on the boundary of its disk of convergence. Let $\sum_{n=0}^{\infty} a_n z^n$ have radius of convergence 1, and let

$$S_n(z) = \sum_{k=0}^{n} a_n z^n; \qquad T_n(z) = \sum_{k=0}^{n} z^k = \frac{z^{n+1} - 1}{z - 1} \qquad (z \neq 1).$$

By partial summation we have

(i) $\quad S_n(z) = \sum_{k=0}^{n-1} (a_k - a_{k+1}) \frac{z^{k+1} - 1}{z - 1} + a_n \frac{z^{n+1} - 1}{z - 1}.$

If $\sum_{k=0}^{\infty} |a_k - a_{k+1}|$ converges and $\lim_{n \to \infty} a_n = 0$, then the estimate

$$\left| \frac{z^{k+1} - 1}{z - 1} \right| \leq \frac{2}{|z - 1|} < \frac{2}{\eta} \qquad (|z - 1| > \eta)$$

and the Weierstrass M-test (1.5) show that the functions S_n converge uniformly on the set

$$L_\eta = \{z \mid |z| \leq 1 \quad \text{and} \quad |z - 1| > \eta\}$$

for every $\eta > 0$. In particular, $\sum_{n=0}^{\infty} a_n z^n$ converges for $|z| = 1$ and $z \neq 1$. We have thus proved the following theorem.

Theorem. If $\sum_{n=0}^{\infty} a_n z^n$ has radius of convergence 1, $\sum_{k=0}^{\infty} |a_k - a_{k+1}|$ converges, and $\lim_{n \to \infty} a_n = 0$, then $\sum_{n=0}^{\infty} a_n z^n$ converges uniformly on $\{z \mid |z| \leq 1 \text{ and } |z - 1| > n\}$ for every $\eta > 0$.

The singular point $z = 1$ can easily be moved to ξ ($|\xi| = 1$). Consider the series $\sum_{n=0}^{\infty} a_n (\bar{\xi} z)^n = \sum_{n=0}^{\infty} (a_n \bar{\xi}^n) z^n$. The series on the left converges if $\bar{\xi} z \neq 1$; that is, if $z \neq \xi$. Hence the power series on the right converges if $z \neq \xi$, $|z| \leq 1$.

The result applies to the case $a_n = n^{-1}$, because $\sum_{n=1}^{\infty} [n^{-1} - (n + 1)^{-1}]$ converges. In fact, the result applies to any series in which the coefficients form a positive sequence decreasing to zero.

EXERCISES

TYPE I

1. (a) Use the equality (2.18.i) to show that $\sum_{k=0}^{\infty} a_k b_k$ converges if
    (i) $(T_n)_{n=1}^{\infty}$ is a bounded sequence
    (ii) $\sum_{k=0}^{\infty} |b_k - b_{k+1}|$ converges
    (iii) $\lim_{n \to \infty} b_n = 0$.
   (b) Show that $\sum_{k=0}^{\infty} a_k b_k$ converges if
    (i) $(T_n)_{n=1}^{\infty}$ is a bounded sequence
    (ii) $(b_k)_{k=0}^{\infty}$ is a positive decreasing sequence
    (iii) $\lim_{n \to \infty} b_n = 0$.

2. Show that the following series converge.
    (a) $\sum_{n=1}^{\infty} i^n / n$
    (b) $\sum_{n=1}^{\infty} i^n / \sqrt{n}$
    (c) $\sum_{n=2}^{\infty} i^n / \log(n)$.
   Notice that the series obtained by replacing i in (a–c) by 1 all diverge.

3. First show that $\sum_{n=1}^{\infty} (\sqrt{n + 1} - \sqrt{n})/n$ converges and then show that $\sum_{n=1}^{\infty} (\sqrt{n + 1} - \sqrt{n})/\sqrt{n}$ diverges. Determine convergence or divergence of $\sum_{n=1}^{\infty} (\sqrt{n + 1} - \sqrt{n})/n^a$, $\frac{1}{2} < a < 1$.

TYPE II

1. Find the set $C = \{z \mid \sum_{n=0}^{\infty} a_n z^n$ converges$\}$ for the following power series.
   (a) $a_n = n^{-\varepsilon} (0 < \varepsilon < \infty)$
   (b) $a_n = [\log (n)]^{-1}, n > 1; a_0 = a_1 = 0$
   (c) $a_n = \log (n), n \geq 1; a_0 = 0$
   (d) $a_n = \log (n)/n, n \geq 1; a_0 = 0.$
   (e) $b_n = \xi^n a_n (|\xi| = 1)$ for each of (a–d)
   (f) Define b_1 on **R** by $b_1(x) = \log (x)$ if $x > 1$, $b_1(x) = \infty$ if $x \leq 1$. Define b_k inductively by $b_k(x) = b_1(b_{k-1}(x))$. Let $a_n = [b_k(n)]^{-1}, k$ fixed.
   (g) Let $a_n = b_n(n)^{-1}, b_n$ as in (f).

2. Show that $\lim_{n \to \infty} [\log n - \sum_{k=1}^{n} k^{-1}]$ exists. Give an upperbound for its value.

3. If $\sum_{n=0}^{\infty} |a_n|$ converges, show that $\sum_{n=0}^{\infty} a_n z^n$ converges uniformly on $\overline{D(1, 0)} = \{z \mid |z| \leq 1\}$.

4. Let $\xi_1, \xi_2, \ldots, \xi_k$ be fixed points on **T** $= \{z \mid |z| = 1\}$. Construct a power series $\sum_{n=0}^{\infty} a_n z^n$ with radius of convergence 1 having convergence set precisely $\overline{D(1, 0)} \setminus \{\xi_1, \ldots, \xi_k\}$.

5. Suppose that $f(z) = \sum_{n=1}^{\infty} a_n z^n$ has radius of convergence $\rho > 0$ and that the series converges at $\xi, |\xi| = \rho$. Show that $\lim_{r \to \rho, r < \rho} f(r\xi) = \sum_{n=1}^{\infty} a_n \xi^n$.

6. Fill in the details of the following alternate proof of (2.18.i).

$$
\begin{aligned}
S_n &= a_0 b_0 + a_1 b_1 + \cdots + a_n b_n \\
&= a_0(b_0 - b_1) + a_0(b_1 - b_2) + \cdots + a_0(b_{n-1} - b_n) + a_0 b_n \\
&\quad + a_1(b_1 - b_2) + a_1(b_2 - b_3) + \cdots + a_1(b_{n-1} - b_n) + a_1 b_n \\
&\quad + a_2(b_2 - b_3) + a_2(b_3 - b_4) + \cdots + a_2(b_{n-1} - b_n) + a_2 b_n \\
&\quad + \cdots + \\
&\quad + a_{n-1}(b_{n-1} - b_n) + a_{n-1} b_n + a_n b_n \\
&= a_0(b_0 - b_1) + (a_0 + a_1)(b_1 - b_2) \\
&\quad + (a_0 + a_1 + a_2)(b_2 - b_3) + \cdots + T_{n-1}(b_{n-1} - b_n) + T_n b_n.
\end{aligned}
$$

CHAPTER 3

Integration

In Chapters 1 and 2, we examined two basic tools of complex analysis, the complex derivative and power series. In this chapter, we introduce the third basic tool, line integrals in the complex plane. These line integrals are integrals of the form $\int_C f(z)\, dz$ where C is a curve in the complex plane and f is a function defined on C. Recall that a curve C is the image of a real interval $[a, b]$ under a continuous mapping α. Thus the integral $\int_C f(z)\, dz$ is equal to $\int_{[a,b]} f(\alpha(t))\, d\alpha(t)$, the integral of the complex-valued function $f \circ \alpha$ with respect to a complex-valued weight function.

As an example, consider the set $C = \{z \mid x = t,\, y = t^3,\, 0 \le t \le 1\}$. C is the image of the weight function α, with $\alpha(t) = t + it^3$. Let $f(z) = (x^2 + y^2) + ixy$, so that $f(\alpha(t)) = t^2 + t^6 + it^4$. Since $d\alpha/dt = (1 + 3it^2)$, it is suggestive to write $dz = d\alpha = (1 + 3it^2)\, dt$ and compute $\int_C f(z)\, dz$ as follows:

$$\int_C f(z)\, dz = \int_0^1 (t^2 + t^6 + it^4)(1 + 3it^2)\, dt$$

$$= \int_0^1 [(t^2 - 2t^6) + i(4t^4 + 3t^8)]\, dt$$

$$= \int_0^1 (t^2 - 2t^6)\, dt + i \int_0^1 (4t^4 + 3t^8)\, dt = \frac{1}{21} + i\,\frac{17}{15}.$$

As a matter of practice one might rephrase the above example less formally: C is $y = x^3,\, 0 \le x \le 1$ and $f(z) = (x^2 + y^2) + ixy$. Then $\alpha(x) = x + ix^3$ and

$$\int_C f(z)\, dz = \int_0^1 (x^2 + x^6 + ix^4)(1 + 3ix^2)\, dx = \frac{1}{21} + i\,\frac{17}{15}.$$

In the first part of the chapter, we develop the integration theory needed for the rest of the book. Calculations of the above type will be justified by this theory. We then define and discuss the elementary properties of the index of a closed path in Sections (3.15–3.17). This material is also essential to much of the subsequent development. We conclude the chapter with an elementary discussion of Fourier series. This material will be used only in starred sections. We include it as a reference so that we can point out connections between various parts of complex analysis and Fourier series.

3.1 DERIVATIVES AND INTEGRALS OF COMPLEX FUNCTIONS OF A REAL VARIABLE

A complex-valued function $f = g + ih$ of a real variable defined on $[a, b]$ has derivative f' on $]a, b[$ if

$$\lim_{\substack{t \to x \\ t \in]a, b[}} \frac{f(t) - f(x)}{t - x} = f'(x)$$

holds for all x in $]a, b[$. It is easy to see that f' exists if and only if g' and h' exist and then $f' = g' + ih'$. Similarly, the integral $\int_{[a, b]} f(x)\, dx$ is defined by

$$\int_{[a, b]} f(x)\, dx = \int_{[a, b]} g(x)\, dx + i \int_{[a, b]} h(x)\, dx,$$

provided that the integrals of g and h exist; for example, it suffices that g and h be continuous. As an example, we have

$$\int_{[0, \pi/6]} e^{it}\, dt = \int_{[0, \pi/6]} [\cos t + i \sin t]\, dt$$

$$= \int_{[0, \pi/6]} \cos t\, dt + i \int_{[0, \pi/6]} \sin t\, dt = \frac{1}{2} + \frac{2 - \sqrt{3}}{2}\, i.$$

The integral $\int_{[a, b]} f(t)\, dt$ is linear in f; that is,

(i) $$\int_{[a, b]} [f(t) + g(t)]\, dt = \int_{[a, b]} f(t)\, dt + \int_{[a, b]} g(t)\, dt$$

and

(ii) $$\int_{[a, b]} \lambda f(t)\, dt = \lambda \int_{[a, b]} f(t)\, dt$$

hold if f and g are complex-valued and λ is a complex number. The reader can easily establish these equalities from the corresponding ones for real-valued functions. The fundamental theorems of calculus, namely,

(iii) $\displaystyle\int_{[a,\,b]} f'(x)\,dx = f(b) - f(a)$

and

(iv) $\displaystyle\frac{d}{dx}\int_{[a,\,x]} f(t)\,dt = f(x)$

are also easily deduced for complex f from the corresponding results for real f. Equality (iv) holds for all functions f continuous on $[a, b]$. Equality (iii) holds if f' exists on $]a, b[$ and $f'(a)$ and $f'(b)$ can be defined so that f' is continuous on $[a, b]$.

Recall that the inequality

(v) $\displaystyle\int_{[a,\,b]} g(t)\,dt \le \int_{[a,\,b]} h(t)\,dt$

holds for real-valued functions g and h if $g(t) \le h(t)$ for all t in $[a, b]$. For complex-valued functions, the analog of this inequality is the inequality

(vi) $\displaystyle\left|\int_{[a,\,b]} f(t)\,dt\right| \le \int_{[a,\,b]} |f(t)|\,dt.$

To prove (vi), let $\theta = \operatorname{Arg}\left[\int_{[a,\,b]} f(t)\,dt\right]$. We have

(vii) $\displaystyle\left|\int_{[a,\,b]} f(t)\,dt\right| = e^{-i\theta}\int_{[a,\,b]} f(t)\,dt = \int_{[a,\,b]} e^{-i\theta}f(t)\,dt$

by (ii). Equality (vii) shows that $\int_{[a,\,b]} e^{-i\theta}f(t)\,dt$ is real. It is thus equal to the integral of its real part, so that we have

$$\left|\int_{[a,\,b]} f(t)\,dt\right| = \int_{[a,\,b]} \operatorname{Re}\left(e^{-i\theta}f(t)\right)\,dt.$$

Since $\operatorname{Re}\left(e^{-i\theta}f(t)\right) \le |e^{-i\theta}f(t)| = |f(t)|$, we obtain (vi) by applying (v).

We leave it to the reader to satisfy himself that the chain rule, integration by parts, and integration by substitution for complex-valued functions of a real variable all follow easily from the real-valued versions. As appropriate, one simply applies the real versions to the real and imaginary parts of the complex-valued functions.

3.2 CURVES AND PATHS

Let α be a continuous complex-valued function on a real interval $I = [a, b]$. As t moves from a to b, one obtains a mental image of α stretching I into a smooth curve in $\mathbf{C}$, with initial point $\alpha(a)$ and terminal point $\alpha(b)$. See Figure

3.1. We define a curve in **C** to be precisely this: a continuous function from a finite closed interval into **C**. Thus a curve is not only a set of points, but also has orientation. To distinguish the point set from the curve α we use a different notation. The point set is denoted α^*; that is, $\alpha^* = \{\alpha(t) \mid t \in I\}$. If α' exists in $]a, b[$ and can be defined at a and b so as to be continuous on $[a, b]$, then we say that α is a *continuously differentiable curve*. If α is a continuous function on $[a, b]$ for which there are points $t_1 < t_2 < \cdots < t_n = b$ in $[a, b]$ such that α' exists on each $]t_{j-1}, t_j[$ $(t_0 = a)$ and α' is continuous on $[t_{j-1}, t_j]$ (that is, for each j there is a function w which is continuous on $[t_{j-1}, t_j]$ and equal to α' on $]t_{j-1}, t_j[$), then α is said to be *piecewise continuously differentiable*, or a *path*. The curves which will be treated in this book will be paths.

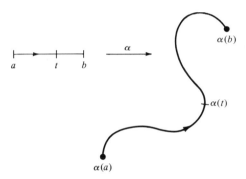

Figure 3.1

The negative, $-\alpha$, of a curve α is defined by

$$(-\alpha)(t) = \alpha(b + a - t).$$

Hence $(-\alpha)(a) = \alpha(b)$ and $(-\alpha)(b) = \alpha(a)$; $-\alpha(t)$ moves from $\alpha(b)$ to $\alpha(a)$.) It is emphasized that the minus sign ("$-$") as used in this context denotes an operation on curves and not an operation on numbers. Notice that $(-\alpha)'(t) = -\alpha'(t)$ if $\alpha'(t)$ exists.

Example. Let $\alpha(t) = e^{it}$, $0 \le t < \pi$. Thus α^* is a semicircle in the upper half-plane. The point $\alpha(t)$ moves counterclockwise as t moves from 0 to π. The negative of α is given by $(-\alpha)(t) = e^{i(\pi - t)}$. Notice that $(-\alpha)^* = \alpha^*$, but that the point $(-\alpha)(t)$ moves clockwise from -1 to 1 as t moves from 0 to π.

A curve α is *closed* if $\alpha(a) = \alpha(b)$; it is *simple* if

$$\alpha(t) \ne \alpha(s) \quad \text{if} \quad t \in \ ]a, b[\text{ and } s \in [a, b]$$

[α does not cross itself, except perhaps $\alpha(a) = \alpha(b)$].

3.3 EXAMPLES

(a) Let $a, b \in \mathbf{C}$, $a \neq b$. The line through the origin and parallel to the line determined by a and b is $\{t(b - a) \mid t \in \mathbf{R}\}$. Hence the line determined by a and b is $\{a + t(b - a) \mid t \in \mathbf{R}\}$. The portion of the line between a and b is $\{a + t(b - a) \mid t \in [0, 1]\}$. Hence if we let $\alpha(t) = a + t(b - a)$ we obtain a path defined on $[0, 1]$ whose image is the line segment from a to b. We often denote α by $\overrightarrow{[a, b]}$. See Figure 3.2. As an example, $\overrightarrow{[0, 1 + i]}$ is $t + it$, $t \in [0, 1]$.

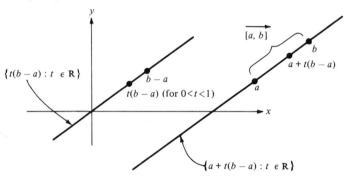

Figure 3.2

(b) If $a \in \mathbf{C}$ and $r > 0$, then let $\alpha(t) = a + re^{it}$, $t \in [-\pi, \pi]$; α is the circle of radius r about a with counterclockwise orientation. We will sometimes denote α as $\overrightarrow{C(r, a)}$. See Figure 3.3.

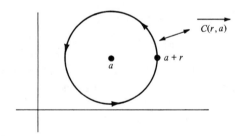

Figure 3.3

(c) If y is an explicit continuous real-valued function of x on $[a, b]$, then we can make the graph of y into a path (in the above sense) by simply letting $\alpha(x) = x + iy(x)$ on $[a, b]$. For example, we can parametrize the parabola $y = x^2$ on $[0, 5]$ to obtain a path by letting

$$\alpha(x) = x + ix^2.$$

(d) Define α on $[-1, 1]$ by $\alpha(x) = x + i|x|$. Hence $\alpha'(x) = 1 - i$ on $]-1, 0[$ and $\alpha'(x) = 1 + i$ on $]0, 1[$. The function α' is discontinuous at 0, but is piecewise continuously differentiable, and hence a path on $[-1, 1]$. See Figure 3.4.

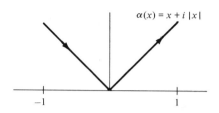

$\alpha(x) = x + i\,|x|$

-1 1

Figure 3.4

Remark. Most readers will have studied vector-valued functions from intervals into $\mathbf{R}^2$; that is, functions of the form $\bar{P}(t) = (x(t), y(t))$, where x and y are real-valued functions of the real variable t. The functions α we are discussing are, of course, the same functions, with emphasis on complex notation.

3.4 INTEGRATION WITH RESPECT TO A COMPLEX WEIGHT FUNCTION AND LINE INTEGRALS

If α is a continuously differentiable function from $[a, b]$ to $\mathbf{C}$ and f is a continuous complex-valued function on $[a, b]$, then the integral of f with respect to the complex-valued weight function α is denoted $\int_{[a, b]} f \, d\alpha$ or $\int_{[a, b]} f(t) \, d\alpha(t)$ and is defined by

(i) $$\int_{[a, b]} f \, d\alpha = \int_{[a, b]} f(t)\alpha'(t) \, dt.$$

If α is a path whose derivative α' has intervals of continuity $[t_{j-1}, t_j]$ as in (3.2), then $\int_{[a, b]} f \, d\alpha$ is defined by

(ii) $$\int_{[a, b]} f \, d\alpha = \sum_{j=1}^{n} \int_{[t_{j-1}, t_j]} f \, d\alpha = \sum_{j=1}^{n} \int_{[t_{j-1}, t_j]} f(t)\alpha'(t) \, dt$$

If f is a continuous complex-valued function on the image α^* of a path α, then $f \circ \alpha$ is continuous on $[a, b]$. The *line integral of f over α* is denoted $\int_{\alpha} f \, dz$ or $\int_{\alpha} f(z) \, dz$ and is defined by

(iii) $$\int_{\alpha} f(z) \, dz = \int_{[a, b]} [f \circ \alpha] \, d\alpha.$$

By (i) we have

(iv) $\displaystyle\int_\alpha f(z)\,dz = \int_{[a,\,b]} f(\alpha(t))\alpha'(t)\,dt$

if α' is continuous on $[a, b]$, and in all cases $\int_\alpha f(z)\,dz$ is a sum of integrals of the form $\int_{[t_{j-1},\,t_j]} f(\alpha(t))\alpha'(t)\,dt$. For example, if $\alpha(t) = e^{it}$, $0 \le t \le \pi$, and $f(z) = 1/z$, then

$$\int_\alpha f(z)\,dz = \int_\alpha \frac{1}{z}\,dz = \int_{[0,\,\pi]} \frac{1}{e^{it}}\,ie^{it}\,dt = i\pi.$$

The basic linearity properties

(v) $\displaystyle\int_\alpha \lambda f(z)\,dz = \lambda \int_\alpha f(z)\,dz \qquad (\lambda \in \mathbf{C})$

and

(vi) $\displaystyle\int_\alpha [f(z) + g(z)]\,dz = \int_\alpha f(z)\,dz + \int_\alpha g(z)\,dz$

follow easily from the linearity of integrals $\int_{[a,\,b]} F(t)\,dt$. Corresponding equalities of course hold for integrals $\int_{[a,\,b]} f\,d\alpha$. Since $(-\alpha)'(t) = -\alpha'(t)$, the equality

(vii) $\displaystyle\int_{-\alpha} f(z)\,dz = -\int_\alpha f(z)\,dz$

is valid. For example, if $\alpha(t) = e^{it}$, then $(-\alpha)(t) = e^{i(\pi-t)}\ 0 \le t \le \pi$ and

$$\int_{-\alpha} \frac{1}{z}\,dz = \int_{[0,\,\pi]} \frac{1}{e^{i(\pi-t)}}\,(-i)e^{i(\pi-t)}\,dt = -i\pi.$$

*3.5 REMARK

By definition, the integral $\int_\alpha f(z)\,dz$ is the limit of approximating sums $\sum_{j=1}^n f(\alpha(t_j'))\alpha'(t_j')[t_j - t_{j-1}]$, where $a = t_0 < t_1 < \cdots < t_n = b$ and $t_{j-1} \le t_j' \le t_j$. The product $\alpha'(t_j')(t_j - t_{j-1})$ is close to $\alpha(t_j) - \alpha(t_{j-1})$ for suitable t_j', as n increases. Using the mean-value theorem, it can be shown that the sums $\sum_{j=1}^n f(\alpha(t_j'))[\alpha(t_j) - \alpha(t_{j-1})]$ also converge to $\int_\alpha f\,dz$. Letting $z_j' = \alpha(t_j')$ and $z_j = \alpha(t_j)$, we thus see that

$$\int_\alpha f(z)\,dz = \lim_{n\to\infty} \sum_{j=1}^n f(z_j')(z_j - z_{j-1})$$

where the z_j are points on the path and z_j' lies on the path between z_{j-1} and z_j.

3.6 THE FUNDAMENTAL THEOREM OF CALCULUS FOR LINE INTEGRALS

Let U be an open set in $\mathbf{C}$, f an analytic function on U such that f' is continuous on U, and α a path in U defined on $[a, b]$. We have

(i) $\displaystyle\int_\alpha f'(z) \, dz = f(\alpha(b)) - f(\alpha(a));$

(ii) $\displaystyle\int_\alpha f'(z) \, dz = 0 \qquad$ if α is closed.

PROOF.   First, suppose that α' is continuous on $[a, b]$. Using (3.1.iii), we have

$$\int_\alpha f'(z) \, dz = \int_{[a, b]} (f' \circ \alpha)(t)\alpha'(t) \, dt$$

$$= \int_{[a, b]} (f \circ \alpha)'(t) \, dt = f(\alpha(b)) - f(\alpha(a)).$$

If α is a path and α' has intervals of continuity $[t_{j-1}, t_j]$, then we have

$$\int_\alpha f'(z) \, dz = \sum_{j=1}^{n} \int_{[t_{j-1}, t_j]} (f' \circ \alpha)(t)\alpha'(t) \, dt$$

$$= \sum_{j=1}^{n} [f(\alpha(t_j)) - f(\alpha(t_{j-1}))] = f(\alpha(b)) - f(\alpha(a)).$$

Equality (ii) is valid because $f(\alpha(b)) = f(\alpha(a))$ if α is closed.

Remarks.   Equality (i) says that the integral of the derivative of an analytic function over a path depends only on the end points of the path; that is, it is independent of the path. Equality (ii) is a weak form of Cauchy's theorem. This theorem asserts that $\int_\alpha f(z) \, dz = 0$ for any closed path provided that f is analytic in an open set U *without* "*holes*" which contains α. Chapter 4 is devoted to this theorem and some of its consequences. It is the central theorem of complex analysis.

3.7 EXAMPLES

To illustrate the definitions and the use of (3.6), we calculate $\int_\alpha z^2 \, dz$ for each of the α in (3.3). Since $z^2 = d(z^3/3)/dz$, we can use (3.6).

(a) $\displaystyle\int_{\overrightarrow{[0, 1+i]}} z^2 \, dz = \int_{\overrightarrow{[0, 1+i]}} \frac{d}{dz}\left(\frac{z^3}{3}\right) dz = \frac{(1 + i)^3}{3} - \frac{0^3}{3} = \frac{-2 + 2i}{3}.$

(b) $\displaystyle\int_{\overrightarrow{C(r, a)}} z^2 \, dz = 0$ by (3.6.ii) $[\overrightarrow{C(r, a)}$ is closed$]$.

(c) Here $\alpha(t) = t + it^2$ on $[0, 5]$. We have

$$\int_\alpha z^2 \, dz = \frac{(5 + i25)^3}{3} = \frac{-3250 + 16250i}{3}.$$

(d) We have

$$\int_\alpha z^2 \, dz = \frac{(1 + i)^3}{3} - \frac{(-1 + i)^3}{3} = \frac{4}{3} i.$$

For contrast, let us next compute $\int |z|^2 \, dz$ over the same paths. Since $|z|^2$ is not analytic, we cannot use (3.6).

(a) $\displaystyle\int_{\overrightarrow{[0, 1+i]}} |z|^2 \, dz = \int_{[0, 1]} |t + it|^2 (1 + i) \, dt.$

$$= (1 + i) \int_{[0, 1]} 2t^2 \, dt = \frac{2}{3} (1 + i).$$

(b) If z is a point on $C(r, a)$, then $z = a + re^{it}$ for some t. Hence

$$|z|^2 = (a + re^{it})(\bar{a} + re^{-it}) = a\bar{a} + are^{-it} + \bar{a}re^{it} + r^2.$$

If $\alpha = \overrightarrow{C(r, a)}$, then $\alpha'(t) = (a + re^{it})' = ire^{it}$ and we have

$$\int_{\overrightarrow{C(r, a)}} |z|^2 \, dz = \int_{[-\pi, \pi]} (a\bar{a} + are^{-it} + \bar{a}re^{it} + r^2)ire^{it} \, dt$$

$$= (a\bar{a} + r^2)ir \int_{[-\pi, \pi]} e^{it} \, dt + i\bar{a}r^2 \int_{[-\pi, \pi]} e^{2it} \, dt + iar^2 \int_{[-\pi, \pi]} dt$$

$$= 2\pi ir^2 a.$$

Notice that in both (a) and (b) we have used (3.1.iii). However, we used (3.6.ii) to determine without further computation that

$$\int_{[-\pi, \pi]} e^{it} \, dt = 0 \quad \text{and} \quad \int_{[-\pi, \pi]} e^{2it} \, dt = 0$$

since we recognized them as the integrals of $\frac{1}{2}(dz^2/dz) = z$ and $\frac{1}{3}(dz^3/dz) = z^2$, respectively, over the closed path $\overrightarrow{C(1, 0)}$. We continue with the examples.

(c) $\displaystyle\int_\alpha |z|^2 \, dz = \int_{[0, 5]} |t + it^2|^2 (1 + i2t) \, dt$

$$= \int_{[0, 5]} [(t^2 + t^4) + i(2t^3 + 2t^5)] \, dt$$

$$= \left(\frac{5^3}{3} + \frac{5^5}{5}\right) + i\left(\frac{5^4}{2} + \frac{5^6}{3}\right)$$

(d) $\displaystyle\int_\alpha |z|^2\, dz = \int_{[-1,\,0]} |t - it|^2 (1 - i)\, dt + \int_{[0,\,1]} |t + it|^2 (1 + i)\, dt$

$$= \int_{[-1,\,1]} 2t^2\, dt = \tfrac{4}{3}.$$

(e) As another example, let $\alpha(t) = t + it$ on $[0, 1]$ and let $f(z) = \bar{z}$.

We have $f(\alpha(t)) = t - it$ and thus

$$\int_\alpha \bar{z}\, dz = \int_{[0,\,1]} (t - it)\alpha'(t)\, dt$$

$$= \int_{[0,\,1]} (t - it)(1 + i)\, dt = \int_{[0,\,1]} 2t\, dt = 1.$$

However, if $\beta(t) = t$ on $[0, 1]$ and $\beta(t) = 1 + (t - 1)i$ on $[1, 2]$, then $f(\beta(t)) = t$ holds on $[0, 1]$ and $f(\beta(t)) = 1 - (t - 1)i$ on $[1, 2]$. Thus we have

$$\int_\beta \bar{z}\, dz = \int_{[0,\,1]} t\, dt + \int_{[1,\,2]} [1 - (t - 1)i]i\, dt$$

$$= \tfrac{1}{2} + \int_{[1,\,2]} (t - 1)\, dt + i \int_{[1,\,2]} 1\, dt = 1 + i.$$

Thus $\int_\beta \bar{z}\, dz \neq \int_\alpha \bar{z}\, dz$, although α and β have the same initial and the same terminal points. Theorem (3.6) does not apply because $\bar{z}$ is not analytic. See Figure 3.5.

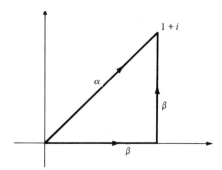

Figure 3.5

EXERCISES

TYPE I

1. A graph $(x, y(x))$ where y is a continuous function of x on $[a, b]$ is the image of a curve α defined on $[a, b]$ by

$$\alpha(x) = x + iy(x).$$

Sketch the image of α for the following examples, and find $\alpha'(x)$. Identify the curve.

(a) $y(x) = x^2$ on $[-1, 1]$
(b) $y(x) = 2\sqrt{1 - x^2}$ on $[0, 1]$
(c) $y(x) = 2\sqrt{1 + x^2}$ on $[0, 1]$
(d) $y(x) = e^x$ on $[0, 1]$.

2. If $\alpha(t) = x(t) + iy(t)$ is a curve on $[a, b]$, we can often identify its image by eliminating t to find an equation satisfied by x and y. As a simple example, if $\alpha(t) = t + it^2$, we have $y(t) = (x(t))^2$, and the image is part of the parabola defined by $y = x^2$. Use this technique to identify the images of the following curves. Find $\alpha'(t)$ in each case.

(a) $\alpha(t) = t + i\sqrt{t}$ on $[0, 1]$
(b) $\alpha(t) = \cos t + i \sin^2 t$ on $[0, \pi]$
(c) $\alpha(t) = e^t + i \log t$ on $[1, 10]$
(d) $\alpha(t) = 1 + ie^t$ on $[1, 5]$
(e) $\alpha(t) = e^t + i$ on $[1, 5]$
(f) $\alpha(t) = t + i$ on $[1, 5]$.

3. The weight functions α of (3.2–3.7) can of course be real-valued. The integrals $\int f \, d\alpha$ are then special cases of Riemann-Stieltjes integrals. Compute the following Riemann-Stieltjes integrals.

(a) $\int_{[1, 2]} x^2 \, d\alpha(x)$, $\alpha(x) = \ln(x)$
(b) $\int_{[-1, 1]} \sin x \, d\alpha(x)$, $\alpha(x) = e^x$
(c) $\int_{[-1, 1]} \cos x \, d\alpha(x)$, $\alpha(x) = \sin x$
(d) $\int_{[-\pi, \pi]} 1 \, d\alpha(x)$, $\alpha(x) = \cos x$
(e) $\int_{[-\pi, \pi]} e^x \, d\alpha(x)$, $\alpha(x) = \sin x$.

4. Compute the line integrals $\int_\alpha f(z) \, dz$ for the given f and α. Identify the graph of α in each case. Use (3.6) whenever possible.

(a) $\alpha(t) = e^{it}$ on $[0, \pi]$
     (i) $f(z) = z$
     (ii) $f(z) = \sin(z^2)$
     (iii) $f(z) = \sin |z|$
     (iv) $f(z) = \bar{z}$.

(b) $\alpha(t) = t + it$ on $[0, 1]$
   (i) $f(z) = 1/(z + 1)^2$
   (ii) $f(z) = 1/(z + 1)$
(c) $\alpha(t) = e^{it}$ on $[-\pi, \pi]$
   (i) $f(z) = 1/z^2$
   (ii) $f(z) = e^z$
   (iii) $f(z) = e^{|z|}$
   (iv) $f(z) = \bar{z}$.

5. Show that $\int_\alpha e^z \, dz = e^w - 1$ for any path α joining 0 and w.

6. Show that $\int_\alpha \sin z \, dz = \cos w$ for any path α joining 0 and w. What is $\int_\alpha \cos z \, dz$?

7. Let w be a point in the open upper half-plane and let α be a path lying in the open upper half-plane (except for 1) joining 1 and w. Show that

$$\int_\alpha \frac{1}{z} \, dz = \text{Log}\,(w).$$

3.8 INDEPENDENCE OF PARAMETRIZATION

Let α be a path on $[a, b]$. Let ϕ be a strictly increasing continuously differentiable function on $[a, b]$ to $[\phi(a), \phi(b)]$, and let ψ denote the inverse of ϕ. Suppose that f is continuous on α^*. We have

$$\int_{[a,\, b]} (f \circ \alpha) \, d\alpha = \int_{[\phi(a),\, \phi(b)]} (f \circ \alpha \circ \psi) \, d(\alpha \circ \psi),$$

that is

$$\int_\alpha f(z) \, dz = \int_{\alpha \circ \psi} f(z) \, dz.$$

PROOF.   Again we calculate, using the appropriate definitions. We suppose first that α is continuously differentiable, and write

$$\int_{[a,\, b]} (f \circ \alpha) \, d\alpha = \int_{[a,\, b]} f(\alpha(t))\alpha'(t) \, dt$$

$$= \int_{[\phi(a),\, \phi(b)]} f(\alpha(\psi(t)))\alpha'(\psi(t))\psi'(t) \, dt$$

$$= \int_{[\phi(a),\, \phi(b)]} [f \circ \alpha \circ \psi](t)[\alpha \circ \psi]'(t) \, dt$$

$$= \int_{[\phi(a),\, \phi(b)]} f \circ (\alpha \circ \psi) \, d(\alpha \circ \psi).$$

The first equality here is a definition, the second is by the change of variables theorem for complex functions of a real variable, the third uses the chain rule in the integrand, and the fourth is a definition.

If α is a path but is not continuously differentiable, then the equality is established by applying the equality to each interval where it is continuously differentiable and then summing.

3.9 DEFINITION OF EQUIVALENT PATHS

Two paths α and β defined on intervals $[a, b]$ and $[c, d]$, respectively, are *equivalent* if there is a strictly increasing continuously differentiable function ϕ from $[a, b]$ onto $[c, d]$ such that $\alpha = \beta \circ \phi$. In these terms, if α and β are equivalent paths, then by (3.8) we have

$$\int_\alpha f \, dz = \int_\beta f \, dz$$

for every continuous function f on $\alpha^*(=\beta^*)$.

3.10 ORIENTED UNIONS OF PATHS

If α and β are paths on $[a, b]$ and $[a_1, b_1]$ such that $\alpha(b) = \beta(a_1)$, then we can define a new path by switching from α to β at $\alpha(b)$. To do this, let $c = b + (b_1 - a_1)$ and map $[b, c]$ to $[a_1, b_1]$ by

$$\phi(t) = t + (a_1 - b)$$

and let $\tilde{\beta} = \beta \circ \phi$ on $[b, c]$. Then $\tilde{\beta}$ is equivalent to β and we define

(i) $\alpha \smile \beta(t) = \begin{cases} \alpha(t) \text{ if } t \in [a, b] \\ \tilde{\beta}(t) \text{ if } t \in [b, c]. \end{cases}$

Thus $\alpha \smile \beta$ is well-defined on the interval of parametrization $[a, c]$. It is not difficult to prove that $\alpha \smile \beta$ is again a path; $\alpha \smile \beta$ is not likely to be differentiable at the point b. By definition [(3.4.iii) and (3.4.ii)], the integral $\int_{\alpha \smile \beta} f(z) \, dz$ is equal to $\int_\alpha f(z) \, dz + \int_{\tilde{\beta}} f(z) \, dz$. Thus by (3.8) we have

(ii) $\displaystyle\int_{\alpha \smile \beta} f(z) \, dz = \int_\alpha f(z) \, dz + \int_\beta f(z) \, dz$

for continuous f.

The definition of $\alpha \smile \beta$ extends easily to $\alpha_1 \smile \alpha_2 \smile \cdots \smile \alpha_m$ where the initial point of α_i is the terminal point of α_{i-1}. Notice that every path is an oriented union of continuously differentiable paths.

As an example, let us compute $\int_\alpha |z|^2 \, dz$ along the path α indicated in Figure 3.6. It would be awkward to parametrize α on one interval by one function, but writing it as the oriented union of two line segments, we have $\alpha = \overrightarrow{[1 + i, 0]} \smile \overrightarrow{[0, 1]}$ and

$$\int_\alpha |z|^2 \, dz = -\int_{[0,\,1]} |t + it|^2 (1 + i) \, dt + \int_{[0,\,1]} t^2 \, dt = -\tfrac{1}{3} - \tfrac{2}{3}i.$$

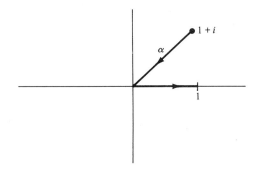

Figure 3.6

3.11 THE VARIATION FUNCTION

Let α be a path defined on $[a, b]$. The *variation function* v_α is defined on $[a, b]$ by

(i) $v_\alpha(t) = \displaystyle\int_{[a,\,t]} |\alpha'(s)| \, ds.$

We have $v_\alpha'(t) = |\alpha'(t)|$ if t is a point of continuity of α'. Thus v_α is also a path (its image is $[0, v_\alpha(b)]$), and so $\int_{[a,\,b]} f \, dv_\alpha$ is well-defined if f is continuous on $[a, b]$ and we have

(ii) $\displaystyle\int_{[a,\,b]} f \, dv_\alpha = \int_{[a,\,b]} f(t)|\alpha'(t)| \, dt.$

If f is a continuous function on α^*, then we define $\int_\alpha f(z)|dz|$ by

(iii) $\displaystyle\int_\alpha f(z)|dz| = \int_{[a,\,b]} f(\alpha(t)) \, dv_\alpha(t)$

and say that $\int_\alpha f(z) \, |dz|$ is the integral of f with respect to arc length.

The following inequalities are important (notation is as above).

(iv) $\left| \int f \, d\alpha \right| \leq \int |f| \, dv_\alpha, f$ continuous on $[a, b]$.

(v) $\left| \int_\alpha f(z) \, dz \right| \leq \int_\alpha |f(z)| \, |dz|, f$ continuous on α^*.

To prove (iv) observe that

$$\left| \int_{[a, b]} f \, d\alpha \right| = \left| \int_{[a, b]} f(t)\alpha'(t) \, dt \right| \leq \int_{[a, b]} |f(t)| \, |\alpha'(t)| \, dt = \int_{[a, b]} |f| \, dv_\alpha.$$

If f is continuous on α^*, then $f \circ \alpha$ is continuous on $[a, b]$, and so (v) follows from (iv).

3.12 ARC LENGTH

Let α be a path defined on $[a, b]$. We define the *length* of α to be the number $v_\alpha(b)$, the total variation of α. Hence

(i) length of $\alpha = \displaystyle\int_{[a, b]} |\alpha'(t)| \, dt.$

For example, if $\alpha = \overrightarrow{[z_1, z_2]}$ then $\alpha' = (z_2 - z_1)$ on $[0, 1]$, and the length of α is (consistently enough) $\int_{[0, 1]} |z_2 - z_1| \, dt = |z_2 - z_1|$. It follows that the formula (i) gives the correct value for the length of any polygonal path. The following theorem reconciles our definition (i) of length with the more intuitive idea of approximating the path by a polygonal path and taking a limit. Recall that a *partition* of an interval $[a, b]$ is a finite ordered set $\Delta = \{t_i\}_{i=0}^m$ where $0 \leq t_{i-1} < t_i \leq b$, $t_0 = a$ and $t_m = b$. The *norm* of the partition is the number $|\Delta| = \max \{t_j - t_{j-1} : 1 \leq j \leq m\}$.

Theorem. Let α be a path on $[a, b]$. Let $(\Delta_n)_{n=1}^\infty$ be a sequence of partitions of $[a, b]$ such that $\lim_{n \to \infty} |\Delta_n| = 0$. Let $\Delta_n = \{t_{j,n} : 0 \leq j \leq m_n\}$. We have

(ii) $v_\alpha(b) = \displaystyle\lim_{n \to \infty} \sum_{j=1}^{m_n} |\alpha(t_{j,n}) - \alpha(t_{j-1, n})|.$

PROOF. We will drop the subscripts n on the t_j of Δ_n. First, suppose that α' is continuous on $[a, b]$. Since $v_\alpha(b) = \int_\alpha |\alpha'(t)| \, dt$, we have $v_\alpha(b) = \lim_{n \to \infty} \sum_{j=1}^{m_n} |\alpha'(s_j)|(t_j - t_{j-1})$ where s_j is any point satisfying $t_{j-1} \leq s_j \leq t_j$. Write $\alpha = x + iy$, x and y real-valued functions of t. For each n, by the mean-value theorem there are points s_j and u_j in $[t_{j-1}, t_j]$ such that

$$x(t_j) - x(t_{j-1}) = x'(s_j)(t_j - t_{j-1}); \quad y(t_j) - y(t_{j-1}) = y'(u_j)(t_j - t_{j-1}).$$

We have

$$\sum_{j=1}^{m_n} |\alpha'(s_j)|(t_j - t_{j-1}) = \sum_{j=1}^{m_n} |(x'(s_j) + iy'(s_j))(t_j - t_{j-1})|$$

$$= \sum_{j=1}^{m_n} \left(|[x(t_j) - x(t_{j-1})] \right.$$

$$+ i[y(t_j) - y(t_{j-1})]$$

$$\left. - i[(y'(u_j) - y'(s_j))(t_j - t_{j-1})]| \right) \quad (1)$$

$$\leq \sum_{j=1}^{m_n} |\alpha(t_j) - \alpha(t_{j-1})|$$

$$+ \sum_{j=1}^{m_n} |y'(u_j) - y'(s_j)|(t_j - t_{j-1}).$$

Let $\varepsilon > 0$. By the uniform continuity of y' (1.13), there is a $\delta > 0$ such that $|y'(v) - y'(w)| \leq \varepsilon/(b - a)$ if $|v - w| < \delta$. There is an n_0 such that $|\Delta_n| < \delta$ if $n \geq n_0$. Thus $|y'(u_j) - y'(s_j)| \leq \varepsilon/(b - a)$ holds if $n \geq n_0$ and by (1) we obtain

$$\sum_{j=1}^{m_n} |\alpha'(s_j)|(t_j - t_{j-1}) \leq \sum_{j=1}^{m_n} |\alpha(t_j) - \alpha(t_{j-1})| + \varepsilon$$

if $n \geq n_0$. Letting n go to ∞, we obtain

$$v_\alpha(b) \leq \lim_{n \to \infty} \sum_{j=1}^{m_n} |\alpha(t_j) - \alpha(t_{j-1})| + \varepsilon.$$

Since this equality holds for all $\varepsilon > 0$, it also holds for $\varepsilon = 0$. On the other hand, using (3.6), for each n we obtain

$$v_\alpha(b) = \int_{[a, b]} |\alpha'(t)| \, dt = \sum_{j=1}^{m_n} \int_{[t_{j-1}, t_j]} |\alpha'(t)| \, dt$$

$$\geq \sum_{j=1}^{m_n} \left| \int_{[t_{j-1}, t_j]} \alpha'(t) \, dt \right|$$

$$= \sum_{j=1}^{m_n} |\alpha(t_j) - \alpha(t_{j-1})|.$$

From this inequality we have

$$\varlimsup_{n \to \infty} \sum_{j=1}^{m_n} |\alpha'(t_j) - \alpha(t_{j-1})| \leq v_\alpha(b).$$

Thus

$$\overline{\lim_{n \to \infty}} \sum_{j=1}^{m_n} |\alpha(t_j) - \alpha(t_{j-1})| \le v_\alpha(b) \le \lim_{n \to \infty} \sum_{j=1}^{m_n} |\alpha(t_j) - \alpha(t_{j-1})|,$$

proving (ii).

If α' is only piecewise continuous on $[a, b]$, (ii) then holds for each of the finitely many intervals of continuity of α'. It follows, after some tedious fussing with partitions, that (ii) also holds on $[a, b]$.

Remark. A complex-valued function α on an interval $[a, b]$ is said to have *finite variation* if the numbers $V(\Delta) = \sum_{j=1}^{m} |\alpha(t_j) - \alpha(t_{j-1})|$ are bounded; that is, there is a constant M such that $V(\Delta) < M$ for all partitions Δ of $[a, b]$. The theorem just proved shows that every path has finite variation. Integrals $\int f \, d\alpha$ can be defined with respect to any finite variation function α. The theory is a little more technical than for paths and is not needed for elementary complex analysis.

3.13 EXAMPLES

(i) Let $\alpha(t) = t + it$ on $[0, 1]$. We have

$$v_\alpha(1) = \int_{[0, 1]} |\alpha'(t)| \, dt = \int_{[0, 1]} |1 + i| \, dt = \sqrt{2};$$

that is, the length of α from $\alpha(0)$ to $\alpha(1)$ is $\sqrt{2}$. This, of course, is not surprising.

Suppose that we want to calculate $\int_\alpha \bar{z} \, |dz|$. Since $|\alpha'(t)| = |1 + i| = \sqrt{2}$, (3.11.ii) gives

$$\int_\alpha \bar{z} \, |dz| = \int_{[0, 1]} (t - it)\sqrt{2} \, dt = \frac{1}{2} - i\frac{1}{2}.$$

(ii) Let $\alpha(t) = re^{it}$ on $[-\pi, \pi]$, so that α^* is a circle of radius r. The circumference of the circle is

$$v_\alpha(\pi) = \int_{[-\pi, \pi]} |rie^{it}| \, dt = \int_{[-\pi, \pi]} r \, dt = 2\pi r.$$

Hence (see 2.11.vi) π is the ratio of the circumference of a circle to its diameter.

(iii) To calculate $\int_\gamma \bar{z} \, dz$ for a parabolic path γ joining 0 and $1 + i$, parametrize the curve $y = x^2$ by $\gamma(t) = t + it^2$ on $[0, 1]$. We then have

$$\int_\gamma \bar{z} \, dz = \int_{[0, 1]} (t - it^2)(1 + i2t) \, dt$$

$$= \int_{[0, 1]} (t + 2t^3) \, dt + i \int_{[0, 1]} (2t^2 - t^2) \, dt = 1 + i\tfrac{1}{3}.$$

The following theorem is very useful.

3.14 THEOREM

Let α be a path defined on $[a, b]$. Let (f_n) be a sequence of continuous functions on α^* which converge uniformly on α^* to the function f. We have

(i) $\quad \lim\limits_{n \to \infty} \int_\alpha f_n(z) \, dz = \int_\alpha f(z) \, dz.$

PROOF.   Let $\varepsilon > 0$. Since the f_n converge uniformly on α^*, there is an n_0 such that

$$|f(z) - f_n(z)| < \frac{\varepsilon}{v_\alpha(b)}$$

if $n \geq n_0$. Using (3.11.v), we have

$$\left| \int_\alpha f(z) \, dz - \int_\alpha f_n(z) \, dz \right| = \left| \int_\alpha [f(z) - f_n(z)] \, dz \right|$$

$$\leq \int_\alpha |f(z) - f_n(z)| \, |dz|$$

$$\leq \int_\alpha \frac{\varepsilon}{v_\alpha(b)} \, |dz| = \varepsilon.$$

EXERCISES

TYPE I

1. Let $w, v \in \mathbf{C}, w \neq v$.
    (a) Define simple paths α, σ, and δ which connect w to v with a straight line, a triangle, and a semicircle, respectively.
    (b) For a fixed positive integer n, let $f(z) = z^n$ and examine each of the four integrals $\int f(z) \, dz$, $\int f(z) \, |dz|$, $\int |f(z)| \, dz$, $\int |f(z)| \, |dz|$ for each of the paths in (a).

2. Let y be a real-valued continuously differentiable function on $[a, b]$. Let $\alpha(t) = t + iy(t)$, $t \in [a, b]$. Show that

$$v_\alpha(t) = \int_a^t \sqrt{1 + \left(\frac{dy}{dx}\right)^2} \, dx.$$

This is the formula for arc length familiar from calculus.

3. Find $v_\alpha(t)$ for the following α. If α is given explicitly, identify the path. If α is described geometrically, parametrize it.
    (a) $\alpha(t) = t + i \sin t$ on $[0, 2\pi]$
    (b) α is "once around" an ellipse centered at 0 with axes of lengths a and b
    (c) $\alpha(t) = e^t + ie^{2t}$, $\quad 0 \leq t \leq 2$.

4. If α is increasing on $[a, b]$, show that $v_\alpha(x) = \alpha(x) - \alpha(a)$.

5. (a) Prove that $\lim_{n \to \infty} (n - n \cos (1/n)) = 0$ by using (3.14) with $f_n(x) = \sin (x/n)$.

   (b) Find $\lim_{n \to \infty} n(\cos (\varepsilon/\sqrt{n}) - \cos (1/\sqrt{n}))$ for $0 < \varepsilon < 1$.

6. Let x and y be real-valued differentiable functions on $[a, b]$. Suppose that $x'(t) \neq 0$ for all t.

   (a) Show that the slope of the tangent to the image of α at $\alpha(t)$ should be defined as $(y'(t))/(x'(t))$.

   (b) Show that $\overrightarrow{[0, \alpha'(t)]}$ has the same slope as the tangent at every t. The unit tangent vector to α at t is defined by

   $$T(t) = \frac{\alpha'(t)}{|\alpha'(t)|}.$$

   In complex notation, it is a complex number of absolute value 1.

   (c) Show that dT/dt is perpendicular to T; that is, $\overrightarrow{[0, T]}$ is perpendicular to $\overrightarrow{[0, dT/dt]}$.

   If T and dT/dt are regarded as free vectors, then $T(t)$ is the unit tangent vector at $\alpha(t)$ and $N(t) = (dT/dt)/(|dT/dt|)$ is the unit normal vector.

TYPE II

1. For paths α and β, write $\alpha \sim \beta$ if $\beta = \alpha \circ \psi$ where ψ is increasing as in (3.8). Show that "$\sim$" is an equivalence relation on the class of all paths.

2. (a) For each positive integer n, define a function f_n on $[0, 1]$ which is zero except on $[1/(n + 1), 1/n]$ and the graph of which is an isosceles triangle based on $[1/(n + 1), 1/n]$ and having area equal to 1.

   (b) Show that $\lim_{n \to \infty} f_n(x) = 0$ for every $x \in [0, 1]$.

   (c) Show that $\lim_{n \to \infty} \int_{[0, 1]} f_n(x) \, dx = 1$.

   This exercise shows that the equality (3.14.i) can fail if the convergence is not uniform.

3. (Re 3.3) If σ is a real-valued continuous function on an interval $[a, b]$ which is differentiable on $[a, b]$, then the mean-value theorem asserts that there is a point $t \in [a, b]$ such that

   $$\frac{\sigma(b) - \sigma(a)}{b - a} = \sigma'(t).$$

   Show that this theorem may fail if σ is replaced by a complex-valued differentiable function α on $[a, b]$ by considering the function

   $$\alpha(t) = t^2 + it^3$$

   on $[0, 1]$.

4. Use partial summation to prove the formula for integration by parts:

$$\int_{[a, b]} f'(x)g(x)\, dx = fg|_a^b - \int_{[a, b]} f(x)g'(x)\, dx.$$

5. Let α be a path of length L, let f be Riemann integrable, and let

$$m = \text{glb}\ \{|f(z)|\colon z \in \alpha^*\}$$
$$M = \text{lub}\ \{|f(z)|\colon z \in \alpha^*\}.$$

(a) Show that

$$m \le \frac{1}{L} \int_\alpha |f|\ |dz| \le M.$$

(b) Give an example where

$$m > \left| \frac{1}{L} \int_\alpha f\, dz \right|.$$

(c) If, in addition, f is continuous, v_α is strictly increasing, and

$$\frac{1}{L} \int_\alpha |f|\ |dz| = M,$$

show that $|f| = M$ on α^*.

6. If a function f of two variables z and w is continuous in z and analytic in w for z in the image of a path and w in some open set, and if $[f(z, w_0 + \Delta w) - f(z, w_0)]/\Delta w$ converges to $df(z, w_0)/dw$ uniformly in z for some fixed w_0, show that

$$\left[\frac{d}{dw} \int_\alpha f(z, w)\, dz \right]_{w_0} = \int_\alpha \left[\frac{d}{dw} f(z, w) \right]_{w_0} dz.$$

Notation for real line integrals. Let α and β be real-valued differentiable functions defined on $[a, b]$. Let $\Gamma = (\alpha, \beta)$, so that $\Gamma\colon [a, b] \to \mathbf{R}^2$. Let f be continuous on the range of Γ to $\mathbf{R}$, say $f = (g, h)$, g and h real-valued. The line integral of f over Γ is defined as the number

$$\int_{[a, b]} g(\alpha(t), \beta(t))\alpha'(t)\, dt + \int_{[a, b]} h(\alpha(t), \beta(t))\beta'(t)\, dt,$$

and is usually denoted

$$\int_\Gamma g(x, y)\, dx + h(x, y)\, dy$$

or

$$\int_\Gamma f \cdot dr$$

where dr is supposed to represent incremental arc length and $\cdot$ is supposed to be the dot (or scalar) product.

7. With notation as above, we also let $\Gamma(t) = \alpha(t) + i\beta(t)$ and $f(z) = f(x, y) = g(x,y) + ih(x, y)$ $(z = x + iy)$. Show that

(a) $\displaystyle\int_\Gamma f(z)\,dz = \int_\Gamma g(x, y)\,dx - h(x, y)\,dy$

$$+ i\int_\Gamma h(x, y)\,dx + g(x, y)\,dy;$$

or in alternate form,

(b) $\displaystyle\int_\Gamma \overline{f(z)}\,dz = \int_\Gamma f \cdot dr + i\int_\Gamma [g(x, y)\,dy - h(x, y)\,dx].$

Equality (a) can also be written $\int_\Gamma f(z)\,dz = \int_\Gamma f(z)\,dx + if(z)\,dy$.

8. Let $f = u + iv$ be analytic on an open set containing the closed path γ. Show that the following relations hold:

$$0 = \int_\gamma u_x\,dx + u_y\,dy = \int_\gamma u_x\,dy - u_y\,dx.$$

Are there any other relations similar to the above which are also true?

9. Let f be analytic on an open set U, and write $f = u + iv$, u and v real. Assume that f' is continuous.

(a) Show that $f' = \partial u/\partial x - i\,\partial u/\partial y$.

(b) Fix $z_0 \in U$. Show that the integral $\int_\alpha ((\partial u/\partial x)\,dx + (\partial u/\partial y)\,dy)$ is independent of the path α in U joining z_0 and z.

Differential forms such as $(\partial u/\partial x)\,dx + (\partial u/\partial y)\,dy$ above are called *exact differentials*. That is, a form $p(x, y)\,dx + q(x, y)\,dy$ with p and q real-valued is *exact* in U if and only if there is a $\mathbb{C}^1$-function u on U such that $\partial u/\partial x = p$ and $\partial u/\partial y = q$ in U. Later we will prove the converse of (b), as a consequence of Cauchy's theorem.

10. Let $f = u + iv$ be a complex-valued function on an open set U. Assume that u and v are $\mathbb{C}^1$-functions. By (7.a), it is reasonable to call the complex differential form $f(z)\,dz$ *exact* if $u\,dx - v\,dy$ and $v\,dy + u\,dx$ are exact. Show that $f(z)\,dz$ is exact if there is an analytic function F on U such that $F' = f$ on U. (Later, as a consequence of Cauchy's theorem, we will prove a converse to this statement.)

In Exercises 11 through 22, we point out some connections between fluid flow problems and analytic functions. Not all necessary physical assumptions are stated, and the student is urged to make a list of additional assumptions needed. He also is urged to describe actual physical situations which the problems idealize.

11. Consider a layer of incompressible fluid flowing parallel to the complex plane in such a way that the velocity of the fluid is independent of depth. Let α be a path lying in the flow. Let $v_x(z)$ and $v_y(z)$ be the x and y components of velocity of the flow at the point $z = x + iy$ (do not confuse the subscripts with partial derivatives). The velocity vector $v = (v_x, v_y)$ can be written as a complex function $v = v_x + iv_y$. Show (by an incremental argument of elementary calculus) that the flow per unit time across α (say V) is given by

(i) $V = \displaystyle\int_\alpha v_x \, dy - v_y \, dx.$

(Volume per unit time across α is Vh, h the depth of the fluid.) In complex notation, we have

(ii) $V = \text{Im} \displaystyle\int_\alpha \overline{v(z)} \, dz.$

12. In addition to the assumptions of the preceding problem, assume that α is closed. Argue from physical considerations that

$$\int_\alpha v_x \, dy - v_y \, dx = 0.$$

Notice that if $\bar{v}$ is the derivative an analytic function, then this equality follows from (3.6).

13. Under the conditions of the preceding two problems, show that

$$\psi(z) = \int_{z_0}^z v_x \, dy - v_y \, dx \qquad (z_0 \text{ fixed})$$

is independent of the path joining z_0 to z.

14. With ψ as in Exercise 13, establish the equalities

$$\frac{\partial \psi}{\partial x} = -v_y \quad \text{and} \quad \frac{\partial \psi}{\partial y} = v_x.$$

15. Show that the path of a particle of fluid lies on a curve determined by

$$\psi(z) = \text{constant}.$$

These paths are called streamlines.

16. (a) Assuming that v_x and v_y have continuous first partials, show that

$$\frac{\partial v_x}{\partial x} = -\frac{\partial v_y}{\partial y}. \tag{$*$}$$

[*Hint:* $\partial^2\psi/(\partial x \, \partial y) = \partial^2\psi/(\partial y \, \partial x)$.]

Remark. $\partial v_x/\partial x = -\partial v_y/\partial y$ is one of the two Cauchy-Riemann equations for $\bar{v} = v_x - iv_y$.

(b) Assuming only that v_x and v_y have bounded partials, use Green's theorem to show that

$$\frac{\partial^2 \psi}{\partial x \, \partial y} = \frac{\partial^2 \psi}{\partial y \, \partial x},$$

and hence that $(*)$ holds.

17. If, in addition to the previous assumptions, the flow is irrotational $(\partial v_y / \partial x = \partial v_x / \partial y)$, then

$$\frac{\partial^2 \psi}{\partial x^2} + \frac{\partial^2 \psi}{\partial y^2} = 0;$$

that is, ψ is harmonic. In this case, show that

$$\Phi(z) = \int_{z_0}^{z} v_x \, dx + v_y \, dy$$

is independent of path.

Remarks.

  (i) $\partial v_y / \partial x = \partial v_x / \partial y$ is the second Cauchy-Riemann equation for $\bar{v} = v_x - iv_y$. Hence with appropriate physical assumptions, $\bar{v}$ is analytic.

  (ii) $\Phi(z) = \operatorname{Re} \int_{z_0}^{z} \overline{v(z)} \, dz.$

  (iii) Since $\partial \Phi / \partial x = v_x$ and $\partial \Phi / \partial y = v_y$, Φ is called the *velocity potential.*

18. Show that the function $\Phi + i\psi$ is analytic. It is called the *complex potential* function for v.

19. (a) Show that Φ is harmonic.
    (b) Show that the streamlines are perpendicular to the curves of constant velocity potential.

    [*Hint:* See Exercise 1, p. 53.]

20. Suppose that the velocity function v is

$$v(z) = \frac{Qiz}{|z|^2} \qquad (Q > 0).$$

Show that the complex potential function $\Phi + i\psi$ is $-Qi \operatorname{Log}(z)$ in any open set not intersecting the nonpositive x-axis. What is happening at 0? Is $v = \operatorname{grad} \Phi$ valid at 0? What about curl (v) at 0?

21. Is there a complex potential function for the fluid flow whose velocity is given by

    (a) iz
    (b) z
    (c) $\bar{z}$?

In Exercises 11 through 21 we have seen that an irrotational steady-state planar flow of an incompressible fluid in a region free of sources and sinks gives rise to a complex potential function $\Phi + i\psi$ which is analytic in this region. The streamlines in the z-plane are given by the curves $\{z \mid \psi(z) = \text{constant}\}$ and the velocity components at a point z are $\partial\Phi/\partial x\,(z)$ and $\partial\Phi/\partial y\,(z)$. When the flow encounters an obstacle, the boundary of the obstacle is evidently a limit of a sequence of streamlines. In the next problem, we reverse the procedure. Given an analytic function $F = \Phi + i\psi$ in a region Ω, we interpret it as a complex potential function for a flow in Ω. If the boundary of Ω is the limit of curves $\psi(z) = \text{constant}$ then the boundary of Ω can be interpreted as a barrier to the flow.

22. For this exercise, let $\sqrt{}$ denote the branch of the square root function defined by

$$\sqrt{z} = \sqrt{|z|}\, e^{i(\theta/2)} \qquad \text{if } z = |z|e^{i\theta} \qquad (0 \le \theta < 2\pi).$$

Thus the function $z \to \sqrt{z}$ is analytic off of the set $\{x + iy \mid y = 0 \text{ and } x \ge 0\}$.

(a) Show that $\sqrt{z^2 - 1}$ is analytic on the set $\Omega = \mathbf{C}\backslash\{x + iy \mid y = 0 \text{ and } |x| \ge 1\}$.

(b) If $z \in \Omega$, then $z + \sqrt{z^2 - 1}$ does not lie on the negative x-axis and is not zero.

(c) Show that $\text{Log}\,(z + \sqrt{z^2 - 1})$ is analytic on Ω.

(d) Show that $\cosh\,(\text{Log}\,(z + \sqrt{z^2 - 1})) = z$, so that $\text{Log}\,(z + \sqrt{z^2 - 1})$ is inverse to $\cosh$ on Ω.

(e) Let $F(z) = \Phi(z) + i\psi(z) = \text{Log}\,(z + \sqrt{z^2 - 1})$ on Ω, Φ and ψ real-valued. Show that the graph of $\psi(z) = \text{constant}$ is the hyperbola

$$\left\{ z = x + iy \,\middle|\, \frac{x^2}{\cos^2\psi} - \frac{y^2}{\sin^2\psi} = 1 \right\}.$$

[*Hint:* See (2.13) and also Exercise 6, p. 76.]

(f) Show that

$$\frac{\partial\Phi}{\partial x}(z) = \text{Re}\,\frac{1}{\sqrt{z^2 - 1}}$$

$$\frac{\partial\Phi}{\partial y}(z) = -\text{Im}\,\frac{1}{\sqrt{z^2 - 1}}$$

[*Hint:* Use $\cosh F(z) = z$ and the Cauchy-Riemann equations.]

Remarks.   If we view F as a complex potential function for a flow in Ω, then the set $\mathbf{C}\backslash\Omega = \{x + iy \mid y = 0, |x| \geq 1\}$ is interpreted physically as a barrier to the flow. The streamlines are given by the hyperbolas in (e) and the x and y components of velocity of the flow at z are given in (f).

3.15 IMPORTANT EXAMPLES

Let $\tau: [-\pi, \pi] \to \mathbf{C}$ be given by the formula $\tau(t) = e^{it}$. Thus, τ is "once around the unit circle, counterclockwise." For functions f continuous on $\mathbf{T} = \tau^*$, we have

$$\int_\tau f(z)\, dz = \int_{[-\pi, \pi]} f(e^{it}) i e^{it}\, dt.$$

In particular, let $f(z) = z^n$, $n \in \mathbf{Z}$. We have

$$\frac{1}{2\pi i} \int_\tau z^n\, dz = \frac{1}{2\pi} \int_{[-\pi, \pi]} e^{i(n+1)t}\, dt.$$

The integral on the right is 1 if $n = -1$ and is 0 for all other n; for if $n \neq -1$, then $z^n = [1/(n + 1)]\, d(z^{n+1})/dz$, and (3.6.ii) applies. Thus we have

$$\frac{1}{2\pi i} \int_\tau z^n\, dz = \begin{cases} 0 & \text{if } n \neq -1 \\ 1 & \text{if } n = -1. \end{cases}$$

If we let $\tau_m(t) = e^{it}$ on $[-m\pi, m\pi]$, then we get

$$\frac{1}{2\pi i} \int_{\tau_m} z^n\, dz = \begin{cases} 0 & \text{if } n \neq -1 \\ m & \text{if } n = -1. \end{cases}$$

Hence, $(1/2\pi i) \int_{\tau_m} (1/z)\, dz$ counts the number of times that τ_m winds around the point 0. This interesting interpretation also holds for other points in $D(1, 0)$. Let $\zeta \in D(1, 0)$, fix m, and let $\zeta = re^{i\theta}$. We have

$$\frac{1}{2\pi i} \int_{\tau_m} \frac{1}{z - \zeta}\, dz = \frac{1}{2\pi} \int_{[-\pi m, \pi m]} \frac{e^{it}}{e^{it} - re^{i\theta}}\, dt$$

$$= \frac{1}{2\pi} \int_{[-\pi m, \pi m]} \frac{1}{1 - re^{i(\theta - t)}}\, dt$$

$$= \frac{1}{2\pi} \int_{[-\pi m, \pi m]} \frac{1}{1 - re^{it}}\, dt.$$

The function $F(\zeta) = (1/2\pi i) \int_{\tau_m} (1/(z - \zeta))\, dz$ is thus constant on the circles $\{|\zeta| = r\}$, so that F is actually a function of r only. Writing

$$\frac{1}{1 - re^{it}} = \sum_{n=0}^{\infty} r^n e^{int},$$

and using uniform convergence of the series to interchange sum and integral (3.14), we see that $F(r) = m$ if $|\zeta| < 1$.

Next, suppose that $|\zeta| > 1$. Again F is constant on each $\{|\zeta| = r\}$, $r > 1$. Fix ζ_0, $|\zeta_0| > 1$. Since ζ_0 is a fixed positive distance from $D(1, 0)$, the difference quotient

$$\frac{1}{\zeta - \zeta_0}\left(\frac{1}{z - \zeta} - \frac{1}{z - \zeta_0}\right)$$

converges uniformly to $(1/(z - \zeta_0))^2$ as ζ goes to ζ_0 through any sequence. Hence differentiation under the integral is valid, by uniform convergence (3.14), and using (3.6) we obtain

$$\frac{d}{d\zeta}\frac{1}{2\pi i}\int_{\tau_m}\frac{1}{z - \zeta}\,dz = \frac{1}{2\pi i}\int_{\tau_m}\frac{d}{d\zeta}\left(\frac{1}{z - \zeta}\right)dz$$

$$= \frac{1}{2\pi i}\int_{\tau_m} -\frac{d}{dz}\left(\frac{1}{z - \zeta}\right)dz = 0.$$

In particular, if we consider F as a function of the real variable r, then we have $F'(r) = 0$. This implies that F is constant on $\{|\zeta| > 1\}$. The constant must be zero, for uniform convergence shows that

$$\lim_{|\zeta| \to \infty} F(\zeta) = 0.$$

In summary, we have

$$\frac{1}{2\pi i}\int_{\tau_m}\frac{1}{z - \zeta}\,dz = \begin{cases} m & \text{if } |\zeta| < 1 \\ 0 & \text{if } |\zeta| > 1. \end{cases}$$

We see that the integral on the left counts the numbers of times that τ_m winds around the point ζ. Notice that

$$\frac{1}{2\pi i}\int_{-\tau_m}\frac{1}{z - \zeta}\,dz = \begin{cases} -m & \text{if } |\zeta| < 1 \\ 0 & \text{if } |\zeta| > 1. \end{cases}$$

3.16 WINDING NUMBER (OR INDEX) OF A CLOSED PATH

Let α be a closed path in $\mathbf{C}$. If $z \notin \alpha^*$, define

$$\text{Ind}_\alpha(z) = \frac{1}{2\pi i}\int_\alpha\frac{1}{\zeta - z}\,d\zeta.$$

In the previous example, $\text{Ind}_{\tau_m}(z) = m$ if $z \in D(1, 0)$ and $\text{Ind}_{\tau_m}(z) = 0$ if $|z| > 1$.

Theorem. The number $\text{Ind}_\alpha(z)$ is always an integer.

PROOF. Define a function ϕ on the interval of parametrization $[a, b]$ of α by

$$\phi(t) = \exp\left(\int_{[a, t]} \frac{\alpha'(s)}{\alpha(s) - z}\, ds\right).$$

Since $\phi'(t) = \phi(t)\alpha'(t)/(\alpha(t) - z)$, the equality

$$\phi'(t) \cdot (\alpha(t) - z) - \phi(t)\alpha'(t) = 0$$

holds. Hence the equality

$$\frac{d}{dt}\left(\frac{\phi(t)}{\alpha(t) - z}\right) = 0$$

is valid in $[a, b]$. It follows that there is a constant K such that

$$\phi(t) = K(\alpha(t) - z), \text{ all } t \text{ in } [a, b].$$

By definition of ϕ, $\phi(a) = 1$. Since $\alpha(b) = \alpha(a)$, $\phi(b) = 1$ also holds; that is,

$$\exp\left(\int_{[a, b]} \frac{\alpha'(s)}{\alpha(s) - z}\, ds\right) = 1.$$

Thus $\int_{[a, b]} (\alpha'(s)/(\alpha(s) - z))\, ds$ equals $2\pi i n$ for some integer n, by (2.11).

We will discuss the index and further substantiate the geometrical interpretation after we have discussed connectedness, in (4.22).

3.17 ORIENTATION OF A CLOSED PATH

If α is a closed path and if $\text{Ind}_\alpha(z) \geq 0$ for all $z \notin \alpha^*$, then we say that α is positively oriented. In Chapter 4 we will make precise the notion of "inside" and "outside" for a large class of closed paths. Then we will be able to say that a closed path is positively oriented if $\text{Ind}_\alpha(z) > 0$ for all z "inside" α^* and negatively oriented if $\text{Ind}_\alpha(z) < 0$ for all z "inside" α^*. It is possible that a path be neither positively nor negatively oriented. We will also show that $\text{Ind}_\alpha(z) = 0$ for all z "outside" α^*. We have seen that for the unit circle the positive orientation is counterclockwise.

EXERCISES

TYPE I

1. Let ∂ be the oriented triangle $[\overrightarrow{1, -1 + i}] \smile [\overrightarrow{-1 + i, -i}] \smile [\overrightarrow{-i, 1}]$. Find $\text{Ind}_\partial(z)$ for $z \notin \partial^*$.
   [*Hint:* First find $\text{Ind}_\partial(0)$.]

2. (a) Define a function α on some $[0, b]$ which goes once counterclockwise around the unit square about 0.
   (b) Calculate $\text{Ind}_\alpha (z)$ for $z \notin \alpha^*$.

3. Let $\alpha(t) = e^{it}$ for $t \in [0, \pi]$ and let $\beta = \alpha \smile \overrightarrow{[-1, 1]}$. Find $\text{Ind}_\beta (i\tfrac{1}{2})$, then find $\text{Ind}_\beta (z)$ for $z \notin \beta^*$.

4. Let

$$\alpha(t) = \begin{cases} \dfrac{e^{it}}{2} & 0 \le t \le 2\pi \\[2mm] 1 + \dfrac{1}{2} e^{-i(t+\pi)} & 2\pi \le t \le 4\pi. \end{cases}$$

Sketch α, find $\text{Ind}_\alpha (0)$ and $\text{Ind}_\alpha (1)$, and show that α is neither positively nor negatively oriented.

*3.18 FOURIER SERIES

If f is a complex-valued function defined on $\mathbf{T} = \{z \mid |z| = 1\}$ such that $|f|$ is Riemann integrable, then define

$$\hat{f}(n) = \frac{1}{2\pi} \int_{[-\pi, \pi]} f(e^{it})e^{-int} \, dt;$$

$\hat{f}(n)$ is called the nth Fourier coefficient for f. (Equivalently, we could consider functions f defined on $[-\pi, \pi]$ with $f(-\pi) = f(\pi)$; or, functions of period 2π on all of $\mathbf{R}$. In this case, $\hat{f}(n) = (1/2\pi) \int_{[-\pi, \pi]} f(t)e^{-int} \, dt$.) In many applications (see exercises at the end of this chapter) it is of interest to recover the function f from the sequence of numbers $\{\hat{f}(n)\}_{n=-\infty}^{\infty}$. The most straightforward approach to doing this is to consider the infinite series

(i) $$\sum_{n=-\infty}^{\infty} \hat{f}(n)e^{int},$$

for if f can be written as a limit of sums of exponentials, say

$$f(e^{it}) = \sum_{n=-\infty}^{\infty} a_n e^{int},$$

then for any k we have

(ii) $$f(e^{it})e^{-ikt} = \sum_{n=-\infty}^{\infty} a_n e^{i(n-k)t}.$$

Ignoring questions of convergence and integrating term-by-term in (ii), we obtain

$$\frac{1}{2\pi} \int_{[-\pi, \pi]} f(e^{it})e^{-ikt} \, dt = \sum_{n=-\infty}^{\infty} a_n \frac{1}{2\pi} \int_{[-\pi, \pi]} e^{i(n-k)t} \, dt.$$

The integrals on the right are 0 unless $n = k$ [see (3.15)], and so we see that $a_n = \hat{f}(n)$. That is, if we attempt to write f as a sum of exponentials $(e^{int})_{n=-\infty}^{\infty}$, then $\hat{f}(n)$ is the leading candidate for the coefficient of e^{int}. Unfortunately, the series (i) (the *Fourier series* for f) may not converge; that is, $\lim_{N\to\infty} \sum_{n=-N}^{N} \hat{f}(n)e^{inx}$ does not exist for all f. A simple hypothesis on f giving convergence appears in the following theorem. To say that f defined on **T** is *continuously differentiable* means that the periodic function F defined on **R** by $F(t) = f(e^{it})$ is continuously differentiable and that $F'(\pi) = F'(-\pi)$. The family of all such functions is denoted $\mathbb{C}^1(\mathbf{T})$; $\mathbb{C}^{(n)}(\mathbf{T})$ denotes the family of n-times continuously differentiable functions, with an obvious inductive definition of "n-times continuously differentiable." Throughout this discussion, $f'(e^{it})$ means $df(e^{it})/dt$.

*3.19 THEOREM

If f, defined on **T**, has a continuous second derivative ($f \in \mathbb{C}^2(\mathbf{T})$) then

$$\sum_{n=-\infty}^{\infty} \hat{f}(n)e^{inx}$$

converges uniformly in x.

Remark. The theorem does not say that $\sum_{n=-\infty}^{\infty} \hat{f}(n)e^{inx}$ converges to $f(e^{ix})$. However, this is true and we will prove it later in this chapter.

PROOF. Integrating by parts twice we obtain

$$\hat{f}(n) = \frac{1}{2\pi}\left[f(e^{it}) \frac{e^{-int}}{-in}\Big|_{-\pi}^{\pi} + f'(e^{it}) \frac{e^{-int}}{n^2}\Big|_{-\pi}^{\pi} - \frac{1}{n^2} \int_{[-\pi, \pi]} f''(e^{it})e^{-int} \, dt \right].$$

By periodicity, the first two terms are zero, so we have

$$|\hat{f}(n)| = \frac{1}{2\pi n^2}\left| \int_{[-\pi, \pi]} f''(e^{it})e^{-int} \, dt \right| \le \frac{1}{2\pi n^2} \int_{[-\pi, \pi]} |f(e^{it})| \, dt.$$

The last inequality follows from the fact that $|e^{-int}| = 1$. The inequality

$$|\hat{f}(n)| \le \frac{1}{2\pi n^2} \int_{[-\pi, \pi]} |f''(e^{it})| \, dt$$

and the dominated convergence theorem for series (1.5) show that

$$\lim_{N \to \infty} \sum_{n=-N}^{N} \hat{f}(n)e^{inx}$$

exists uniformly in x.

*3.20 MEAN SQUARE ERROR

We next examine another method of dealing with the sequence of numbers $\hat{f}(n)$. We suppose that f is Riemann integrable as a function on $[-\pi, \pi]$. We again make use of the equality

$$\int_{[-\pi, \pi]} e^{in\theta}\, d\theta = \begin{cases} 2\pi & \text{if } n = 0 \\ 0 & \text{if } n \neq 0; \end{cases}$$

see (3.15). An expression of the form

$$S_N = \sum_{n=-N}^{N} a_n e^{inx}$$

where each a_n is a complex number and one of $\{a_N, a_{-N}\}$ is not zero is called a *trigonometric polynomial of degree N*. Clearly S_N can be viewed either as a function on $\mathbf{T}$ (a function of e^{ix}), or as a periodic function on $\mathbf{R}$. If S_N is a trigonometric polynomial of degree at most N and f is a function defined on $\mathbf{T}$, we call

$$E_N = \frac{1}{2\pi} \int_{[-\pi, \pi]} |f(e^{ix}) - S_N(x)|^2 \, dx$$

the *mean square error* in the approximation of f by S_N.

(i) *Theorem.* If f is Riemann integrable on $\mathbf{T}$, then, for each integer N the unique trigonometric polynomial of degree less than or equal to N which minimizes E_N is

$$S_N(x) = \sum_{n=-N}^{N} \hat{f}(n)e^{inx}.$$

PROOF.   If S_N has coefficients $(a_n)_{n=-N}^{N}$, then we have

$$E_N = \frac{1}{2\pi} \int_{[-\pi, \pi]} |f(e^{ix}) - S_N(x)|^2 \, dx$$

$$= \frac{1}{2\pi} \int_{[-\pi, \pi]} (f(e^{ix}) - S_N(x))(\overline{f(e^{ix}) - S_N(x)}) \, dx$$

$$= \frac{1}{2\pi} \int_{[-\pi, \pi]} (|f(e^{ix})|^2 + S_N(x)\overline{S_N(x)} - \overline{S_N(x)}f(e^{ix}) - S_N(x)\overline{f(e^{ix})}) \, dx.$$

Again using the equality $\int_{[-\pi,\,\pi]} e^{ikx}\,dx = 0$ if $k \neq 0$, we have

$$\frac{1}{2\pi}\int_{[-\pi,\,\pi]} S_N(x)\overline{S_N(x)}\,dx = \frac{1}{2\pi}\sum_{n=-N}^{N}\sum_{m=-N}^{N} a_n\bar{a}_m \int_{[-\pi,\,\pi]} e^{inx}e^{-imx}\,dx$$

$$= \sum_{n=-N}^{N} |a_n|^2;$$

$$-\frac{1}{2\pi}\int_{[-\pi,\,\pi]} \overline{S_N(x)}f(e^{ix})\,dx = -\frac{1}{2\pi}\sum_{n=-N}^{N} \bar{a}_n \int_{[-\pi,\,\pi]} f(e^{ix})e^{-inx}\,dx$$

$$= -\sum_{n=-N}^{N} \bar{a}_n\hat{f}(n);$$

and

$$-\frac{1}{2\pi}\int_{[-\pi,\,\pi]} S_N(x)\overline{f(e^{ix})}\,dx = -\frac{1}{2\pi}\sum_{n=-N}^{N} a_n \int_{[-\pi,\,\pi]} \overline{f(e^{ix})}e^{inx}\,dx$$

$$= -\sum_{n=-N}^{N} a_n\overline{\hat{f}(n)}.$$

Hence

$$E_N = \frac{1}{2\pi}\int_{[-\pi,\,\pi]} |f(e^{ix})|^2\,dx + \sum_{n=-N}^{N} (|a_n|^2 - \bar{a}_n\hat{f}(n) - a_n\overline{\hat{f}(n)}).$$

But

$$|a_n|^2 - \bar{a}_n\hat{f}(n) - a_n\overline{\hat{f}(n)} = (a_n - \hat{f}(n))\overline{(a_n - \hat{f}(n))} - \hat{f}(n)\overline{\hat{f}(n)}$$

$$= |a_n - \hat{f}(n)|^2 - |\hat{f}(n)|^2,$$

so that

(ii) $$E_N = \frac{1}{2\pi}\int_{[-\pi,\,\pi]} |f(e^{ix})|^2\,dx + \sum_{n=-N}^{N} |a_n - \hat{f}(n)|^2 - \sum_{n=-N}^{N} |\hat{f}(n)|^2.$$

From (ii) it is clear that the unique choice for $\{a_n\}$ which makes E_N a minimum for each N is $a_n = \hat{f}(n)$, $n = 0, \pm 1, \pm 2, \pm 3, \ldots$. This minimum is

$$\frac{1}{2\pi}\int_{[-\pi,\,\pi]} |f(e^{ix})|^2\,dx - \sum_{n=-N}^{N} |\hat{f}(n)|^2.$$

(iii) *Corollary.*

$$\sum_{n=-\infty}^{\infty} |\hat{f}(n)|^2 \leq \frac{1}{2\pi}\int_{[-\pi,\,\pi]} |f(e^{ix})|^2\,dx.$$

EXERCISES

TYPE I

1. If f is real-valued on **T**, show that
   (a) $\hat{f}(n) = \overline{\hat{f}(-n)}$
   (b) $\hat{\bar{f}}(n) = \overline{\hat{f}(n)}$

2. The Fourier series of a function f can be written

   $$\sum_{n=-\infty}^{\infty} \hat{f}(n)e^{inx} = \sum_{n=-\infty}^{\infty} \hat{f}(n)(\cos(nx) + i\sin(nx))$$

   $$= \sum_{n=0}^{\infty} a_n \cos(nx) + b_n \sin(nx).$$

   Find a_n and b_n in terms of $\hat{f}(n)$ and $\hat{f}(-n)$.

   [*Remark.* The coefficients a_n and b_n are also called Fourier coefficients of f.]

3. If f (considered as a function defined on $[-\pi, \pi]$) is odd, that is, $-f(x) = f(-x)$ and if $f(x) = \sum_{n=0}^{\infty} (a_n \cos(nx) + b_n \sin(nx))$, show that $a_n = 0$ for all n and that $b_n = (2/\pi) \int_{[0,\pi]} f(x) \sin(nx) \, dx$.

4. What can you say about the Fourier coefficients of a function defined on $[-\pi, \pi]$ which is even, that is, $f(x) = f(-x)$?

5. Let $f(x) = x$. Compute the Fourier coefficients for f by two methods.
   (a) Integrate by parts.
   (b) Observe that $ze^{inz}/in + e^{inz}/n^2$ is a primative for ze^{inz} and use (3.11).

6. Compute the Fourier coefficients for x^2 by two methods.
   [*Hint:* $d(z^2 e^{inz}/in + 2ze^{inz}/n^2 - 2e^{inz}/in^3)/dz = z^2 e^{inz}$.]

7. Find the Fourier series for the periodic function described below, left.

8. Find the Fourier series for the saw-tooth function described below, right.

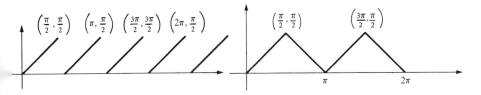

*3.21 FEJER'S THEOREM

The results (3.18–3.20) are of a preliminary nature. We know that the Fourier series of a $\mathfrak{C}^2$ function converges, but we do not know the limit. We know the best trigonometric polynomial mean-square approximation to a continuous function, but we do not know that the approximation converges in mean square to the function. Motivated by these results, we thus formulate three important questions.

1. If $f \in \mathbb{C}(\mathbf{T})$, does $\lim_{N \to \infty} \int_{[-\pi, \pi]} |f(e^{ix}) - \sum_{n=-N}^{N} \hat{f}(n)e^{inx}|^2 \, dx = 0$?
2. If $f \in \mathbb{C}(\mathbf{T})$, does $\sum_{n=-\infty}^{\infty} |\hat{f}(n)|^2 = (1/2\pi) \int_{[-\pi, \pi]} |f(e^{ix})|^2 \, dx$?
3. If $f \in \mathbb{C}^2(\mathbf{T})$, does $\lim_{N \to \infty} \sum_{n=-N}^{N} \hat{f}(n)e^{inx} = f(e^{ix})$ hold for all x?

We shall use a theorem of Fejér (which we prove next) to give an affirmative answer to all three questions. For a fixed $f \in \mathbb{C}(\mathbf{T})$, let

$$S_K(x) = \sum_{n=-K}^{K} \hat{f}(n)e^{inx}$$

and

$$\sigma_N(x) = \frac{1}{N}(S_0(x) + S_1(x) + \cdots + S_{N-1}(x)).$$

The function σ_N is the Nth arithmetic mean of the partial sums of the Fourier series for f. Expanding the right side of the equality defining $\sigma_N(x)$, we obtain

$$\sigma_N(x) = \frac{1}{N}\left(\hat{f}(0) + \sum_{n=-1}^{1} \hat{f}(n)e^{inx} + \cdots + \sum_{n=-N+1}^{N-1} \hat{f}(n)e^{inx}\right)$$

$$= \frac{1}{N}(N\hat{f}(0) + (N-1)(\hat{f}(1)e^{ix} + \hat{f}(-1)e^{-ix}) + \cdots$$

$$+ \hat{f}(N-1)e^{i(N-1)x} + \hat{f}(-N+1)e^{-i(N-1)x})$$

$$= \frac{1}{N}\sum_{n=-N+1}^{N-1} (N - |n|)\hat{f}(n)e^{inx}$$

$$= \frac{1}{N}\sum_{n=-N+1}^{N-1} (N - |n|)\left(\frac{1}{2\pi}\int_{[-\pi, \pi]} f(e^{it})e^{-int} \, dt\right)e^{inx}$$

$$= \sum_{n=-N+1}^{N-1} \frac{(N - |n|)}{N}\frac{1}{2\pi}\int_{[-\pi, \pi]} f(e^{it})e^{in(x-t)} \, dt$$

$$= \frac{1}{2\pi}\int_{[-\pi, \pi]} f(e^{it})\left(\sum_{n=-N+1}^{N-1} \frac{N - |n|}{N} e^{in(x-t)}\right) dt.$$

The function of t with values

$$(i) \quad \sum_{n=-N+1}^{N-1} \left(\frac{N - |n|}{N}\right) e^{int}$$

is called Fejér's kernel; we denote it by $K_N(t)$. With this notation, we have proved above that

$$\sigma_N(x) = \frac{1}{2\pi} \int_{[-\pi, \pi]} f(e^{it}) K_N(x - t) \, dt.$$

(ii) *Lemma.* We have $K_N(0) = N$. If $\theta \in [-\pi, \pi]$ and $\theta \neq 0$, then we have

$$K_N(\theta) = \frac{1}{N} \frac{1 - \cos N\theta}{1 - \cos \theta} = \frac{1}{N} \frac{\sin^2 [(N/2)\theta]}{\sin^2 (\theta/2)}.$$

PROOF. We have

$$(N + 1)K_{N+1}(\theta) = \sum_{n=-N}^{N} (N + 1 - |n|)e^{in\theta} = \sum_{n=-N}^{N} e^{in\theta} + NK_N(\theta).$$

First we examine the *Dirichlet kernel* $D_N(\theta) = \sum_{n=-N}^{N} e^{in\theta}$. We have

$$D_N(\theta) = \sum_{n=0}^{N} e^{in\theta} + \sum_{n=0}^{N} e^{-in\theta} - 1 = \frac{1 - e^{i(N+1)\theta}}{1 - e^{i\theta}} + \frac{1 - e^{-i(N+1)\theta}}{1 - e^{-i\theta}} - 1,$$

if $\theta \neq 0$. After some elementary manipulation using the formula $e^{iy} = \cos y + i \sin y$, we obtain

$$D_N(\theta) = \frac{\cos N\theta - \cos(N + 1)\theta}{1 - \cos \theta}, \quad \theta \in [-\pi, \pi], \theta \neq 0; \ D_N(0) = 2N + 1.$$

The functions K_N thus satisfy the recursive relation

$$(N + 1)K_{N+1}(\theta) = \frac{\cos N\theta - \cos (N + 1)\theta}{1 - \cos \theta} + NK_N(\theta),$$

if $\theta \neq 0$. Since $K_1(\theta) = 1$, it follows that

$$K_N(\theta) = \frac{1}{N} \frac{1 - \cos N\theta}{1 - \cos \theta}, \quad \theta \in [-\pi, \pi], \theta \neq 0.$$

The second expression for K_N in the statement of the lemma follows from the identity $\cos (\theta) = 1 - 2 \sin^2 (\theta/2)$. Evaluating (i) at 0 gives $K_N(0) = N$.

(iii) *Lemma.* 1. $K_N(\theta) \geq 0$.

      2. $(1/2\pi) \int_{[-\pi, \pi]} K_N(\theta) \, d\theta = 1$.

      3. For each $\delta > 0$, $\lim_{N \to \infty} (\text{lub } \{K_N(t): \delta \leq |t| \leq \pi\}) = 0$.

PROOF. Assertion (1) is trivial from the preceding lemma. For assertion (2), we have

$$\frac{1}{2\pi} \int_{[-\pi, \pi]} K_N(\theta) \, d\theta = \frac{1}{2\pi} \int_{[-\pi, \pi]} \frac{1}{N} \sum_{n=-N+1}^{N-1} ((N - |n|)e^{in\theta}) \, d\theta = 1.$$

To prove assertion (3), note that $\sin^2(t/2) \geq \sin^2(\delta/2)$ if $\pi \geq |t| \geq \delta$ holds. Since $|\sin^2(N\theta/2)| \leq 1$, we have $0 \leq K_N(\theta) \leq (1/N)(1/\sin^2 (\delta/2))$. Hence

$$\text{lub } \{K_N(t) \mid \delta \leq |t| \leq \pi\} \leq \frac{1}{N} \frac{1}{\sin^2 \delta/2},$$

and assertion (3) follows.

(iv) *Theorem.* (Fejér) If f is continuous on **T**, then σ_N converges to f uniformly on **T** as $N \to \infty$.

PROOF.   We must estimate the difference

$$f(e^{ix}) - \sigma_N(x) = f(e^{ix}) - \frac{1}{2\pi} \int_{[-\pi, \pi]} f(e^{it})K_N(x - t) \, dt.$$

By assertion (2) of the previous lemma, we have

$$f(e^{ix}) = \frac{1}{2\pi} \int_{[-\pi, \pi]} K_N(y)f(e^{ix}) \, dy.$$

By a change of variables,

$$\frac{1}{2\pi} \int_{[-\pi, \pi]} f(e^{it})K_N(x - t) \, dt = \frac{1}{2\pi} \int_{[x-\pi, x+\pi]} f(e^{i(x-y)})K_N(y) \, dy$$

$$= \frac{1}{2\pi} \int_{[-\pi, \pi]} f(e^{i(x-y)})K_N(y) \, dy.$$

The last equality holds because the integrand has period 2π, as a function on **R**; this reflects the fact that we are integrating around the unit circle. We may write

$$f(e^{ix}) - \sigma_N(x) = \frac{1}{2\pi} \int_{[-\pi, \pi]} K_N(y)f(e^{ix}) \, dy - \frac{1}{2\pi} \int_{[-\pi, \pi]} f(e^{i(x-y)})K_N(y) \, dy$$

$$= \frac{1}{2\pi} \int_{[-\pi, \pi]} K_N(y) [f(e^{ix}) - f(e^{i(x-y)})] \, dy.$$

Given $\varepsilon > 0$, by uniform continuity (1.13) of f there is a $\delta > 0$ such that $|f(e^{ix}) - f(e^{i(x-y)})| < \varepsilon/2$ holds for $|y| < \delta$. The last displayed integral above can be written as $\int_{[-\delta, \delta]} + \int_{|y| \geq \delta}$. By (1) and (2) of the previous lemma, the integral over $[-\delta, \delta]$ is dominated by

$$\frac{1}{2\pi} \int_{[-\delta, \delta]} K_N(y)|f(e^{ix}) - f(e^{i(x-y)})| \, dy \leq \frac{1}{2\pi} \int_{[-\pi, \pi]} K_N(y) \frac{\varepsilon}{2} \, dy = \frac{\varepsilon}{2}.$$

Next, we estimate $\int_{|y| \geq \delta}$. Let $M = \text{lub} \{|f(e^{ix})|\}$. Using (3) of the previous lemma, choose N_0 such that $|K_N(y)| < \varepsilon/(4M)$ for all $N > N_0$ and $\delta \leq |y| \leq \pi$. The integral over $\{|y| > \delta\}$ is then dominated by

$$\frac{1}{2\pi} \int_{|y| \geq \delta} \frac{\varepsilon}{4M} |f(e^{ix}) - f(e^{i(x-y)})| \, dy \leq \frac{\varepsilon}{2\pi} \int_{|y| \geq \delta} dy < \frac{\varepsilon}{2}.$$

Hence for $N > N_0$ and all x we have

$$|f(e^{ix}) - \sigma_N(x)| \leq \frac{\varepsilon}{2} + \frac{\varepsilon}{2} = \varepsilon.$$

The last statement completes the proof of Fejér's theorem.

(v) *Remark.* If f is only Riemann integrable in $[-\pi, \pi]$ (not necessarily continuous on **T**), then the proof shows that

$$\lim_{N \to \infty} \sigma_N(x) = f(e^{ix})$$

if e^{ix} is a point of continuity of f. Of course, there is no uniform convergence result in this case.

*3.22 COROLLARIES TO FEJER'S THEOREM

We are now in a position to answer the questions raised at the beginning of section (3.21). Since

$$\sigma_{N+1}(x) = \sum_{n=-N}^{N} \frac{N + 1 - |n|}{N + 1} \hat{f}(n) e^{inx}$$

is a trigonometric polynomial of degree at most N, the inequality

$$\int_{[-\pi, \pi]} \left| f(e^{ix}) - \sum_{n=-N}^{N} \hat{f}(n) e^{inx} \right|^2 dx \leq \int_{[-\pi, \pi]} |f(e^{ix}) - \sigma_{N+1}(x)|^2 \, dx$$

holds (3.20.i). Since uniform convergence on $[-\pi, \pi]$ implies mean-square convergence, Fejér's theorem settles question 1.

We have seen in (3.20.ii) that

$$\int_{[-\pi, \pi]} \left| f(e^{ix}) - \sum_{n=-N}^{N} \hat{f}(n) e^{inx} \right|^2 dx = \int_{[-\pi, \pi]} |f(e^{ix})|^2 \, dx - \sum_{n=-N}^{N} |\hat{f}(n)|^2;$$

this equality combined with the affirmative answer to question 1 answers question 2 affirmatively.

To answer question 3 we use the following lemma.

Lemma. If the sequence $(z_n)_{n=1}^{\infty}$ converges to z, then the sequence $(w_N)_{N=1}^{\infty}$ defined by $w_N = (z_1 + z_2 + \cdots + z_N)/N$ also converges to z.

PROOF.   Given $\varepsilon > 0$, choose N_0 so that $N \geq N_0$ implies that $|z_n - z| < \varepsilon/2$. There is an $N_1 > N_0$ such that

$$\left| \frac{(z_1 - z) + \cdots + (z_{N_0} - z)}{N} \right| \leq \frac{\varepsilon}{2}$$

if $N \geq N_1$. Hence for $N \geq N_1$, we have

$$|w_N - z| = \left| \frac{z_1 + \cdots + z_N}{N} - z \right|$$

$$\leq \left| \frac{(z_1 - z) + \cdots + (z_{N_0} - z)}{N} \right| + \left| \frac{(z_{N_0+1} - z) + \cdots + (z_N - z)}{N} \right|$$

$$< \varepsilon.$$

An affirmative answer to question 3 is now immediate.

*3.23 MEAN-SQUARE APPROXIMATION FOR DISCONTINUOUS FUNCTIONS

The inequality

(i) $$\int_{[-\pi, \pi]} \left| f(e^{ix}) - \sum_{n=-N}^{N} \hat{f}(n) e^{inx} \right|^2 dx \leq \int_{[-\pi, \pi]} |f(e^{ix}) - \sigma_{N+1}(x)|^2 \, dx$$

established in (3.20.i) is valid for any (bounded) Riemann integrable function f. We have observed that

$$\lim_{n \to \infty} \sigma_N(x) = f(e^{ix})$$

at every point of continuity of f. If we can write

(ii) $$\lim_{N \to \infty} \int_{[-\pi, \pi]} |f(e^{ix}) - \sigma_{N+1}(x)|^2 \, dx$$

$$= \int_{[-\pi, \pi]} \lim_{N \to \infty} |f(e^{ix}) - \sigma_{N+1}(x)|^2 \, dx$$

and f does not have too many discontinuities, then apparently the right side of (i) will go to zero as $N \to \infty$ and we will have mean-square convergence of the Fourier series of f to f. As a matter of fact, if f is Riemann integrable, then it does not have too many discontinuities and the interchange of limits in (ii) is valid. Showing this in general would carry us too far afield, and we will not do it. However, if f is piecewise continuous, then the integrals may be written as sums of integrals over subintervals where f is continuous. On these subintervals σ_N converges uniformly to f and the interchange of limits in (ii) is thus valid on each subinterval. Hence the right side in (i) goes to zero, and we have a proof of mean-square convergence for piecewise continuous functions.

Mean-square convergence is actually valid for much more general functions, the square-integrable measurable functions. Study of Fourier series for such functions requires a knowledge of Lebesgue integration. The theory of Fourier series acquires a much more complete form in this context.

EXERCISES

TYPE II

1. Suppose that f is a real-valued continuous function on $\mathbf{T} = \{|z| = 1\}$ such that $\hat{f}(n) = 0$ for all $n < 0$. Show that f is a constant.

   Discussion: Vibrating string. Consider a taut string of length l and of uniform mass density ρ units per unit length. Suppose that the string is perfectly elastic and offers no resistance to bending. Suppose that the initial tension is so great that the effects of gravity are negligible. Finally, suppose that the motion of the string is small vertical vibrations in the x, y plane (no motion along the x-axis). Let P and Q be two points on the string. Let the tension at P and Q be T_1 and T_2, respectively. Let α and β be the angles that T_1 and T_2 respectively make with the x-axis.

   Since there is no horizontal motion, the horizontal components of tension are equal, say

   $$T = \text{constant} = T_1 \cos(\alpha) = T_2 \cos(\beta).$$

The vertical components are $T_1 \sin(\alpha)$ and $T_2 \sin(\beta)$. For a point x such that $0 < x < l$ and $t \geq 0$, let $u(t, x)$ denote the y-component of the position of the x-component of the string at the time t. If the x components of P and Q are x and $x + x$, respectively, then $\rho \, \Delta x$ is the mass of the string segment between P and Q and by Newton's law we have

$$u_{tt}(\bar{x}, t)\rho \, \Delta x = T_1 \sin \alpha + T_2 \sin \beta; \quad x \leq \bar{x} < x + \Delta x. \tag{1}$$

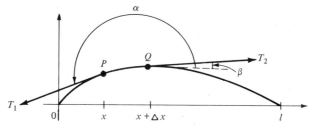

We divide this equality by $T \, \Delta x$ and obtain

$$\rho u_{tt}(\bar{x}, t) = \frac{1}{\Delta x} (\tan(\alpha) + \tan(\beta)) \tag{2}$$

since $\tan(\alpha) = -u_x(x, t)$ and $\tan(\beta) = u_x(x + \Delta x, t)$, (2) may be written

$$\rho u_{tt}(\bar{x}, t) = \frac{u_x(x + \Delta x, t) - u_x(x, t)}{\Delta x}.$$

As Δx goes to zero, $\bar{x}$ goes to x and we obtain the partial differential equation

$$\rho u_{tt}(x, t) = u_{xx}(x, t) \tag{3}$$

for the position of the string. In Exercises 2–8 we outline a method of solving (3) using Fourier series. It is convenient to write the constant ρ as a square. Hence we consider the equation

$$a^2 u_{tt}(x, t) - u_{xx}(x, t). \tag{4}$$

2. Suppose that $u(x, t) = X(x)T(t)$, that u satisfies (4), and that $\partial X / \partial t = \partial T / \partial x = 0$. Show that there is a constant k so that the equations

$$X'' - kX = 0 \tag{5}$$

and

$$T'' - \frac{k}{a^2} = 0 \tag{6}$$

are satisfied.

3. We want to determine choices for k so that the pair of equations (5) and (6) can have a solution. Given that the general solution of $X'' - kX = 0$ is $X = Ae^{\sqrt{k}x} + Be^{-\sqrt{k}x}$ and that $X(0) = X(l) = 0$, show that Re $(\sqrt{kl}) = 0$ and hence that for cach $n = 0, \pm 1, \pm 2, \ldots, k$ could satisfy $k = -(n\pi/l)^2$.

4. We cannot tell $(-n)^2$ from $(n)^2$ so we may as well assume that $n \geq 0$. Show that $X_n(x) = \sin(n\pi x/l)$ is a solution of (5) for each positive integer n and that the Fourier series for any solution X is $X(x) = \sum_{n=1}^{\infty} A_n \sin(n\pi x/l)$ for appropriate choice of A_n.

5. Show that any solution T of (6) has a Fourier series

$$T(t) = \sum_{n=0}^{\infty} C_n \cos\left[\left(\frac{n\pi}{la}\right)t\right] + D_n \sin\left[\left(\frac{n\pi}{la}\right)t\right].$$

6. Show that any solution $u(x, t)$ of (4) of the form XT, as in Exercise 2, may be written

$$u(x, t) = \sum_{n=1}^{\infty} \left[a_n \cos\left(\frac{n\pi t}{la}\right) + b_n \sin\left(\frac{n\pi t}{la}\right)\right] \sin\left(\frac{n\pi x}{l}\right)$$

for appropriate choice of a_n and b_n.

7. If $f(x) = u(x, 0)$, $(0 \le x \le l)$ gives the initial position of the string and if $g(x) = u_t(x, 0)$, $(0 \le x \le l)$ gives the initial velocity, show that

$$a_n = \frac{2}{l} \int_{[0, l]} f(x) \sin \left(\frac{n\pi x}{l}\right) dx$$

and

$$b_n = \frac{a}{n\pi} \int_{[0, l]} g(x) \sin \left(\frac{n\pi x}{l}\right) dx.$$

8. Find the series expansion for $u(x, t)$ if the initial position is $f(x) = l/100 - (1/25l)(x - l/2)^2$ and the initial velocity is zero.

CHAPTER 4

Cauchy's Theorem

Discussion. If f is an analytic function on an open set U and f' is continuous on U, then the equality

$$\int_\alpha f'(z)\,dz = 0$$

holds for every closed path α such that $\alpha^* \subset U$, by (3.6) (a simple calculation). Stated somewhat differently, if f is an analytic function on an open set U which has a primitive F in U ($F' = f$ in U), then the equality

(i) $\quad\displaystyle\int_\alpha f(z)\,dz = 0$

holds for every closed path α in U. In particular, if f is an entire function with a power series $f(z) = \sum_{n=0}^{\infty} a_n z^n$ valid in the entire complex plane, then (i) holds for every closed path α in $\mathbf{C}$. Thus the equality holds for every polynomial f. The equality (i) for appropriate f and α is of central importance in complex analysis, as we will see. Roughly speaking, (i) holds if f is analytic on the "inside" of α. In view of the above discussion, we can prove (i) for an analytic function f and a closed path α by finding a primitive for f in an open set that contains α^*. Not every function has a primitive. For example, if α is the unit circle ($\alpha(t) = e^{it}$), then the equality $\int_\alpha z^{-1}\,dz = 2\pi i$ shows that the function $f(z) = z^{-1}$ has no primitive in every open set U containing $\alpha^* = \mathbf{T}$. This is reasonable, for the supposed primitive would be Log, and Log is not analytic on any open set containing $\mathbf{T}$.

The situation in the real case is distinctly different. If f is differentiable on $[a, b]$ (or even only continuous), then F defined by

$$F(t) = \int_{[a,\,t]} f(s)\,ds$$

is a primitive for f. In defining $F(t)$, there is only one way to go from a; straight to t. However, in the complex case, one can choose any number of paths. Suppose we attempt to imitate the "real" construction of a primitive in the complex case. Let f be analytic on an open set U, and pick a fixed point $a \in U$. Define F by

$$F(z) = \int_\alpha f(\zeta)\, d\zeta,$$

where α is a path in U with initial point a and terminal point z. What α do we choose? The answer to this question is not totally obvious since, for example, if $U = \mathbf{C}\backslash\{0\}$ and $f(z) = z^{-1}$, then F is dependent on the path α as well as the point z. The difficulty in this example is in part the nature of the set U. The function $1/z$ does have a primitive in $V = \{\mathrm{Re}\,(z) > 0\}$, namely, $\mathrm{Log}\,(z)$. We thus see that there is going to be some restriction on the set U in order that functions analytic on U have primitives, and (i) holds. In fact, it is sufficient that the "inside" of α be contained in U whenever α is a closed path in U, as we will see in (4.28).

Our first objective, however, is to prove Cauchy's theorem for a class of open sets which obviously have no holes; namely, convex sets. A subset C of the complex plane $\mathbf{C}$ is said to be *convex* if, given any two points in C, C contains the straight line segment connecting these two points. If C is convex, $a \in C$, and f is continuous on C, we can define

$$F(z) = \int_{\overrightarrow{[a,\,z]}} f(\zeta)\, d\zeta \qquad \text{for each } z \in C;$$

F is the most reasonable candidate available for a primitive to f. We will first prove Cauchy's theorem for a triangle whose interior is contained in an open set in which f is analytic. It will easily follow that F, as defined above, is indeed a primitive for f, provided that f is analytic.

4.1 CAUCHY'S THEOREM FOR A TRIANGLE

If a, b, and c are not collinear, then the oriented triangle ∂ formed by a, b, and c is the directed union of the directed line segments from a to b, b to c, and c to a. Thus, $\partial = l_1 \smile l_2 \smile l_3$, where the segments l_i are parametrized on $[0, 1]$ by

$$l_1(t) = a + t(b - a)$$
$$l_2(t) = b + t(c - b)$$
$$l_3(t) = c + t(a - c).$$

See also Figure 4.1.

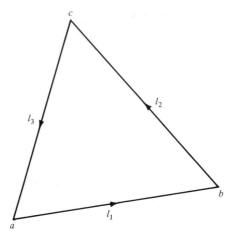

Figure 4.1

It is geometrically clear that the triangle ∂ separates the plane into two disjoint open sets. We will call the bounded one $\triangle$. We can also describe $\triangle$ as follows. Consider the infinitely extended line through b and c. It plainly divides the complex plane into two open half-planes. Let H_a be the half-plane containing a. Define H_b and H_c similarly. Then $\triangle = H_a \cap H_b \cap H_c$ is the triangle excluding the boundary ∂^*. With this notation we state and prove the Cauchy-Goursat theorem. (See Figure 4.1.)

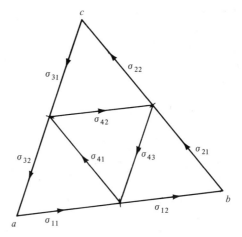

Figure 4.2

Theorem. Let U be an open set containing $\overline{\Delta}$ ($\overline{\Delta}$ is the triangle Δ and its boundary) and let w be a point in U. Suppose that f is continuous on U and analytic on $U \backslash \{w\}$. The equality

$$\int_\partial f(z) \, dz = 0$$

is valid.

Remark. The hypothesis of the theorem does not require that f be analytic at the point w. This is for technical convenience in the proof of Cauchy's integral formula (4.4). Indeed we will see in (4.6) that f is analytic at w.

PROOF. Suppose first that $w \in U \backslash \overline{\Delta}$. Bisect each of the three line segments and so form four new triangles with oriented curves as boundaries, as indicated in Figure 4.2 (the line segments σ_{ij} can of course be parametrized). Let

$$\sigma_1 = \sigma_{11} \smile \sigma_{41} \smile \sigma_{32}$$

$$\sigma_2 = \sigma_{12} \smile \sigma_{21} \smile \sigma_{43}$$

$$\sigma_3 = \sigma_{22} \smile \sigma_{31} \smile \sigma_{42}$$

$$\sigma_4 = -\sigma_{41} \smile -\sigma_{43} \smile -\sigma_{42}.$$

In the sum $\sum_{j=1}^4 \int_{\sigma_j} f(z) \, dz$ the integrals over the interior line segments cancel and we have

$$\int_\partial f(z) \, dz = \sum_{j=1}^4 \int_{\sigma_j} f(z) \, dz,$$

and so the inequality $|\int_{\sigma_j} f(z) \, dz| \geq \frac{1}{4}|\int_\partial f(z) \, dz|$ holds for at least one σ_j. Let ∂_1 denote a σ_j for which the inequality is valid, and let Δ_1 be the corresponding triangle. Apply the same process to Δ_1 to obtain a ∂_2 and Δ_2. Inductively, define paths $(\partial_n)_{n=1}^\infty$ which bound triangles $(\Delta_n)_{n=1}^\infty$ and satisfy

$$\left| \int_{\partial_n} f(z) \, dz \right| \leq 4 \left| \int_{\partial_{n+1}} f(z) \, dz \right| \tag{1}$$

$$l(\partial_n) = 2l(\partial_{n+1}) \qquad [l(\partial_n) \text{ is the perimeter of } \partial_n] \tag{2}$$

$$\overline{\Delta}_{n+1} \subset \overline{\Delta}_n. \tag{3}$$

Since the set $\bar{\Delta}$ is compact, it follows from (1.11.v) that the set $S = \cap_{n=1}^{\infty} \bar{\Delta}_n$ is nonvoid. The set S cannot contain two distinct points (because $\lim_{n\to\infty} l(\partial_n) = 0$), and so in fact there is a point z_0 such that $\cap_{n=1}^{\infty} \bar{\Delta}_n = \{z_0\}$. If $z_0 \notin \partial_n^*$, then we estimate as follows.

$$\left| \int_{\partial_n} f(z)\, dz \right| = \left| \int_{\partial_n} [f(z) - f(z_0)]\, dz \right|$$

$$= \left| \int_{\partial_n} [z - z_0] \cdot \frac{f(z) - f(z_0)}{z - z_0}\, dz \right|$$

$$= \left| \int_{\partial_n} [z - z_0] \left[\frac{f(z) - f(z_0)}{z - z_0} - f'(z_0) \right] dz \right|$$

$$\leq l(\partial_n) \int_{\partial_n} \left| \frac{f(z) - f(z_0)}{z - z_0} - f'(z_0) \right| |dz|.$$

The reader should verify the equalities $\int_{\partial_n} f(z_0)\, dz = \int_{\partial_n} (z - z_0) f'(z_0)\, dz = 0$ used above. If $z_0 \in \partial_n^*$, then the difference quotient $(f(z) - f(z_0))/(z - z_0)$ in the second line is replaced with $f'(z_0)$ at $z = z_0$ and the estimates are unaffected. Let $\varepsilon > 0$. There is a $\delta > 0$ such that

$$\left| \frac{f(z) - f(z_0)}{z - z_0} - f'(z_0) \right| < \varepsilon \text{ if } |z - z_0| < \delta.$$

For $l(\partial_n) < \delta$, the inequality

$$\left| \int_{\partial_n} f(z)\, dz \right| \leq [l(\partial_n)]^2 \varepsilon$$

is thus valid. Using this inequality and (1) and (2) we have

$$\left| \int_{\partial} f(z)\, dz \right| \leq 4^n [l(\partial_n)]^2 \varepsilon = 4^n [2^{-n} l(\partial)]^2 \varepsilon = l(\partial)^2 \varepsilon.$$

Since ε is arbitrary, we have proved that $\int_{\partial} f(z)\, dz = 0$ holds.

If the exceptional point w is on ∂^*, then shrink Δ by η to a smaller triangle with vertices a', b', c' and oriented boundary ∂_η, as in Figure 4.3. Let l_η denote the oriented segment from a' to b'. We have

$$\int_{l_1} f(z)\, dz - \int_{l_\eta} f(z)\, dz = \int_{[0,1]} (f(l_1(t))(b - a) - f(l_\eta(t))(b' - a'))\, dt.$$

By the uniform continuity of f on $\overline{\triangle}$ (1.12), we see that $\lim_{\eta \to 0} \int_{l_\eta} f(z)\, dz = \int_{l_1} f(z)\, dz$. It follows that $\lim_{\eta \to 0} \int_{\partial_\eta} f(z)\, dz = \int_\partial f(z)\, dz$ holds. Since $\int_{\partial_\eta} f(z)\, dz = 0$ for each η, we have $\int_\partial f(z)\, dz = 0$.

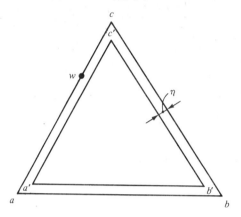

Figure 4.3

If the exceptional point w is in $\triangle$, then triangulate $\triangle$ at w, as in Figure 4.4. We have

$$\int_\partial f(z)\, dz = \sum_{j=1}^{3} \int_{\partial_j} f(z)\, dz = 0.$$

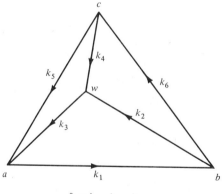

$$\partial_1 = k_1 \smile k_2 \smile k_3$$
$$\partial_2 = k_6 \smile k_4 \smile - k_2$$
$$\partial_3 = k_5 \smile - k_3 \smile - k_4$$

Figure 4.4

4.2 CAUCHY-GOURSAT THEOREM FOR A CONVEX SET

Let U be an open convex subset of $\mathbf{C}$, and let $w \in U$. Suppose that f is continuous on U and analytic on $U \backslash \{w\}$. The equality

(i) $\qquad \int_\gamma f(z) \, dz = 0$

holds for every closed path γ contained in U.

PROOF.   Fix $a \in U$ and define F on U by

$$F(z) = \int_{\overrightarrow{[a, z]}} f(\zeta) \, d\zeta.$$

By (4.1), we have

$$F(z) - F(z_0) = \int_{\overrightarrow{[a, z]}} f(\zeta) \, d\zeta - \int_{\overrightarrow{[a, z_0]}} f(\zeta) \, d\zeta = \int_{\overrightarrow{[z_0, z]}} f(\zeta) \, d\zeta.$$

Thus the equality

$$\frac{F(z) - F(z_0)}{z - z_0} - f(z_0) = \frac{1}{z - z_0} \int_{\overrightarrow{[z_0, z]}} [f(\zeta) - f(z_0)] \, d\zeta. \qquad (1)$$

holds. For a given $\varepsilon > 0$, there is a $\delta > 0$ such that $|f(\zeta) - f(z_0)| < \varepsilon$ if $|\zeta - z_0| < \delta$. By (1), we obtain

$$\left| \frac{F(z) - F(z_0)}{z - z_0} - f(z_0) \right| < \varepsilon$$

if $|z - z_0| < \delta$. Thus, the equality $F'(z) = f(z)$ holds for all $z \in U$. We have

$$\int_\gamma f(z) \, dz = \int_\gamma F'(z) \, dz = F(\gamma(b)) - F(\gamma(c)) = 0,$$

where γ is parametrized on $[c, b]$.

4.3 REMARKS

(i) This version of the Cauchy-Goursat theorem can easily be used to give the equality (i) for many curves in many nonconvex regions. Consider for example the shaded region U in Figure 4.5(a) and the closed path γ in U, and suppose that f is analytic on U. Since U is not convex, (4.2) does not apply directly. However, if we split α^* at the dotted line as in Figure 4.5(b), then each of the paths α and β lies in a convex region in which f is analytic, and so the integrals of f over these paths are zero. The segment l is a portion of α while $-l$ is a portion of β, and so in the sum $\int_\alpha f(z)\,dz + \int_\beta f(z)\,dz$ the integrals over l and $-l$ cancel giving us

$$\int_\gamma f(z)\,dz = \int_\alpha f(z)\,dz + \int_\beta f(z)\,dz = 0.$$

We will not attempt to formulate a general theorem based on this technique, but we will use the technique occasionally.

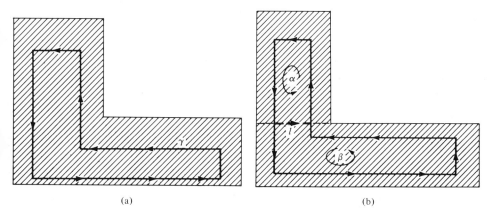

(a)   (b)

Figure 4.5

(ii) *The index.* Cauchy's theorem is often useful in calculating the index of a closed path. As an example, consider $\operatorname{Ind}_\partial (z)$ for z on the inside of a triangle ∂ which has counterclockwise orientation. See Figure 4.6(a). Let C be the circle through a, b, and c. Let a_1, a_2, and a_3 be the arcs of the circle opposite l_1, l_2, and l_3 (respectively) with orientations opposite to l_i (see Figure 4.6(b)). For each i, the function $1/(\zeta - z)$ is analytic on an open convex set containing $l_i \smile a_i$. Hence by (4.2) we have

$$\int_{l_i} \frac{1}{\zeta - z}\,d\zeta + \int_{a_i} \frac{1}{\zeta - z}\,d\zeta = 0.$$

Summing from 1 to 3 and letting $\overrightarrow{C}$ denote C with clockwise orientation, we obtain

$$\int_\partial \frac{1}{\zeta - z}\,d\zeta + \int_{\overrightarrow{C}} \frac{1}{\zeta - z}\,d\zeta = 0.$$

Since we know that

$$\int_{\overrightarrow{C}} \frac{1}{\zeta - z}\,d\zeta = -2\pi i,$$

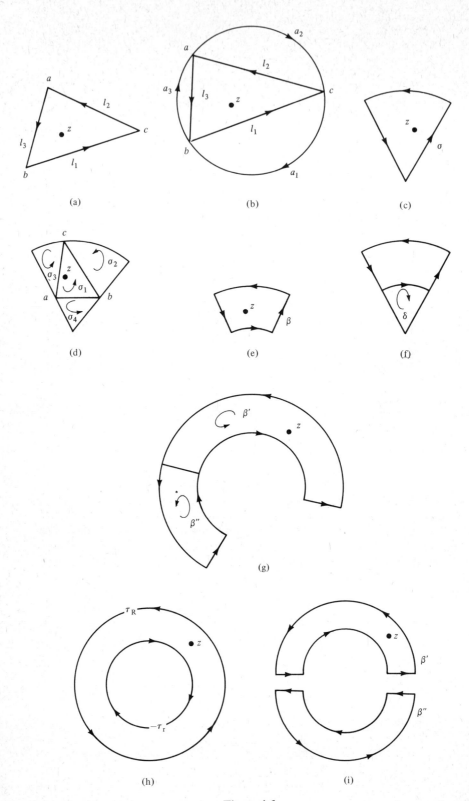

(a)   (b)   (c)

(d)   (e)   (f)

(g)

(h)   (i)

Figure 4.6

130

by (3.15), we have

$$\int_{\partial} \frac{1}{\zeta - z} \, d\zeta = 2\pi i.$$

Next consider a simple closed path σ which is the boundary of a sector of a circle. Suppose that the central angle of the sector is less than or equal to π. See Figure 4.6(c). Let z be in the sector. Draw a segment $[a, b]$ below z from one side of σ to the other and enclose z in a triangle having $[a, b]$ as one side and opposite vertex on the arc of the circle. See Figure 4.6(d). The sum of the integrals $\int_{\sigma_i} (1/(\zeta - z)) \, dz$ $(1 \leq i \leq 4)$ is equal to $\int_{\sigma} (1/(\zeta - z)) \, d\zeta$. As we have seen, the integral over σ_1 is $2\pi i$. The integrals over the other σ_i are all zero because each σ_i^* is contained in a convex open set in which $1/(\zeta - z)$ is analytic. Hence we have $\text{Ind}_{\sigma} (z) = 1$.

Next we truncate the sector in Figure 4.6(c) by swinging a circular arc from one side to the other, using the same center. Hence we obtain the path β pictured in Figure 4.6(e). It encloses a portion of an annulus. If z is inside β, we have

$$\int_{\beta} \frac{1}{\zeta - z} \, d\zeta = \int_{\sigma} \frac{1}{\zeta - z} \, d\zeta + \int_{\delta} \frac{1}{\zeta - z} \, d\zeta$$

where δ is pictured in Figure 4.6(f) and σ is shown in Figure 4.6(c). The integral over δ is zero because $1/(\zeta - z)$ is analytic on a convex region containing δ^*. Hence $\text{Ind}_{\beta} (z)$ is 1.

If the central angle of β is larger than π, then we can enclose the given point z in a path β' of the same type and having central angle π. We have $\text{Ind}_{\beta'} (z) = 1$, and the integral $\int_{\beta''} 1/(\zeta - z) \, d\zeta$ is zero. (See Figure 4.6(g).) It follows that $\text{Ind}_{\beta} (z) = 1$.

We can go a step further and consider the entire annulus; that is, we can let the central angle be 2π. Then β becomes two concentric circles. For convenience, suppose that the circles are centered at 0. Let $0 < r < R < \infty$ and let z satisfy $r < |z| < R$. (See Figure 4.6(h).) Denote the positively oriented circles by τ_r and τ_R. Draw line segments between the circles as indicated in Figure 4.6(i), forming two curves β' and β'' each having central angle π. Performing the proper cancellations, we obtain

$$\int_{-\tau_r} \frac{1}{\zeta - z} \, d\zeta + \int_{\tau_R} \frac{1}{\zeta - z} \, d\zeta = \int_{\beta'} \frac{1}{\zeta - z} \, d\zeta + \int_{\beta''} \frac{1}{\zeta - z} \, d\zeta.$$

The first integral on the right is $2\pi i$. The second is zero because β'' is contained in a convex set in which $1/(\zeta - z)$ is analytic. Hence we have

$$\int_{-\tau_r} \frac{1}{\zeta - z} \, d\zeta + \int_{\tau_R} \frac{1}{\zeta - z} \, d\zeta = 2\pi i.$$

Notice that all the closed paths considered here are positively oriented, according to the definition in (3.17). In each case, as the path is traversed the inside of the curve lies to the left. This interpretation holds for all positively oriented closed paths, but we will not attempt to prove it. We will return to the index again later in this chapter. Notice that the formula given in (4.4.i) contains $\text{Ind}_\gamma(z)$ as a factor. It is thus important to have methods of finding the index. The methods of this section are quite effective if the paths are not too complicated.

4.4 CAUCHY INTEGRAL REPRESENTATION

Let U be an open convex set and suppose that f is analytic on U. If γ is a closed path in U, then

(i) $f(z)\,\text{Ind}_\gamma(z) = \dfrac{1}{2\pi i}\displaystyle\int_\gamma \dfrac{f(\zeta)}{\zeta - z}\,d\zeta$

for all $z \in U$.

PROOF.   For fixed z, the function g defined by

$$g(\zeta) = \begin{cases} \dfrac{f(\zeta) - f(z)}{\zeta - z} & \text{if } \zeta \neq z \\[2mm] f'(z) & \text{if } \zeta = z \end{cases}$$

is analytic on $U \setminus \{z\}$ and continuous on U. By (4.2), the equality

$$\int_\gamma g(\zeta)\,d\zeta = 0$$

is valid; that is, we have

$$\int_\gamma \frac{f(\zeta)}{\zeta - z}\,d\zeta = \int_\gamma \frac{f(z)}{\zeta - z}\,d\zeta.$$

This proves (i) in view of the definition of the index (3.16).

For an interpretation of (i) when $z \notin U$, see Exercise 14, p. 152.

4.5 COROLLARY

Suppose that f is analytic on the open set U. Then $f^{(n)}$ exists and is analytic in U for all positive integers n. If $a \in U$ and $\overline{D(r, a)} \subset U$, then the equality

(i) $f^{(n)}(z) = \dfrac{n!}{2\pi i}\displaystyle\int_\tau \dfrac{f(\zeta)}{(\zeta - z)^{n+1}}\,d\zeta,$

holds for $z \in D(r, a)$, where $\tau(t) = a + re^{it}$, $t \in [-\pi, \pi]$.

PROOF. We saw in Section (3.15) that $\mathrm{Ind}_\tau(z) = 1$ if $|z - a| < r$. By (4.4.i), we have

$$\frac{f(w) - f(z)}{w - z} = \frac{1}{w - z}\frac{1}{2\pi i}\int_\tau\left[\frac{f(\zeta)}{\zeta - w} - \frac{f(\zeta)}{\zeta - z}\right]d\zeta$$

$$= \frac{1}{2\pi i}\int_\tau\frac{f(\zeta)}{(\zeta - w)(\zeta - z)}\,d\zeta.$$

If $(w_n)_{n=1}^\infty$ is a sequence in $D(r, a)$ converging to z, then

$$\lim_{n \to \infty}\frac{f(\zeta)}{(\zeta - w_n)/(\zeta - z)} = \frac{f(\zeta)}{(\zeta - z)^2}$$

uniformly for $\zeta \in \tau^*$. (The reader should verify that the convergence is uniform.) By (3.14), we conclude that

(ii) $f'(z) = \dfrac{1}{2\pi i}\displaystyle\int_\tau\frac{f(\zeta)}{(\zeta - z)^2}\,d\zeta.$

Fix z_0 in $D(r, a)$. Since

$$\frac{1}{(z - z_0)}\left[\frac{1}{(\zeta - z)^2} - \frac{1}{(\zeta - z_0)^2}\right]$$

converges uniformly on τ^* to $2/[(\zeta - z_0)^3]$ as z goes to z_0, it follows from (ii) that $f''(z_0)$ exists and equals

$$\frac{1}{2\pi i}\int_\tau\frac{2f(\zeta)}{(\zeta - z_0)^3}\,d\zeta.$$

The analyticity of $f^{(n)}$ and the equality (i) can be established by induction.

4.6 COROLLARY

If f is continuous on an open set U, $w \in U$, and f is analytic on $U\backslash\{w\}$, then f is in fact analytic on U.

PROOF. There is a disk $D(\varepsilon, w)$ which is contained in U. Define F on $D(\varepsilon, w)$ by

$$F(z) = \int_{\overrightarrow{[w, z]}}f(\zeta)\,d\zeta.$$

Using (4.2), we see (as in the proof of (4.2)) that F is analytic on $D(\varepsilon, w)$ and that $F' = f$ holds on $D(\varepsilon, w)$. By (4.5), F' is analytic on $D(\varepsilon, w)$. Hence f is analytic on $D(\varepsilon, w)$ and hence on all of U.

There follow a number of important corollaries of Cauchy's theorem and Cauchy's integral formula.

4.7 CAUCHY'S ESTIMATE

If f is analytic on the open set U and $\overline{D(a, r)} \subset U$, then

$$|f^{(n)}(a)| \le \frac{n!\, M}{r^n} ,$$

where $M = \text{lub } \{|f(\zeta)| \mid |\zeta - a| = r\}$.

The proof is an obvious application of (4.5.i) and is left to the reader.

4.8 LOCAL POWER SERIES (TAYLOR EXPANSION)

Let U be open and f analytic on U. For $a \in U$, let $\rho = \text{lub } \{r \mid D(r, a) \subset U\}$, and suppose that $0 < \mu < \rho$. The equality

(i) $f(z) = \displaystyle\sum_{n=0}^{\infty} \frac{f^{(n)}(a)}{n!} (z - a)^n$

is valid on $D(\rho, a)$, and the series converges uniformly on $D(\mu, a)$.

PROOF.   Suppose that $\mu < r < \rho$ and let $\tau(t) = a + re^{it}$, $t \in [-\pi, \pi]$. We simply calculate, using (4.4) and a geometric series. For $z \in D(r, a)$, we have

$$f(z) = \frac{1}{2\pi i} \int_\tau \frac{f(\zeta)}{\zeta - z}\, d\zeta = \frac{1}{2\pi i} \int_\tau \frac{f(\zeta)}{\zeta - a} \frac{1}{1 - \dfrac{(z - a)}{(\zeta - a)}}\, d\zeta$$

$$= \frac{1}{2\pi i} \int_\tau \frac{f(\zeta)}{(\zeta - a)} \sum_{k=0}^{\infty} \left[\frac{(z - a)}{(\zeta - a)}\right]^k d\zeta.$$

Since $|(z - a)/(\zeta - a)| \le \mu/r$ for $\zeta \in \tau^*$ and $z \in D(\mu, a)$, the series in the integrand converges uniformly on τ^* for each $z \in D(\mu, a)$. By (3.14), the last integral above is thus equal to

$$\sum_{k=0}^{\infty} \left[\frac{1}{2\pi i} \int_\tau \frac{f(\zeta)}{(\zeta - a)^{k+1}}\, d\zeta\right] (z - a)^k;$$

the series converges uniformly for $z \in D(\mu, a)$. The bracketed expression is equal to $f^{(n)}(a)/n!$, by (4.5).

4.9 EXAMPLES

The equality (4.8.i) provides a straightforward way of calculating Taylor expansions. We give some examples.

(i) Let $f(z) = e^z$. Then we have $f^{(n)}(z) = e^z$, and so for each fixed $a \in C$, the expansion

$$e^z = \sum_{n=0}^{\infty} \frac{e^a}{n!} (z - a)^n$$

is valid for all $z \in C$. As a special case, we have $e^z = \sum_{n=0}^{\infty} (-1/n!) (z - \pi i)^n$.

(ii) The function $f(z) = 1/z$ satisfies

$$f^{(n)}(z) = \frac{(-1)^n n!}{z^{n+1}} \quad \text{if } z \neq 0.$$

Hence the expansion of f about $a \neq 0$ is

$$f(z) = \frac{1}{a} + \sum_{n=1}^{\infty} \frac{(-1)^n}{a^{n+1}} (z - a)^n.$$

By (4.8), the expansion is valid if $|z - a| < |a|$. As a special case, we have

$$1/z = 1 + \sum_{n=1}^{\infty} (-1)^n (z - 1)^n, \quad |z - 1| < 1.$$

(iii) If $f(z) = \text{Log}\, (z + 1)$, then $f'(z) = 1/(z + 1)$ holds if $|z| < 1$. In general, we have

$$f^{(n)}(z) = \frac{(-1)^{n+1}(n - 1)!}{(z + 1)^{n+1}},$$

and so

$$\text{Log}\, (z + 1) = \sum_{n=1}^{\infty} \frac{(-1)^{n+1}}{n} z^n \quad (|z| < 1)$$

is the Taylor expansion of $\text{Log}\, (z + 1)$ about 0. Notice that the series diverges at $z = -1$ but converges (conditionally) at every other point of $\{|z| = 1\}$; see (2.18) and also exercises, (p. 59).

Sometimes there is a more convenient way of finding the Taylor expansion of a function than by direct calculation of the coefficients. We give some illustrations.

(iv) We can find the Taylor expansion of $1/z^2$ about $a \neq 0$ by using $-1/z^2 = d(1/z)/dz$, (ii) above, and (2.4). We have

$$\frac{1}{z^2} = \sum_{n=1}^{\infty} \frac{n(-1)^{n+1}}{a^{n+1}} (z - a)^{n-1} \quad (|z - a| < |a|).$$

(v) To find the expansion of $1/z$ about 1, we can also write

$$\frac{1}{z} = \frac{1}{1 - (1 - z)} = \sum_{n=0}^{\infty} (1 - z)^n = \sum_{n=0}^{\infty} (-1)^n (z - 1)^n \qquad |z - 1| < 1.$$

It is often convenient to spot a geometric series, as in this example.

There are many examples of Taylor expansions in the exercises. First, we give four more important consequences of Cauchy's theorem.

4.10 THE CONVERSE OF CAUCHY'S THEOREM (4.1): (MORERA'S THEOREM)

The definition of the primitive for f in the proof of (4.2) will work when the equality $\int_{\partial} f \, dz = 0$ holds for oriented triangles ∂. A precise statement follows.

Theorem. If f is a continuous function on an open set U with the property that $\int_{\partial} f(z) \, dz = 0$ for the oriented boundary ∂ of every triangle $\bar{\Delta} \subset U$, then f is analytic on U.

PROOF.   Fix $a \in U$ and suppose that $\overline{D(r, a)} \subset U$. Define F on $D(r, a)$ by

$$F(z) = \int_{\overrightarrow{[a, z]}} f(\zeta) \, d\zeta.$$

As in the proof of (4.2), we have $F'(z) = f(z)$ on $D(r, a)$. By (4.5), f is analytic on $D(r, a)$. Since a is arbitrary, f is analytic on U.

Remark. Corollary (4.5) is a consequence of Cauchy's theorem (4.2), which is a consequence of (4.1). Thus (4.1) was used to prove its own converse.

4.11 ANALYTICITY OF UNIFORM LIMITS

Let $(f_n)_{n=1}^{\infty}$ be a sequence of analytic functions on an open set U such that $\lim_{n \to \infty} f_n = f$ uniformly on every disk contained in U. Then f is analytic on U.

PROOF.   Let $\bar{\Delta}$ be a closed triangle contained in U, with oriented boundary ∂. By Cauchy's theorem, we have

$$\int_{\partial} f_n(z) \, dz = 0$$

for each n. By uniform convergence of f_n and (3.14), it follows that

$$\int_\partial f(z)\, dz = 0.$$

By (4.10) f is analytic on U.

4.12 THE MAXIMUM MODULUS THEOREM FOR ANALYTIC FUNCTIONS

Let f be analytic on an open set U. If $|f|$ has a local maximum at z_0 in U, then f is constant on every disk in U centered at z_0.

PROOF. Suppose that $|f|$ has a local maximum at z_0 and let $D(r, z_0) \subset U$. We must show that f is constant on $D(r, z_0)$. There is a μ with $0 < \mu \le r$ such that $|f(\xi)| \le |f(z_0)|$ for $\xi \in D(\mu, z_0)$. If $\rho \le \mu$, let $\tau_\rho(t) = z_0 + \rho e^{it}$ on $[-\pi, \pi]$. By (4.4), we have

$$|f(z_0)| = \left| \frac{1}{2\pi i} \int_{\tau_\rho} \frac{f(\xi)}{\xi - z_0}\, d\xi \right| \le \frac{1}{2\pi} \int_{\tau_\rho} \frac{|f(\xi)|}{\rho} |d\xi|$$

$$= \frac{1}{2\pi} \int_{[-\pi, \pi]} |f(z_0 + \rho e^{it})|\, dt;$$

that is, $|f(z_0)|$ is less than or equal to the average of $|f(\zeta)|$ over τ_ρ^*. Since $|f(\xi)| \le |f(z_0)|$, it follows that $|f(z_0)| = |f(\xi)|$ for $\xi \in \tau_\rho^*$. Since this holds for each $\rho \le \mu$, we conclude that $|f|$ is constant on $D(\mu, z_0)$. If $|f(z)| = 0$ for all $z \in D(\mu, z_0)$, then $f(z) = 0$ on $D(\mu, z_0)$. If $|f(z)| \ne 0$, then we have $\overline{f(z)} = c/f(z)$ for some constant c. Thus both f and $\bar{f}$ are analytic in $D(\mu, z_0)$. It is an easy exercise to show that this can happen only if f is a constant on $D(\mu, z_0)$. Finally f is constant on $D(r, z_0)$ by the identity theorem for power series (2.7), and (4.8).

Later in this chapter we will introduce the notion of a connected open set. It will then be apparent that the modulus of a function analytic on a connected open set can have a local maximum on the set if and only if the function is a constant on the set in question.

*4.13 COROLLARY: (FUNDAMENTAL THEOREM OF ALGEBRA)

Let $P(z) = a_0 + a_1 z + \cdots + z^n$ be a polynomial with complex coefficients, $n \ge 1$. The equation $P(z) = 0$ has a solution in the complex numbers.

PROOF. If $P(z) \ne 0$ for all $z \in \mathbb{C}$, then $1/P$ is analytic on $\mathbb{C}$. Since $\lim_{|z| \to \infty} |P(z)| = \infty$, $|P|$ must have an absolute minimum, which is, of course, also a local minimum. This local minimum is a local maximum of $|1/P|$.

Remark. The fundamental theorem was also proved in (1.15). The above proof uses (4.12) and hence ultimately (4.2). Notice that (4.2) is not particularly difficult to prove for polynomials, for it is clear that polynomials have primitives and so (3.6) applies.

EXERCISES

TYPE I

1. Show that $\int_{\gamma} f(z)\,dz = 0$, if γ is the closed path in the shaded open set U indicated in Figure 4.7, and if f is analytic on U. Base your proofs on (4.2).

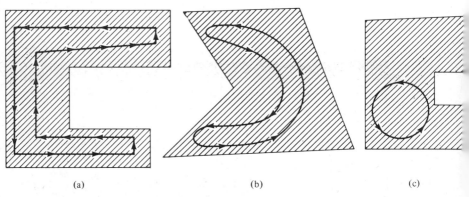

(a)   (b)   (c)

Figure 4.7

2. Find the Taylor expansion of f about a for the following f and a. Give the domain of validity in each case.
   (a) $f(z) = \exp(z)$, $a = 1$
   (b) $f(z) = \cos(z)$, $a = \pi/2$
   (c) $f(z) = \text{Log}(z)$, $a = 1$
   (d) $f(z) = 1/(1 - z)$; $a = 0$, $a = 2$, $a = i$
   (e) $f(z) = \sinh(z)$, $a = 0$
   (f) $f(z) = \cosh(z)$, $a = 0$
   (g) $f(z) = [\sin(z)]^2$, $a = 0$
   (h) $f(z) = [\cos(z)]^2$, $a = 0$.

3. Find the Taylor expansion of f about a for the following f and a. Give the domain of validity in each case.
   (a) $f(z) = 1/(1 + z^2)$, $a = 0$
      [*Hint:* Write $1/(1 + z^2) = 1/(1 - (-z^2)) = \sum_{k=0}^{\infty} (-z^2)^k$. For amusement write $1/(1 + z^2) = 1/(1 - iz)(1 + iz)$, express each of $1/(1 - iz)$ and $1/(1 + iz)$ as geometric series, and take the Cauchy product of the two geometric series.]

(b) $f(z) = 1/(1 - z^2)$, $a = 0$

(c) $f(z) = 1/[(1 - z)^2]$, $a = 0, 2, i$

[*Hint:* Differentiate the series obtained in Exercise 2(d), or take the Cauchy product of the series with itself.]

(d) $f(z) = 1/[(1 - z)^3]$, $a = 0, 2, i$

(e) $f(z) = z/(1 - z)$, $a = 0$

[*Hint:* Multiply the series for $1/(1 - z)$ by z.]

(f) $f(z) = 1/(z - 2)$, $a = 1$

[*Hint:* Write $z - 2 = -(1 - (z - 1))$ and use geometric series.]

(g) $f(z) = e^{z^2}$, $a = 0$

(h) $f(z) = e^{2z}$, $a = 0$

(i) $f(z) = 1/z^2$, $a = -1$.

4. Fix θ satisfying $-\pi < \theta \leq \pi$. Let $h_\theta = \{re^{i\theta} \mid r \geq 0\}$; h_θ is the ray from the origin in the direction θ. Define an analytic function f on $\mathbb{C}\backslash h_\theta$ which satisfies $e^{f(z)} = z$ for $z \in \mathbb{C}\backslash h_\theta$. How does f relate to Log? Find the power series expansion of f about $e^{-i\theta}$

5. Write down the first five terms of the Taylor expansion of f about a for the f and a given.

(a) $f(z) = \tan(z)$, $a = 0$

(b) $f(z) = \exp[z/(1 - z)]$, $a = 0$

(c) $f(z) = \exp(z \sin z)$, $a = 0$.

6. (Binomial expansion.) Suppose that $0 < \alpha < 1$ and let $a \in \mathbb{C}\backslash\{0\}$. Let $g(z) = F_\alpha(z + a)$.

(a) Show that g is discontinuous on $L = \{-a + t : t \leq 0\}$ and analytic on $U = \mathbb{C}\backslash L$.

(b) Show that

$$g^{(n)}(z) = \alpha(\alpha - 1)\cdots(\alpha - n + 1)F_\alpha(z + a)\,\frac{1}{(z + a)^n}, \qquad z \in U.$$

(c) With the definition of $\binom{\alpha}{h}$ given in (2.17) show that the Taylor expansion of g centered at 0 is

$$F_\alpha(z + a) = \sum_{n=0}^{\infty}\left[\binom{\alpha}{h}F_\alpha(a)a^{-n}\right]z^n.$$

(d) Show that $D(|a|, 0) \subset U$ if and only if $\text{Re}(a) \geq 0$.

(e) If $\text{Re}(a) \geq 0$, then by (d) and (4.8) the expansion in (c) is valid in $D(|a|, 0)$. By (2.17) the disk of convergence of the series is $D(|a|, 0)$ for all a. If $\text{Re}(a) < 0$, show that the equality in (d) does not hold for some z in $D(|a|, 0)$. Sketch the largest disk centered at 0 in which the equality is valid if $\text{Re}(a) < 0$.

(f) Since $F_\alpha(a)a^{-n} = F_{\alpha-n}(a)$ the equality in (d) can be written

$$F_\alpha(z + a) = \sum_{n=0}^{\infty} \left[\binom{\alpha}{n} F_{\alpha-n}(a) \right] z^n.$$

7. Find the Taylor expansion of

$$g(z) = (z^2 + z + i)/(z - 1)(z - i)$$

about the point 0.

8. Suppose that $f(z) = \sum_{n=0}^{\infty} a_n z^n$, where the series has radius of convergence $r > 0$. Without using integration or the Cauchy theorem, write a primitive F of f, valid on $D(r, 0)$.

9. A function f on $D = D(\rho, 0)$ is said to be *even* if $f(z) = f(-z)$ for all $z \in D$, and odd if $f(z) = -f(-z)$, all $z \in D$. If f is analytic on D with Taylor series $f(z) = \sum_{n=0}^{\infty} a_n z^n$, show that f is even if and only if $a_n = 0$ for odd n and that f is odd if and only if $a_n = 0$ for all even n.

10. Suppose that f is analytic on $D = D(1, 0)$ and that there is a sequence $(z_m)_{m=1}^{\infty}$ such that $z_m \neq 0$ for all m, $\lim_{m \to \infty} z_m = 0$, and $f(z_m) = f(-z_m)$. Show that f is even.

11. Let f be analytic on $D(\rho, 0)$, $\rho > 0$. Let m be a fixed positive integer. Suppose that $f(\zeta z) = f(z)$ for every m^{th} root of unity ζ and all $z \in D(\rho, 0)$. Determine which coefficients in the Taylor expansion of f about 0 must be zero.

12. Find $\text{Ind}_\gamma (z)$ for the γ and z indicated in Figure 4.8.

TYPE II

1. Let U be an open convex set and suppose f is analytic on U. Let $Z(f) = \{z \in U \mid f(z) = 0\}$. Prove that $Z(f)$ has a cluster point in U if and only if $Z(f) = U$; that is, f is zero every place on U.

   [*Hint:* Use (4.8) and (2.7).]

2. Let $f(z) = \sum_{n=0}^{\infty} a_n z^n$ have radius of convergence $r > 0$. Give a proof of the Cauchy integral representation (4.4) for f which does not use Cauchy's theorem.

3. (Liouville's Theorem) A bounded analytic function f on $\mathbf{C}$ is constant. [*Hint:* Use (4.7). See (5.24) for a different proof.]

4. Prove the analyticity of uniform limits of analytic functions (4.11) by direct use of Cauchy's integral formula (4.4) for a circle.

5. Suppose that f is analytic on $D(R, 0)$ with Taylor expansion $f(z) = \sum_{n=0}^{\infty} a_n z^n$, $a_0 \neq 0$ on $D(R, 0)$. Show that there is an $r > 0$ such that $1/f$ is analytic on $D(r, 0)$. If $\sum_{n=0}^{\infty} b_n z^n$ is the Taylor expansion of $1/f$ on

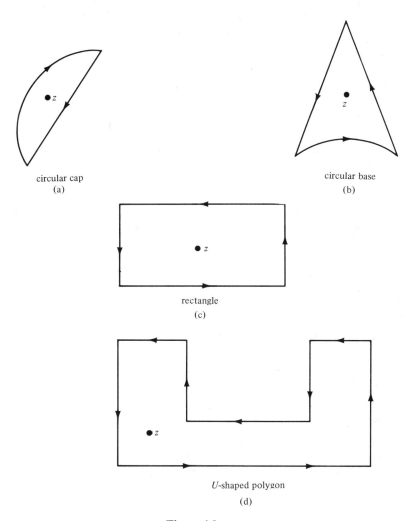

circular cap
(a)

circular base
(b)

rectangle
(c)

U-shaped polygon
(d)

Figure 4.8

$D(r, 0)$, show that the coefficients (a_n) and (b_n) satisfy the recursive relation

$$a_0 b_0 = 1$$
$$a_0 b_1 + a_1 b_0 = 0$$
$$\vdots$$
$$a_0 b_n + \cdots + a_n b_0 = 0.$$
$$\vdots$$

$(*)$

Given the a's, the b's are thus uniquely determined by the infinitely many equations in $(*)$.

6. Apply the recursive relations in Exercise 5 to find the Taylor expansion of $1/[\cos(z)]$ about 0, then find the Taylor expansion of $\tan(z)$ about 0 by using Cauchy products.

7. (Bernoulli numbers.)   The function $f(z) = z/(e^z - 1)$ is analytic on $D(2\pi, 0)$. We have

$$\frac{z}{e^z - 1} = \frac{1}{1 + z/2! + z^2/3! + \cdots} = \sum_{n=0}^{\infty} \frac{1}{n!} B_n z^n \qquad (|z| < 2\pi)$$

for suitably defined B_n. The numbers B_n are called the *Bernoulli numbers*. Use Exercise 5 to find $\{B_n\}_{n=0}^{8}$.

8. Suppose that f and g are analytic on a disk containing 0 and that $f(0) = 0$. Find a formula for the coefficients in the Taylor expansion of $g \circ f$ in terms of those of f and g.

9. Let U be an open convex set in $\mathbf{C}$ and suppose that f is analytic on U. Let γ be a closed path in U and suppose that $|f(z)| \geq M > 0$ for all $z \in \gamma^*$.
   (a) If $f(z) \neq 0$ for all z such that $\text{Ind}_\alpha(z) \neq 0$, show that $|f(z)| \geq M$ for all z such that $\text{Ind}_\alpha(z) \neq 0$.
   (b) Give an example where $f(z) \neq 0$ for some z with $\text{Ind}_\alpha(z) \neq 0$ and $|f(z)| < M$.

10. Let γ be a simple closed path in $\mathbf{C}$. Suppose that γ^* is the boundary of open sets B (bounded) and U (unbounded) such that $B \cap U = \phi$ and $\mathbf{C} \backslash \gamma^* = B \cup U$, and suppose that γ has counterclockwise orientation with respect to B. Let V be an open set containing $B \cup \gamma^*$ and suppose that P and Q are real-valued functions in $\mathbf{C}^1(V)$. Recall that Green's theorem says that

$$\int_\gamma P \, dx + Q \, dy = \iint_B \left(\frac{\partial Q}{\partial x} - \frac{\partial P}{\partial y} \right) dx \, dy. \qquad (*)$$

   (a) Show that $(*)$ also holds if P and Q are complex-valued $\mathbf{C}^1$ functions on V.
   (b) Let $f \in \mathbf{C}^1(V)$. Show that $\int_\gamma f(z) \, dz = \int_\gamma f \, dx + if \, dy$.
   (c) Show that

$$\int_\gamma f(z) \, dz = i \iint_B \left[\frac{\partial f}{\partial x} + i \frac{\partial f}{\partial y} \right] dx \, dy.$$

(d) Suppose that f is analytic on V. Use Green's theorem (c) for f to prove Cauchy's theorem for γ.

Remarks.

(1) The basic formula (∗) is difficult to prove for general situations. If one considers an arbitrary simple closed path and wishes to base a proof of Cauchy's theorem for γ on (∗), then the sets B and U are difficult to describe and deep topological questions arise. The point we wish to make here is simply that any valid proof of (∗) for a pair of γ and B yields a valid proof of Cauchy's theorem for functions with continuous first derivatives.

(2) The formula (∗) can be extended to regions B bounded by a collection of curves, and the corresponding results also yield proofs of Cauchy theorems for these situations.

(3) Recall that the bracketed expression in (c) is the definition of $2\partial f/\partial \bar{z}$. Hence the result in (c) can be written

$$\int_{\gamma} f(z)\,dz = 2i \iint_{B} \frac{\partial f}{\partial \bar{z}}\,dx\,dy.$$

(4) Versions of Green's theorem are contained in most books on advanced calculus or functions of several variables.

∗4.14 THE CAUCHY KERNEL AND CAUCHY INTEGRAL

If f is analytic on $D(r, 0)$, $r > 1$, then the Cauchy integral representation (4.4) of f on $D(1, 0)$ is

(i) $$f(z) = \frac{1}{2\pi i} \int_{\tau} \frac{f(\zeta)}{\zeta - z}\,d\zeta = \frac{1}{2\pi} \int_{[-\pi, \pi]} f(e^{it}) \frac{e^{it}}{e^{it} - z}\,dt,$$

where $\tau(t) = e^{it}$ on $[-\pi, \pi]$. The function $1/(2\pi i(\zeta - z))$ (or $e^{it}/(2\pi(e^{it} - z))$ if we use the interval of parametrization) serves as a *kernel* for the *Cauchy integral of f*. We define the Cauchy integral of any continuous function f on $T = \{z \mid |z| = 1\}$ by the equality

(ii) $$Cf(z) = \frac{1}{2\pi} \int_{[-\pi, \pi]} f(e^{it}) \frac{e^{it}}{e^{it} - z}\,dt.$$

Thus, if f is the restriction to $\mathbf{T}$ of a function f which is analytic on some $D(r, 0)$, $r > 1$, we know that $Cf = f$ on $\mathbf{T}$.

(iii) *Power Series for Cf.* Let f be continuous on **T**. The following calculation is justified [for $z \in D(1, 0)$] by uniform convergence of the integrand and (3.14).

$$Cf(z) = \frac{1}{2\pi} \int_{[-\pi, \pi]} f(e^{it}) \frac{e^{it}}{e^{it} - z} \, dt = \frac{1}{2\pi} \int_{[-\pi, \pi]} f(e^{it}) \frac{1}{[1 - (z/e^{it})]} \, dt$$

$$= \frac{1}{2\pi} \int_{[-\pi, \pi]} f(e^{it}) \sum_{n=0}^{\infty} (z/e^{it})^n \, dt = \sum_{n=0}^{\infty} \left[\frac{1}{2\pi} \int_{[-\pi, \pi]} f(e^{it}) e^{-int} \, dt \right] z^n$$

$$= \sum_{n=0}^{\infty} \hat{f}(n) z^n,$$

Recall that $\hat{f}(n)$ is the nth *Fourier coefficient* of f, defined for all $n \in \mathbf{Z}$ by

$$\hat{f}(n) = \frac{1}{2\pi} \int_{[-\pi, \pi]} f(e^{it}) e^{-int} \, dt;$$

see (3.18). The power series $\sum_{n=0}^{\infty} \hat{f}(n) z^n$ thus has radius of convergence not less than 1. The Cauchy integral of an arbitrary continuous function on **T** thus has a power series valid on $D(1, 0)$ for which the coefficients are precisely the Fourier coefficients of f. In particular, if f is analytic on $D(r, 0)$ $(r > 1)$, then $\hat{f}(n) = f^{(n)}(0)/n!$, for $n \geq 0$, by (4.8) and (2.7). By (2.4), we have the following theorem.

(iv) *Theorem.* If f is continuous on T, Cf is analytic on $D(1, 0)$.

Of course, (iv) could also be proved by direct consideration of difference quotients.

(v) *A Boundary Value Problem.* If f is continuous on **T**, we pose the following problem.

Problem. Find a function F on $\overline{D(1, 0)}$ such that F is analytic on $D(1, 0)$ and $F = f$ on **T**.

In view of the above discussion, the obvious function to try for F is Cf. But suppose that $(a_n)_{n=-\infty}^{\infty}$ is a two-sided complex sequence for which $\sum_{n=-\infty}^{\infty} |a_n|$ converges. The formula

$$f(e^{it}) = \sum_{n=-\infty}^{\infty} a_n e^{int}$$

then defines a function f continuous on **T**, and it is easy to show that $\hat{f}(n) = a_n$ for $n = 0, \pm 1, \pm 2, \ldots$ [use (3.14) to justify the technique in (3.18)]. We therefore obtain

$$Cf(z) = \sum_{n=0}^{\infty} a_n z^n \quad (z \in D(1, 0)) \quad \text{and} \quad f(z) = \sum_{n=-\infty}^{\infty} a_n z^n \quad (|z| = 1).$$

Hence the equality $\lim_{\zeta \to z, |\zeta| < 1} Cf(\zeta) = f(z)$ appears unlikely if there are negative n such that $a_n = \hat{f}(n) \neq 0$. Thus the Cauchy integral appears a likely candidate to solve the problem only if $\hat{f}(n) = 0$ for $n < 0$. We use Fejér's theorem on Fourier series (3.21.iv) to prove that the Cauchy integral does solve the problem.

(vi) *Theorem.* The problem (v) has a solution if and only if $\hat{f}(n) = 0$ for $n < 0$. The unique solution is

$$F(z) = \begin{cases} Cf(z) & \text{if } z \in D(1, 0) \\ f(z) & \text{if } |z| = 1. \end{cases} \tag{*}$$

PROOF. If a solution F exists, it must be the displayed function, for it is easy to verify that letting $r \to 1$ in the Cauchy representation

$$F(z) = \frac{1}{2\pi i} \int_{\tau_r} \frac{F(\zeta)}{\zeta - z}\, d\zeta \qquad (|z| < r < 1)$$

gives (*). To prove the rest of the assertion, we use the Fejér means of f. See (3.21) for notation and results. We consider σ_n as a function on $\mathbf{T}$ and write its values as $\sigma_n(e^{it})$.

Suppose first that $\hat{f}(n) = 0$ for $n < 0$, so that σ_N is given by

$$\sigma_N(e^{ix}) = \frac{1}{N} \sum_{n=0}^{N-1} (N - n)\hat{f}(n)e^{inx}.$$

Extend σ_N to all of $\overline{D(1, 0)}$ by simply defining

$$\sigma_N(z) = \frac{1}{N} \sum_{n=0}^{N-1} (N - n)\hat{f}(n)z^n \qquad (|z| \leq 1).$$

Thus each σ_N is analytic on $D(1, 0)$ and continuous on $\overline{D(1, 0)}$. For each N and m, $|\sigma_N - \sigma_m|$ has a maximum on $\overline{D(1, 0)}$, by (1.14). Since each $\sigma_N - \sigma_m$ is analytic in $D(1, 0)$, the maximum must occur on $\mathbf{T}$ by (4.12); that is,

$$\text{lub } \{|\sigma_N(z) - \sigma_m(z)| \mid z \in \overline{D(1, 0)}\}$$
$$= \text{lub } \{|\sigma_N(e^{it}) - \sigma_m(e^{it})| \mid t \in [-\pi, \pi]\}. \tag{**}$$

By Fejér's theorem, σ_N converges uniformly on $\mathbf{T}$ to f, and so by (**) and the Cauchy criterion σ_N converges uniformly on $\overline{D(1, 0)}$. For each z, $\sigma_N(z)$ is the Nth average of the partial sums of the series $\sum_{k=0}^{\infty} \hat{f}(k)z^k$; that is,

$$\sigma_N(z) = \frac{1}{N} \sum_{n=0}^{N-1} S_n(z),$$

where

$$S_n(z) = \sum_{k=0}^{n} \hat{f}(k)z^k.$$

By the lemma in (3.22), we thus obtain

$$\lim_{N \to \infty} \sigma_N(z) = \sum_{n=0}^{\infty} \hat{f}(n)z^n = F(z) \qquad (|z| < 1).$$

Since σ_N converges uniformly to F on $\overline{D(1, 0)}$, F is continuous. [We can also conclude the proof by appealing to (4.11). The sequence $(\sigma_N)_{N=1}^{\infty}$ converges uniformly, and so the limit must be analytic on $D(1, 0)$. Since $\lim \sigma_N = f$ on **T**, a solution exists.]

Suppose next that the function F as defined in (∗) is a solution to our problem. Since the function $e^{it} \to F(re^{it})$ has n^{th} Fourier coefficient equal to $\hat{f}(n)r^n$ if $n \geq 0$ and equal to 0 if $n < 0$, by the affirmative answer to (3.21.2) we have

$$\int_{[-\pi, \pi]} |F(re^{it})|^2 \, dt = \sum_{n=0}^{\infty} |r^n \hat{f}(n)|^2 \qquad (r < 1).$$

As $r \to 1$, the integral on the left converges to $\int_{[-\pi, \pi]} |f(e^{it})|^2 \, dt$ by the uniform continuity of F; and this integral is equal to $\sum_{n=-\infty}^{\infty} |\hat{f}(n)|^2$ by (3.21.2). As $r \to 1$, the series on the right converges to $\sum_{n=0}^{\infty} |\hat{f}(n)|^2$. Thus the equality

$$\sum_{n=0}^{\infty} |\hat{f}(n)|^2 = \sum_{n=-\infty}^{\infty} |\hat{f}(n)|^2$$

holds, and so we have proved that $\hat{f}(n) = 0$ if $n < 0$.

EXERCISES

TYPE II

1. Let f be continuous on the unit circle **T**. Prove that Cf is analytic on $D(1, 0)$ by direct consideration of difference quotients.

4.15 CONNECTEDNESS

Disconnected sets are easier to describe than connected sets. A subset S of **C** is *disconnected* if there are open sets U and V such that $U \cap V = \varnothing$, $S \subset U \cup V$, $U \cap S \neq \varnothing$, and $V \cap S \neq \varnothing$. A set is thus disconnected if there are disjoint open sets each having nonempty intersection with S and whose union contains S. A set is *connected* if it is not disconnected. In Figure 4.9, the shaded open set in (a) is disconnected and that in (b) is connected. In Figure 4.9(c), the set shaded vertically is connected and the set shaded horizontally is disconnected. As one would expect, curves are connected. We begin by proving this fact.

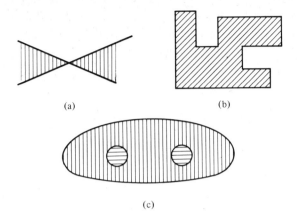

(a)   (b)

(c)

Figure 4.9

(i) *Theorem.*   Let α be a curve defined on $[a, b]$ to **C**. Then α^* is connected.

PROOF.   Suppose that there are disjoint open sets U and V such that $\alpha^* \subset U \cup V$, $U \cap \alpha^* \neq \varnothing$, and $V \cap \alpha^* \neq \varnothing$.   Suppose that $\alpha(a) \in U$.   Let $D(\varepsilon, \alpha(a)) \subset U$.   Since α is continuous, there is a $\delta > 0$ such that $\alpha[a, a + \delta] \subset D(\varepsilon, \alpha(a))$.   Let $c = \text{lub } \{t \in [a, b] \mid \alpha([a, t]) \subset U\}$.   Since $\alpha[a, a + \delta] \subset U$, we have $c > a$. If $c = b$, then we have $\alpha([a, b]) \subset U$, contradicting the fact that $V \cap \alpha^* \neq \varnothing$. Hence we have $a < c < b$. The point $\alpha(c)$ must be in U or in V. If $\alpha(c) \in U$, then there is an $\varepsilon > 0$ such that $D(\varepsilon, \alpha(c)) \subset U$. Since α is continuous, there is a $\delta > 0$ such that $\alpha(t) \in D(\varepsilon, \alpha(c))$ if $|t - c| \leq \delta$. In particular, we have $\alpha(t) \in U$ if $t \in [a, c + \delta]$; this contradicts the definition of c. Thus $\alpha(c)$ cannot be in U. If $\alpha(c) \in V$, then again there is $\delta > 0$ such that $\alpha(t) \in V$ if $|c - t| \leq \delta$. In particular, we have $\alpha(c - \delta) \in V$, again contradicting the definition of c. Hence sets U and V with the stated properties cannot exist, so α^* is connected.

Recall that z and w are said to be joined by a polygonal path if there are oriented line segments $(l_i)_{i=1}^n$ such that the initial point of l_{i+1} is the terminal point l_i, the initial point of l_1 is z, and the terminal point of l_n is w. The equivalence given in the following theorem is very useful.

(ii) *Theorem.*   An open set S in **C** is connected if and only if every pair of points in S can be joined by a polygonal path lying in S.

PROOF.   Suppose first that S is disconnected.   Since S is open, there are disjoint nonvoid open sets U and V such that $S = U \cup V$. Let $l = l_1 \smile l_2 \smile \cdots \smile l_n$ be a polygonal path from $z \in U$ to $w \in V$. There is a smallest k such that the initial point of l_k is in U but l_k^* is not (entirely) contained in U.

We will show that l_k^* cannot be contained (entirely) in S. If $l_k^* \subset S$, then we have

$$l_k^* = (l_k^* \cap U) \cup (l_k^* \cap V). \qquad (1)$$

Since l_k^* is the image of a curve, l_k^* is connected by (i). Since $l_k^* \cap U$ and $l_k^* \cap V$ are not void, (1) is thus a contradiction. We have thus proved that there is no polygonal path lying in S and joining z and w.

Next, suppose that S is an open connected set. Fix $z \in S$ and let $U(z) = \{w \in S \mid w$ is joined to z by a polygonal path in $S\}$. It is easy to verify that $U(z)$ is open. To see that $S \backslash U(z)$ is also open, let $w \in S \backslash U(z)$ and let $D(r, w) \subset S$. If $D(r, w)$ contains a point v which can be connected to z with a polygonal path, then this polygonal path together with the segment from v to w connects z to w. Thus the disk $D(r, w)$ contains no points of $U(z)$, and hence $S \backslash U(z)$ is open. Since S is connected and $U(z)$ is not void, $S \backslash U(z)$ is void. Since z is arbitrary, any two points in S can be joined by a polygonal path which lies entirely in S.

4.16 CONNECTED COMPONENTS

Let S be an open set in $\mathbf{C}$. Define a relation $\sim$ on pairs of elements of S by letting $z \sim w$ if there is a polygonal path in S with initial point z and terminal point w. The relation $\sim$ is an equivalence relation on S. The *connected components* of S are the equivalence classes of $\sim$; that is, the sets

$$C(z) = \{w \mid w \sim z\}.$$

For any z_1 and z_2, either $C(z_1) = C(z_2)$ holds or $C(z_1) \cap C(z_2) = \varnothing$ holds. If $\mathbf{C}$ is the family of distinct components of S, then in summary we have

(i) $S = \cup_{\mathbf{C}} C$;

(ii) if $C \in \mathbf{C}$ and $D \in \mathbf{C}$, then either $C = D$ or $C \cap D = \varnothing$;

(iii) each $C \in \mathbf{C}$ is open;

(iv) each $C \in \mathbf{C}$ is connected;

(v) if U is a connected subset of S, then U is contained in some $C \in \mathbf{C}$.

In Figure 4.9(a) the connected components of the open shaded set are the two triangular shaped open sets. In Figure 4.9(c), the horizontally shaded open set has two components, the vertically shaded open set one component and two holes.

4.17 THEOREM

If K is a compact subset of $\mathbf{C}$, then one and only one of the connected components of $\mathbf{C} \backslash K$ is unbounded, that is, contains numbers of arbitrarily large magnitude.

PROOF. There is an R such that $K \subset \overline{D(R, 0)}$. The set $\mathbf{C} \backslash \overline{D(R, 0)}$ is a connected open set in $\mathbf{C} \backslash K$, and so by (4.16.v) it is contained in a connected component C of $\mathbf{C} \backslash K$. This component C is obviously unbounded. Every other component of $\mathbf{C} \backslash K$ is contained in $\overline{D(R, 0)}$, and so is bounded.

4.18 IDENTITY THEOREM FOR ANALYTIC FUNCTIONS

Let U be an open set, and let f be analytic on U. If $w \in U$ is a cluster point of the set $Z(f) = \{z \in U \mid f(z) = 0\}$, then f is identically zero in the connected component $C(w)$ of U.

PROOF. Since f has a local power series at w (4.8), there is a $\delta > 0$ such that $D(\delta, w) \subset U$ and $f(z) = 0$ for $z \in D(\delta, w)$. To see this, refer to (2.7). Define

$V = \{v \in C(w) \mid$ there is an $\varepsilon > 0$ such that
$$f(z) = 0 \text{ for } z \in D(\varepsilon, v) \text{ and } D(\varepsilon, v) \subset U\}.$$

It is immediately clear that V is open. If $x \in C(w) \backslash V$, then x cannot be a cluster point of $Z(f)$, for if it were, the power series expansion (4.8) of f about x would vanish identically (by (2.7)) and x would be in V. Hence there is an $\varepsilon > 0$ such that $f(\zeta) \neq 0$ for $\zeta \in D(\varepsilon, x) \backslash \{x\}$. In particular, $D(\varepsilon, x) \subset C(w) \backslash V$. Thus $C(w) \backslash V$ is also open. Since V is not void, it follows that $V = C(w)$.

(i) *Corollary.* If f and g are analytic functions on an open set U and $w \in U$ is a cluster point of the set $\{z \mid f(z) = g(z)\}$, then $f = g$ on $C(w)$.

4.19 THEOREM

Let f be analytic on an open set U. If $f'(z) = 0$ for all $z \in U$, then f is constant on each connected component of U.

FIRST PROOF. Since $f^{(n)}(z) = 0$ on U for all $n \geq 1$, f is locally constant on U (by 4.8). By (4.18.i), f is constant on the components of U.

SECOND PROOF. Fix z in a component C of U. If $[z, w]$ is in C, then

$$F(t) = f((1 - t)z + tw)$$

defines a differentiable function on $[0, 1]$ for which $F'(t) = 0$. Thus F is constant and $f(z) = f(w)$. Since two points in C can be connected by a polygonal path, f is constant on C.

4.20 COROLLARY

Let f be real-valued and analytic on an open set U. Then f is constant on each connected component of U.

PROOF.   The difference quotients

$$\frac{f(z + \varepsilon) - f(z)}{\varepsilon} \quad \text{and} \quad \frac{f(z + i\varepsilon) - f(z)}{i\varepsilon}$$

each converge to $f'(z)$ as ε goes to 0 through real values. The first quotient is real-valued and the second is pure-imaginary or zero. Thus the equality $f'(z) = 0$ holds. By (4.19), f is constant on the components of U.

4.21 THEOREM

Let α be a path and suppose that ϕ is continuous on α^*. Define F on $\mathbb{C}\backslash\alpha^*$ by

$$F(z) = \int_\alpha \frac{\phi(\zeta)}{\zeta - z}\, d\zeta.$$

Then F is analytic on $\mathbb{C}\backslash\alpha^*$.

PROOF.   If $z \notin \alpha^*$, then $\phi(\zeta)/(\zeta - z)$ is continuous on α^*, so F is well-defined. Let z_0 be in $\mathbb{C}\backslash\alpha^*$. The difference quotient for F at z_0 is

$$\frac{F(z) - F(z_0)}{z - z_0} = \int_\alpha \phi(\zeta)\left[\frac{1}{(\zeta - z)(\zeta - z_0)}\right] d\zeta.$$

Let $\lim_{n\to\infty} z_n = z_0$ $(z_n \neq z_0)$, $z_n \in \mathbb{C}\backslash\alpha^*$. Since $\mathbb{C}\backslash\alpha^*$ is open, there is an $\varepsilon > 0$ such that $D(2\varepsilon, z_0) \cap \alpha^* = \varnothing$. There is an n_0 such that $|z_n - z_0| < \varepsilon$ if $n \geq n_0$. We have $|\zeta - z_n| \geq \varepsilon$ for all $n \geq n_0$ and all $\zeta \in \alpha^*$. Let $M = \text{lub }\{|\phi(\zeta)| \mid \zeta \in \alpha^*\}$; M is finite by (1.14). We have

$$\left| \phi(\zeta)\frac{1}{(\zeta - z_n)(\zeta - z_0)} - \phi(\zeta)\frac{1}{(\zeta - z_0)^2} \right|$$

$$\leq \frac{|\phi(\zeta)|}{|\zeta - z_0|}\left|\frac{1}{(\zeta - z_n)} - \frac{1}{(\zeta - z_0)}\right| \leq \frac{M}{\varepsilon}\left[\frac{|z_n - z_0|}{\varepsilon^2}\right]$$

for all $\zeta \in \alpha^*$. The sequence of functions of ζ with values

$$\left(\frac{\phi(\zeta)}{(\zeta - z_n)(\zeta - z_0)}\right)_{n=1}^\infty$$

thus converges uniformly on α^* to $\phi(\zeta)/(\zeta - z_0)^2$. By (3.14), we have

$$\lim_{n\to\infty} \frac{F(z_n) - F(z_0)}{z_n - z_0} = \int_\alpha \frac{\phi(\zeta)}{(\zeta - z_0)^2}\, d\zeta.$$

Since this equality holds for all sequences z_n converging to z_0, we conclude that $F'(z_0)$ exists and is equal to $\int_\alpha (\phi(\zeta)/(\zeta - z_0)^2)\, d\zeta$.

4.22 THEOREM ON THE INDEX

Let γ be a closed path in $\mathbf{C}$. The function Ind_γ is constant on each connected component of $U = \mathbf{C}\backslash\gamma^*$, and is zero on the unbounded component.

FIRST PROOF.   By (3.16), $\mathrm{Ind}_\gamma (z)$ is an integer for all $z \in U$. By (4.21) Ind_γ is analytic. By (4.20), it must be the same integer on each component of U.

Let C be the unbounded component of U, let $k = \mathrm{Ind}_\gamma (z)$ for $z \in C$, and let $(z_n)_{n=1}^\infty$ be a sequence in C such that $\lim_{n\to\infty} |z_n| = \infty$. For given $\delta > 0$, there is an n_0 such that $|\zeta - z_n| > 1/\delta$ for all $\zeta \in \gamma^*$ and all $n \geq n_0$; this is because γ^* is bounded. Hence the sequence $[1/(\zeta - z_n)]_{n=1}^\infty$ converges to 0 uniformly on γ^*. Applying (3.14), we obtain

$$k = \lim_{n\to\infty} \frac{1}{2\pi i} \int_\gamma \frac{1}{\zeta - z_n} \, d\zeta = \frac{1}{2\pi i} \int_\gamma \lim_{n\to\infty} \frac{1}{\zeta - z_n} \, d\zeta = 0.$$

SECOND PROOF.   Fix $z_0 \in \mathbf{C}\backslash\gamma^*$. Using the proof of (4.21) to take a derivative under an integral sign, we have

(i) $\displaystyle \mathrm{Ind}'_\gamma (z_0) = \frac{1}{2\pi i} \int_\gamma \left[\frac{d}{dz}\left(\frac{1}{\zeta - z} \right) \right]_{z_0} d\zeta.$

The equality

(ii) $\displaystyle \left[\frac{d}{dz}\left(\frac{1}{\zeta - z} \right) \right]_{z_0} = -\left[\frac{d}{d\xi}\left(\frac{1}{\xi - z_0} \right) \right]_\zeta$

holds if $\zeta \neq z_0$. Using (ii) in (i), we obtain

(iii) $\displaystyle \mathrm{Ind}'_\gamma (z_0) = \frac{1}{2\pi i} \int_\gamma \left[\frac{d}{d\xi}\left(\frac{1}{\xi - z_0} \right) \right]_\zeta d\zeta.$

The value of the integral in (iii) is 0 by (3.6). Hence we have $\mathrm{Ind}'_\gamma (z) = 0$ for $z \in \mathbf{C}\backslash\alpha^*$. By (4.19), Ind_γ is constant on each component. One sees that the index is 0 in the unbounded component as in the first proof.

Remark.   The second proof is independent of (3.16), but does not yield the fact that Ind_γ is always an integer.

EXERCISES

TYPE I

1. Sketch each of the following open sets and determine if it is connected.
    (a) $\{z \mid |\mathrm{Im}\, z| > 1\}$
    (b) $\{z \mid |z| > 1\}$
    (c) $\{z \mid |\mathrm{Re}\, z| < |\mathrm{Im}\, z|\}$
    (d) $\{z \mid \tfrac{1}{2} < |z| < 1\}.$

2. Find the connected components of the following sets.
   (a) $\{z \mid |z| \neq 1\}$
   (b) $\{z \mid \mathrm{Re}\, z \neq 0\}$
   (c) $\{z \mid \mathrm{Im}\, z \neq 0\}$
   (d) $\{z \mid |\mathrm{Re}\, z| > 1\}$
   (e) $\{z \mid \mathrm{Re}\, z > 1\}$
   (f) The complement of a path shaped as a figure eight.

3. Prove that a convex open set is connected.

4. Give an example of connected open sets U and V such that $U \cap V \neq \phi$ and $U \cap V$ is not connected.

5. Let U and V be open connected sets such that $U \cap V \neq \phi$. Prove that $U \cup V$ is connected.

6. Let $U = D(\frac{1}{2}, \frac{1}{2}) \cup D(\frac{1}{2}, -\frac{1}{2})$. Define an analytic function f on U such that $f' = 0$ on U but f is not constant.

7. Let S be a connected subset of $\mathbf{C}$. If f is a continuous function from S into $\mathbf{C}$, show that $f(S)$ is connected.

8. Prove that every connected subset of $\mathbf{R}$ is an interval.

9. Let α be a continuous real-valued function defined on a real-interval $[a, b]$. Show that $\alpha([a, b])$ is an interval.

10. (Intermediate value theorem.) Let f be a real-valued continuous function on $[a, b]$, $a < b$. If $f(a) < f(b)$, and $f(a) < y < f(b)$, prove that there is an $x \in [a, b]$ such that $f(x) = y$.

11. Let U and V be disjoint open connected sets in $\mathbf{C}$ such that $E = \overline{U \cup V}$ is connected. Suppose that f is continuous on E and analytic on $U \cup V$.
    (a) If $f' = 0$ on $U \cup V$, show that f is constant on E.
    (b) Give an example in which $f' = 0$ on U but f is not constant on E.

12. Suppose that f is analytic on $D(1, 0)$ and $f(1/n) = e^{1/n}$. Find f. What if $f'(1/n) = e^{1/n}$?

13. Can there be an analytic function f on $D(2, 0)$ such that $f(1/n) = e^{1/n}$ and $f(1 - 1/n) = e^{1/n}$? On $D(1, 0) \cup D(1, 1)$; on $D(\frac{1}{2}, 0) \cup D(\frac{1}{2}, 1)$? On $D(1, 0)$?

14. Let f be analytic on a convex open set U, and let γ be a closed path in U. Prove that

$$\int_\gamma \frac{f(\zeta)}{\zeta - z}\, d\zeta = 0$$

if z is in the unbounded component of $\mathbf{C}\backslash\gamma^*$.

TYPE II

1. Let S be a subset of $\mathbf{C}$. Show that S is disconnected if and only if there are nonvoid open sets U and V such that $U \cap V \cap S = \varnothing$ and $S \subset U \cup V$.

2. Let C_j be the circle of diameter $1/2^j$ centered at $1/2^j + 1/2^{j+1}$. Without writing an explicit formula, show that there is a path α such that $\alpha^* = \cup_{j=1}^{\infty} C_j$. Sketch α^*. Observe that $\mathbf{C}\backslash\alpha^*$ has infinitely many components and that α has total variation (arc length) π.

4.23 SIMPLE CONNECTEDNESS (OR SIMPLE CONNECTIVITY)

Let S be an open proper subset of $\mathbf{C}$. Then S is said to be *simply connected* if it is connected and if for every $z \in \mathbf{C}\backslash S$ there is an unbounded connected set in $\mathbf{C}\backslash S$ containing z. The complex plane $\mathbf{C}$ is taken to be simply connected by definition.

Intuitively, simply connected sets have no "holes." (If z were in a hole of U, then every unbounded connected set containing z would have to pass through U.) In Figure 4.10, we give four examples. The reader should verify the assertions made there. In the case of the annulus one can use (4.24) below. There are many equivalent definitions of simple connectedness. We have chosen one which is easy to apply, and which has much intuitive appeal. If f is analytic on a simply connected set U, then $\int_\gamma f(z)\,dz = 0$ for every closed path γ in U. The proof of this fact is the principal content of (4.24–4.28); the formal result is presented in theorem (4.28).

Before proceeding to the proof, we should remark that most of the remainder of the book is heavily dependent on Cauchy's theorem for a simple closed path in an open convex set and its immediate consequences; that is, on Sections (4.2–4.12). Cauchy's theorem for simply connected sets is needed only occasionally in the remainder of the book, and then only for the added generality it offers. Thus, some readers might prefer to skip the rest of Chapter 4, at least on a first reading.

*4.24 LEMMA

Let U be a simply connected subset of $\mathbf{C}$ and let γ be a closed path in U. Then every bounded component of $\mathbf{C}\backslash\gamma^*$ is contained in U.

Remark. The converse is also valid, that is, if the bounded components of the complement of every closed path γ lying in U are also in U, then U is simply connected. We will not need this result, and will not prove it. It does, however, provide a good characterization of simple connectivity.

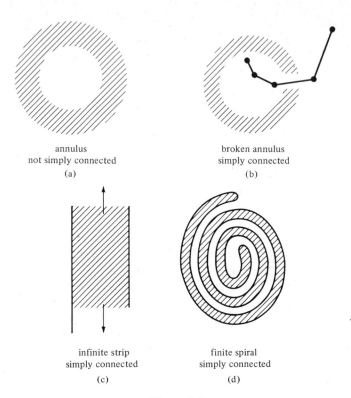

annulus
not simply connected
(a)

broken annulus
simply connected
(b)

infinite strip
simply connected
(c)

finite spiral
simply connected
(d)

Figure 4.10

PROOF. If $U = \mathbf{C}$, there is nothing to prove. Suppose that $U \neq \mathbf{C}$. Let $z \in \mathbf{C} \backslash [\gamma^* \cup U]$. Let C be an unbounded connected set in $\mathbf{C} \backslash U$ which contains z. The connected set C is contained in $\mathbf{C} \backslash \gamma^*$. By (4.16.v), C lies in a connected component A of $\mathbf{C} \backslash \gamma^*$. Since C is unbounded and there is only one unbounded component of $\mathbf{C} \backslash \gamma^*$, A must be this unbounded component. In particular, we have $z \in A$.

We will use (4.24) to prove the general Cauchy theorem (4.28) below. We break the proof into a number of lemmas. The main idea of our approach is to use polygonal paths to define primitives for analytic functions in simply connected regions. To do this we must show that the integrals defining the primitives are independent of the polygonal path over which we integrate. The motivation is thus the same as for convex sets, but instead of Cauchy's theorem for a triangle we will require Cauchy's theorem for closed polygonal paths. Actually we will not need arbitrary polygonal paths but only ones with special properties. Hence we make the definition in (4.25). Recall that a polygonal path is a finite oriented union of oriented line segments. See (3.10).

*4.25 DEFINITIONS

We will call a closed path γ *intersection-finite* if the set

$$K(\gamma) = \{t \in [a, b] \mid \gamma(t) = \gamma(s), \text{ some } s \neq t\}$$

is finite; we then let $n(\gamma)$ denote the number of elements in $K(\gamma)$. [We introduce this notion for technical reasons only; we will not use it after (4.28).] For any closed path γ, let $A(\gamma)$ denote the unbounded connected component of $\mathbb{C} \backslash \gamma^*$ and $B(\gamma)$ denote the union of the bounded components of $\mathbb{C} \backslash \gamma^*$.

*4.26 LEMMA

Let γ be an intersection-finite closed path. There are simple closed paths $(\gamma_m)_{m=1}^M$ such that

(i) $\cup_{m=1}^M \gamma_m^* = \gamma^*$;

(ii) $\int_\gamma f(z)\, dz = \sum_{m=1}^M \int_{\gamma_m} f(z)\, dz$ if f is continuous on γ^*;

(iii) $\cup_{m=1}^M B(\gamma_m) = B(\gamma)$.

If γ is polygonal, so are the γ_m.

PROOF. We give a proof by induction on $n(\gamma)$. If $n(\gamma) = 2$, then γ is simple and the assertions (i–iii) hold with $M = 1$ and $\gamma_1 = \gamma$.

Suppose that suitable γ_m can be found for all γ satisfying $n(\gamma) < k$, $k > 2$. Let γ satisfy $n(\gamma) = k$, and suppose that γ is defined on $[0, 1]$. Let t_0 be a point in $]0, 1[$ for which there are points x in $]0, 1[$ satisfying $\gamma(t_0) = \gamma(x)$ and $x \neq t_0$; t_0 exists because $k > 2$. Define $\tilde{\gamma}$ on $[0, 1]$ by letting

$$\tilde{\gamma}(t) = \begin{cases} \gamma(t + t_0) & \text{if } t \in [0, 1 - t_0] \\ \gamma(t + t_0 - 1) & \text{if } t \in [1 - t_0, 1]. \end{cases}$$

Then $\tilde{\gamma}$ is a closed path for which

(a) $\tilde{\gamma}^* = \gamma^*$,

(b) $n(\tilde{\gamma}) = \eta(\gamma)$,

(c) $\int_\gamma f(z)\, dz = \int_{\tilde{\gamma}} f(z)\, dz$ if f is continuous on γ^*,

and

(d) there is an $s \in]0, 1[$ for which $\tilde{\gamma}(0) = \tilde{\gamma}(s) = \tilde{\gamma}(1)$

hold. Let $0 = s_0 < s_1 < s_2 < \cdots < s_q = 1$ be the points in $[0, 1]$ for which $\tilde{\gamma}(s_j) = \tilde{\gamma}(0)$. Let α be $\tilde{\gamma}$ restricted to $[0, s_1]$ and let β be $\tilde{\gamma}$ restricted to $[s_1, 1]$.

Clearly $K(\alpha) \subset K(\gamma)$ and $K(\beta) \subset K(\gamma)$. But s_2 is not in $K(\alpha)$ and s_1 is not in $K(\beta)$, so we conclude that $n(\alpha) < k$ and $n(\beta) < k$. From the inductive hypothesis and the equality

$$\int_{\tilde{\gamma}} f(z)\, dz = \int_{\alpha} f(z)\, dz + \int_{\beta} f(z)\, dz$$

for continuous f, we conclude that there are γ_i for γ satisfying (i) and (ii). It is clear that α and β are polygonal if γ is polygonal, so the inductive hypothesis gives polygonal γ_i if γ is polygonal.

To prove (iii), let R be such that $\gamma^* \subset D(R, 0)$ and let $w \in \mathbf{C}\backslash D(R, 0)$. Thus $w \in A(\gamma)$ and $w \in A(\gamma_m)$ for all m. If $z \in A(\gamma)$, then there is a polygonal path in $A(\gamma)$ from z to w. This path lies in $\mathbf{C}\backslash\gamma_m^*$ for each m, and so $z \in A(\gamma_m)$ for each m. Hence we have

$$A(\gamma) \subset \bigcap_{m=1}^{M} A(\gamma_m).$$

Since $\cap_{m=1}^{M} A(\gamma_m)$ is a connected open set in $\mathbf{C}\backslash\gamma^*$ which contains $A(\gamma)$, it is equal to $A(\gamma)$ by (4.16.v).

The proof of (4.26) is thus complete.

The next lemma will allow us to define primitives for analytic functions in simply connected regions. The proof is somewhat technical, although entirely elementary, and the notation is often cumbersome. The idea, however, is simple. We know the result for triangles and we prove it by induction for polygonal paths having more than three vertices. The inductive step consists of splitting a polygon with n vertices into two polygons with fewer than n vertices by drawing a line segment between two vertices of the original polygon. The inductive hypothesis then applies to the two new polygons, and in the two integrations the integrals over the new line segment are taken in opposite direction and so cancel, leaving the integral over the original polygon. The reader should sketch a few simple examples. The proof becomes technical because the geometry of a polygon can be messy. The connecting line segment referred to above must lie in the simply connected region, so we cannot arbitrarily choose the vertices to connect.

*4.27 LEMMA

Let α be an intersection-finite closed polygonal path contained in a simply connected set U. If f is analytic on U, then we have

$$\int_{\alpha} f(z)\, dz = 0.$$

PROOF.   By (4.26), it suffices to prove the equality for simple α.   Hence suppose that α is a simple closed polygonal path, and let

$$\alpha = \overrightarrow{[p_0, p_1]} \smile \overrightarrow{[p_1, p_2]} \smile \cdots \smile \overrightarrow{[p_{n-1}, p_0]}.$$

For notational convenience, let $p_n = p_0$ and $p_{n+1} = p_1$. By reparametrizing without altering the value of $\int_\alpha f(z)\, dz$, we suppose that no three points p_k, p_{k+1}, p_{k+2} $(0 \le k \le n - 1)$ are collinear.   Under this assumption, we call each p_k a vertex of α.   The proof is by induction on the number of vertices of α.

If α has three vertices, it is a triangle.   The bounded component $B(\alpha)$ of $\mathbf{C}\backslash\alpha^*$ is the interior of the triangle, which is included in U as U is simply connected.   Hence (4.2) applies to prove that $\int_\alpha f(z)\, dz = 0$.

Suppose that the assertion is valid for all simple closed polygonal paths in U having fewer than n vertices.   Let α have n vertices, say $\{q_j\}_{j=0}^{n-1}$.   Among these vertices there is one for which Re (q_j) is maximal; call it q_k.   Parametrize α on $[0, 1]$ so that $\alpha(0) = q_k$, and then relabel the vertices $p_0 = q_k$, $p_1 = q_{k+1}, \ldots p_{n-1}$.   This can be done without altering $\int f(z)\, dz$; see (3.8).   Hence we have

$$\alpha = \overrightarrow{[p_0, p_1]} \smile \overrightarrow{[p_1, p_2]} \smile \cdots \smile \overrightarrow{[p_{n-1}, p_0]},$$

where no three points p_k, p_{k+1}, p_{k+2} are collinear, and $\mathrm{Re}(p_0)$ is maximal among $\{\mathrm{Re}(p_j) \mid 0 \le j \le n - 1\}$.   (See Figure 4.11.)   Let Δ be the open triangle determined by p_0, p_1, p_{n-1}.   We consider the two possibilities for the set $\tilde{\Delta} = \Delta \cup\,]p_{n-1}, p_1[$.   The first possibility is that $\tilde{\Delta}$ contains no points of α^*.   In this case, let

$$\beta = \overrightarrow{[p_1, p_2]} \smile \cdots \smile \overrightarrow{[p_{n-2}, p_{n-1}]} \smile \overrightarrow{[p_{n-1}, p_1]}$$

and

$$\partial = \overrightarrow{[p_{n-1}, p_0]} \smile \overrightarrow{[p_0, p_1]} \smile \overrightarrow{[p_1, p_{n-1}]}.$$

We will prove that the segment $]p_1, p_{n-1}[$ is contained in $B(\alpha)$.   The triangle Δ must be entirely contained in either $A(\alpha)$ or $B(\alpha)$, for Δ is a connected open set in $\mathbf{C}\backslash\alpha^*$.   Since β^* is compact, there is a disk D about p_0 which does not intersect β^*.   Points in $\Delta \cap D$ are in $A(\beta)$ because they can be connected to (for example) $p_0 + 1$ by a polygonal path lying in $\mathbf{C}\backslash\beta^*$, and $p_0 + 1$ is clearly in $A(\beta)$.   Let $z_0 \in \Delta \cap D$.   In the sum

$$\frac{1}{2\pi i} \int_\beta \frac{1}{z - z_0}\, dz + \frac{1}{2\pi i} \int_\partial \frac{1}{z - z_0}\, dz$$

the integrals over $\overrightarrow{[p_{n-1}, p_1]}$ and $\overrightarrow{[p_1, p_{n-1}]}$ cancel.   Hence the sum is equal to the integral over α, and we have $\mathrm{Ind}_\alpha (z_0) = \mathrm{Ind}_\beta (z_0) + \mathrm{Ind}_\partial (z_0)$.   Since

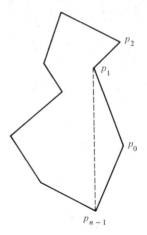

Figure 4.11

z_0 is in the interior of ∂, $\text{Ind}_\partial (z_0)$ is ± 1 by (4.3.ii). Since $z_0 \in A(\beta)$, $\text{Ind}_\beta (z_0)$ is 0. Hence $\text{Ind}_\alpha (z_0) \neq 0$, and so by (4.22) we have $z_0 \in B(\alpha)$. Thus we have $\Delta \subset B(\alpha)$. It follows that $]p_1, p_{n-1}[\subset B(\alpha)$. Since U is simply connected, $B(\alpha)$ is contained in U by (4.24). Hence we have $\partial^* \subset U$ and $\beta^* \subset U$. The closed paths β and ∂ have fewer than n vertices, so the inductive hypothesis applies to each of them, and the integral of f over each is zero. We have

$$\int_\alpha f(z)\, dz = \int_\beta f(z)\, dz + \int_\partial f(z)\, dz = 0.$$

The second possibility is that $\tilde{\Delta}$ contains some points of α^*. If $[p_1, p_{n-1}] \subset \alpha^*$, then since α is a simple path, α must be a triangle. Otherwise $\tilde{\Delta}$ must contain vertices of α. Let p_j be a vertex in $\tilde{\Delta}$ which is such that no other vertex is closer to $[p_{n-1}, p_0]$. (See Figure 4.12.) The line segment $[p_0, p_j]$ cannot intersect α^* and hence it must lie entirely in $A(\alpha)$ or entirely in $B(\alpha)$. We will show that it lies in $B(\alpha)$. There is a disk D centered at p_0 which intersects α^* only on $[p_0, p_1[$ and $[p_0, p_{n-1}[$. Let Δ be the triangle determined by the points p_j, p_0, and p_{n-1} and let $z_0 \in \Delta \cap D$. (See Figure 4.13.) Let

$$\omega = \overrightarrow{[p_0, p_1]} \smile \cdots \smile \overrightarrow{[p_{n-1}, p_j]} \smile \overrightarrow{[p_j, p_0]}$$

and

$$\delta = \overrightarrow{[p_0, p_j]} \smile \overrightarrow{[p_j, p_{n-1}]} \smile \overrightarrow{[p_{n-1}, p_0]}.$$

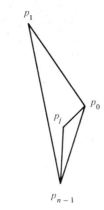

Figure 4.12

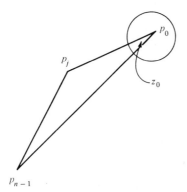

Figure 4.13

There is a polygonal path from z_0 to $p_0 + 1$ which does not intersect ω^*. Hence we have $z_0 \in A(\omega)$ and thus $\text{Ind}_\omega (z_0) = 0$. Since we have

$$\text{Ind}_\alpha (z_0) = \text{Ind}_\omega (z_0) + \text{Ind}_\delta (z_0),$$

we conclude that $\text{Ind}_\alpha (z_0) \neq 0$; hence, we have $z_0 \in B(\alpha)$. Clearly there is a line segment in $B(\alpha)$ from z_0 to $[p_0, p_j]$, so we have $]p_0, p_j[\subset B(\alpha)$. Let

$$\gamma = [\overrightarrow{p_0, p_1}] \smile [\overrightarrow{p_1, p_2}] \smile \cdots \smile [\overrightarrow{p_{j-1}, p_j}] \smile [\overrightarrow{p_j, p_0}]$$

and let

$$\sigma = [\overrightarrow{p_0, p_j}] \smile [\overrightarrow{p_j, p_{j+1}}] \smile \cdots \smile [\overrightarrow{p_{n-1}, p_0}].$$

Applying (4.24) again, we have $]p_0, p_j[\subset B(\alpha) \subset U$, and hence γ^* and σ^* are in U. Both γ and σ have fewer than n vertices. The inductive hypothesis thus applies to show that

$$\int_\gamma f(z)\, dz = \int_\sigma f(z)\, dz = 0.$$

The integrals over $\overrightarrow{[p_0, p_j]}$ and $\overrightarrow{[p_j, p_0]}$ in the sum

$$\int_\gamma f(z)\, dz + \int_\sigma f(z)\, dz$$

cancel, showing that the sum is $\int_\alpha f(z)\, dz$. Hence we have $\int_\alpha f(z)\, dz = 0$, and the proof is complete.

*4.28 CAUCHY'S THEOREM FOR A SIMPLY CONNECTED REGION

Let U be simply connected and suppose that f is analytic on U. If γ is a closed path in U, then we have

(i) $\displaystyle\int_\gamma f(z)\, dz = 0.$

PROOF.   Fix $z_0 \in U$. For each $z \in U$, there are polygonal paths in U from z_0 to z which have the property that all their line segments have either constant real part or constant imaginary part. Call these polygonal paths *rectangular*. If α and β are two rectangular paths in U from z_0 to z, then $\sigma = \alpha \smile -\beta$ is a closed polygonal path lying in U. (See Figure 4.14.) If σ is intersection-finite then we have $\int_\sigma f(z)\, dz = 0$ by (4.27). If σ is not intersection-finite then there are segments of σ which overlap other segments of σ. Let $\overrightarrow{[a, b]}$ be such a segment. Suppose that $[a, b]$ is horizontal. There is a polygonal path $\vec{l}$ from a to b which intersects σ^* only on vertical segments and at a and b and which has the property that $\overrightarrow{[a, b]}$ is contained in a convex subset of U. (See Figure 4.15.) By (4.3), we have

$$\int_{\overrightarrow{[a, b]}} f(z)\, dz = \int_{\vec{l}} f(z)\, dt.$$

Let σ_1 be the closed polygonal path obtained from σ by replacing $\overrightarrow{[a, b]}$ by $\vec{l}$. Then we have $\int_{\sigma_1} f(z)\, dz = \int_\sigma f(z)\, dz$. If $[a, b]$ is vertical, it can be replaced with an $\vec{l}$ in a similar way to give a σ_1. If σ_1 is not intersection-finite then we select an interval which overlaps other segments of σ_1 and construct a

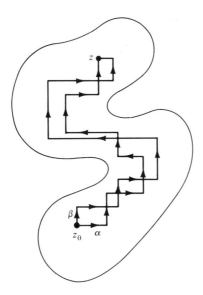

Figure 4.14

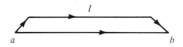

Figure 4.15

σ_2 from σ_1 as above. Applying this technique a finite number of times, we obtain an intersection-finite closed polygonal path $\tilde{\sigma}$ in U for which

$$\int_{\tilde{\sigma}} f(z)\, dz = \int_{\sigma} f(z)\, dz.$$

By (4.27), we have $\int_{\tilde{\sigma}} f(z)\, dz = 0$, and hence $\int_{\sigma} f(z)\, dz = 0$ in all cases. Thus for every pair of rectangular paths in U from z_0 to z we have

$$\int_{\alpha} f(z)\, dz = \int_{\beta} f(z)\, dz.$$

Hence the formula

$$F(z) = \int_{\alpha} f(\zeta)\, d\zeta$$

where α is a rectangular path in U from z_0 to z defines a function in U. We will prove that F is a primitive for f in U. To this end, fix $z_1 \in U$ and let $D(z_1, r) \subset U$. Let α be a rectangular path in U from z_0 to z_1. Let

$z \in D(z_1, r)$. If $[z_1, z]$ is collinear with the final segment of α, let $\beta = \alpha \smile \overrightarrow{[z_1, z]}$. Otherwise, let

$$\beta = \alpha \smile \overrightarrow{[z_1, z_1 + \operatorname{Re}(z - z_1)]} \smile \overrightarrow{[z_1 + \operatorname{Re}(z - z_1), z]}.$$

(See Figure 4.16.) Then β is a rectangular path from z_0 to z. We have

$$F(z) - F(z_1) = \int_{\overrightarrow{[z_1, z]}} f(\zeta)\, d\zeta.$$

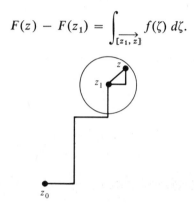

Figure 4.16

Hence we have

$$\left| \frac{F(z) - F(z_1)}{z - z_1} - f(z_1) \right| = \frac{1}{|z - z_1|} \left| \int_{\overrightarrow{[z_1, z]}} [f(\zeta) - f(z_1)]\, d\zeta \right|. \quad (*)$$

Let $\varepsilon > 0$. By the continuity of f, there is a $\delta > 0$ such that $|f(\zeta) - f(z_1)| < \varepsilon$ if $|\zeta - z_1| < \delta$. If $z \in D(z_1, r)$ satisfies $|z - z_1| < \delta$, then $|f(\zeta) - f(z_1)| < \delta$ holds for all ζ on $[z_1, z]$. Hence the integral on the right in $(*)$ is less than $\varepsilon |z - z_1|$, and so the expression on the left is less than ε. Thus we have $F'(z_1) = f(z_1)$.

By (3.6), we obtain (i) for a closed path γ in U.

*4.29 CAUCHY'S INTEGRAL FORMULA

Let f be analytic on a simply connected set U and let γ be a closed path in U. We have

(i) $f(z) \operatorname{Ind}_\gamma(z) = \dfrac{1}{2\pi i} \displaystyle\int_\gamma \frac{f(\zeta)}{\zeta - z}\, d\zeta$

for all $z \in U \backslash \gamma^*$.

PROOF. The function G defined on U by

$$G(\zeta) = \begin{cases} \dfrac{f(\zeta) - f(z)}{\zeta - z} & \zeta \ne z \\[2ex] f'(z) & \zeta = z \end{cases}$$

is analytic on $U \backslash z$ and continuous on U. By (4.6), G is analytic on U. By (4.28), we have $\int_\gamma G(\zeta)\,d\zeta = 0$; that is,

(ii) $\displaystyle \int_\gamma \frac{f(\zeta)}{\zeta - z}\,d\zeta = f(z) \int_\gamma \frac{1}{\zeta - z}\,d\zeta.$

Equality (i) follows from (ii).

It is clearly desirable for applications of (4.29) to have general conditions under which $\text{Ind}_\gamma(z)$ is ± 1. In (4.3.ii) we calculated Ind_α for various elementary paths α by dissecting the path in ad hoc ways and applying Cauchy's theorem. Showing that Ind_γ is ± 1 for general classes of paths is principally a topological problem. The rest of the chapter is devoted to this problem and to a discussion of the related concept of orientation. First we consider simple polygonal paths. We have already completed most of the work [in the proof of (4.27)] for the proof of the following theorem.

*4.30 THEOREM

Let α be a simple closed polygonal path. The set $\mathbf{C} \backslash \alpha^*$ has exactly two connected components, a bounded one $B(\alpha)$ and an unbounded one $A(\alpha)$. The number $\text{Ind}_\gamma(z)$ is either 1 or -1 for all $z \in B(\alpha)$.

PROOF. As in the proof of (4.27), we use induction on the number of vertices of α. If α has three vertices, then it is a triangle and (4.3.ii) applies. Suppose that $\mathbf{C} \backslash \sigma^*$ has just one bounded component for every simple closed polygonal path σ having fewer than n vertices. Let α have n vertices. Let the notation for α be as in the proof of (4.27). Consider first the case in which $\tilde{\Delta}$ contains no points of α^*, and let β and ∂ be the paths defined as in the proof of (4.27). The set $A(\beta) \cap A(\partial)$ is an unbounded connected set in $\mathbf{C} \backslash \alpha^*$ which contains $p_0 + 1$. Let $z' \in A(\alpha)$. There is a polygonal path ρ in $A(\alpha)$ from z' to $p_0 + 1$. Since we have $]p_1, p_{n-1}[\subset B(\alpha)$, ρ^* is contained in $\mathbf{C} \backslash \beta^*$ and in $\mathbf{C} \backslash \partial^*$. Hence we have $z' \in A(\beta) \cap A(\partial)$ and we have proved that $A(\alpha) \subset A(\beta) \cap A(\partial)$. By (4.16.iv), we have $A(\alpha) = A(\beta) \cap A(\partial)$. It follows that

$$B(\alpha) = B(\beta) \cup B(\partial) \cup]p_1, p_{n-1}[= B(\beta) \cup \tilde{\Delta} \qquad (1)$$

holds. We will use (1) to show that $B(\alpha)$ is connected. Let $z_1 \in]p_1, p_{n-1}[$. Since $z_1 \in B(\alpha)$ and $B(\alpha)$ is open, there is a disk $D(r, z_1) \subset B(\alpha)$. Let z_2 be in $D(r, z_1)$ but not in $\tilde{\Delta}$. We have

$$\text{Ind}_\alpha(z_2) = \text{Ind}_\beta(z_2) + \text{Ind}_\partial(z_2).$$

We know from (4.27) that $\Delta \subset B(\alpha)$ and that $\text{Ind}_\alpha (z) = \pm 1$ for $z \in \Delta$. Since z_2 is in the same component of $\mathbf{C}\backslash\alpha^*$ as Δ, we have $\text{Ind}_\alpha (z_2) = \pm 1$. Since $\text{Ind}_\partial (z_2)$ is zero, we have $\text{Ind}_\beta (z_2) = \pm 1$. In particular, z_2 is in $B(\beta)$. By the inductive hypothesis, $B(\beta)$ is the bounded component of $\mathbf{C}\backslash\beta^*$. Let ζ be in $B(\beta)$ and let $\xi \in \tilde{\Delta}$. There is a polygonal path in $\mathbf{C}\backslash\beta^*$ from ζ to z_2 and there is a polygonal path from z_2 to ξ which lies in $\mathbf{C}\backslash\alpha^*$. Putting these together, we obtain a polygonal path in $\mathbf{C}\backslash\alpha^*$ from ζ to ξ. By (1), we conclude that $B(\alpha)$ is connected. We now know that $\Delta \subset B(\alpha)$, that $\text{Ind}_\alpha (z) = \pm 1$ for $z \in \Delta$, and that $B(\alpha)$ is connected. Hence by (4.22), we have $\text{Ind}_\alpha (z) = \pm 1$ for $z \in B(\alpha)$.

In the case that $\tilde{\Delta}$ contains points of α^*, construct p_j as in the proof of (4.27). We proved that $]p_0, p_j[\subset B(\alpha)$. Arguing as above, we see that

$$B(\alpha) = B(\gamma) \cup B(\sigma) \cup]p_0, p_j[. \tag{2}$$

Let $w_1 \in]p_0, p_j[$. There is a disk $D(\varepsilon, w_1) \subset B(\alpha)$. Let w_2 be in $D(\varepsilon, w_1) \cap \Delta(p_{n-1}, p_j, p_0)$ but not on $]p_0, p_j[$ and let w_3 be in $D(\varepsilon, w_1) \cap \Delta(p_1, p_j, p_0)$ but not on $]p_0, p_j[$. There is a polygonal path from w_2 to $p_0 + 1$ which intersects α^* only on $[p_{n-1}, p_0]$; hence we have $w_2 \in A(\gamma)$. (See Figure 4.17 also.) Similarly, there is a polygonal path in $\mathbf{C}\backslash\sigma^*$ from w_3 to $p_0 + 1$; hence, $w_3 \in A(\sigma)$. We have

$$\text{Ind}_\alpha (w_i) = \text{Ind}_\gamma (w_i) + \text{Ind}_\sigma (w_i) \tag{3}$$

for $i = 2, 3$. Since both w_2 and w_3 are in the same component of $\mathbf{C}\backslash\alpha^*$ as z_0 in the proof of (4.27), the left side of (3) is ± 1 for $i = 2, 3$. We conclude that $\text{Ind}_\gamma (w_3) = \pm 1$ and that $\text{Ind}_\sigma (w_2) = \pm 1$. Hence we have $w_3 \in B(\gamma)$ and $w_2 \in B(\sigma)$. Let $\zeta \in B(\gamma)$ and $\xi \in B(\sigma)$. Since σ and γ have fewer than n vertices, $B(\sigma)$ and $B(\gamma)$ are the bounded components of $\mathbf{C}\backslash\sigma^*$ and $B(\alpha)$, respectively. Hence there is a polygonal path from ζ to w_3 in $\mathbf{C}\backslash\gamma^*$ and a polygonal path from ξ to w_2 in $\mathbf{C}\backslash\sigma^*$. Since $]p_0, p_j[\subset B(\alpha)$, it follows that there is a polygonal path from ζ to ξ lying in $B(\alpha)$. Hence $B(\alpha)$ is connected. We have already seen that $\text{Ind}_\alpha (z) = \pm 1$ for $z \in \Delta(p_{n-1}, p_j, p_0)$, so the proof is complete.

*4.31 INDEX HEURISTICS; THE KEYHOLE ARGUMENT

Suppose that α is a "well-behaved" simple closed path, as pictured in Figure 4.18. Let z_0 be in the bounded component of $\mathbf{C}\backslash\alpha^*$. Referring to Figure 4.18, let α_1 be the portion of α from $\alpha(0)$ to w and let α_2 be the portion from v to $\alpha(0)$. The point z_0 is in the unbounded component of the complement of the path

$$\eta_\varepsilon = \alpha_1 \smile l_1 \smile \sigma \smile l_2 \smile \alpha_2.$$

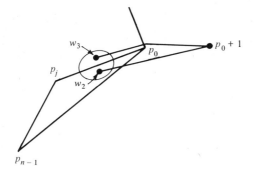

Figure 4.17

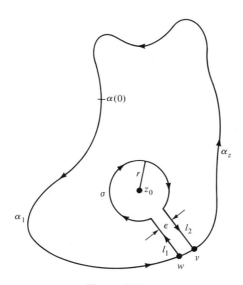

Figure 4.18

Hence we have

$$0 = \int_{\eta_\varepsilon} \frac{1}{z - z_0} \, dz = \left[\int_{\alpha_1} + \int_{\alpha_2} + \int_{l_1} + \int_{l_2} + \int_{\sigma} \right] \left(\frac{1}{z - z_0} \right) dz.$$

As ε goes to zero, the integrals over l_1 and l_2 cancel, the sum $\int_{\alpha_1} + \int_{\alpha_2}$ goes to $\int_\alpha$, and the integral over σ goes to

$$\int_{\overleftarrow{C(z_0, \, r)}} \frac{1}{z - z_0} \, dz = -2\pi i.$$

Hence it appears that $\text{Ind}_\alpha (z_0) = 1$. This argument is difficult to make precise in general situations because the behavior of the curve between w and v may be unruly. For particular nice curves, the technique can be applied on an ad hoc basis. In (4.33), we prove a general theorem based on the technique. Another one is outlined in the exercises. We should remark at this point that it is an important topological theorem (the *Jordan curve theorem*) that the complement of a simple closed path has exactly two components, and that α^* is the boundary of each. Although the validity of this theorem may seem obvious, the proof is actually very difficult. Indeed, Jordan's original proof in about 1887 was incorrect and it was not until 1905 that Veblen gave a correct proof.

*4.32 ORIENTATION

For circles, triangles, squares, and other familiar paths we have seen that $\text{Ind}_\alpha (z) = 1$ if the path α is oriented so that the bounded component of α is on the left of α and that $\text{Ind}_\alpha (z) = -1$ if the bounded component is on the right of α. Hence in accordance with definition (3.17), these paths are positively oriented if the bounded component stays on the left. This phenomenon persists for all simple closed paths α, but for complicated paths it is difficult to make "lies to the left" precise. However, for well-behaved paths which have easily identified components, it is often easy to establish orientation. For example, if the keyhole argument of (4.31) can be applied to the path, then that argument shows that the orientation of α will be opposite to that of the circle $C(r, z_0)$ used in (4.31). In particular, if it is geometrically clear that the bounded component $B(\alpha)$ of α is to the left of α (as in Figure 4.18), then the index of z_0 for the circle is -1 and $\text{Ind}_\alpha = 1$ on the bounded component.

The orientation of a simple closed polygonal path α can be determined as follows. Among all the vertices p of α for which $\text{Re}(p)$ is maximal, let p_0 be the one for which $\text{Im}(p_0)$ is also maximal. Write α as $\alpha = \overrightarrow{[p_0, p_1]} \smile \overrightarrow{[p_1, p_2]} \smile \cdots \smile \overrightarrow{[p_{n-1}, p_0]}$. We have seen in (4.27) and (4.30) that the triangle $\Delta(p_0, p_1, p_{n-1})$ contains points in $B(\alpha)$. Applying the keyhole argument to one of these points b, we can determine Ind_α. (See also Figure 4.19.)

Another technique for finding Ind_α for a simple closed path α is based on polygonal approximation. Suppose first that α is so well-behaved that it is clear that $\mathbb{C} \backslash \alpha^*$ has exactly one bounded component $B(\alpha)$. Let $b \in B(\alpha)$. Suppose that α is parametrized on $[0, 1]$ and suppose that

$$\Gamma = \{0 = t_0, t_1, \ldots, t_n = 1\}$$

is the nth regular partition of $[0, 1]$. Let

$$\alpha_n = \overrightarrow{[\alpha(0), \alpha(t_1)]} \smile \overrightarrow{[\alpha(t_1), \alpha(t_2)]} \smile \cdots \smile \overrightarrow{[\alpha(t_{n-1}), \alpha(t_n)]}.$$

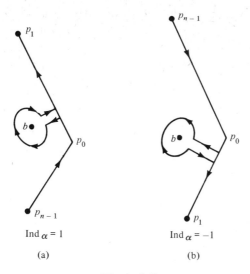

Ind $\alpha = 1$ Ind $\alpha = -1$

(a)   (b)

Figure 4.19

It can be established that $\lim_{n \to \infty} \alpha_n(t) = \alpha(t)$, that $b \in \alpha_n^*$ for all sufficiently large n, and that

$$\lim_{n \to \infty} \int_{\alpha_n} \frac{1}{z - b} \, dz = \int_{\alpha} \frac{1}{z - b} \, dz.$$

Since Ind is integer valued, it follows that there is an n_0 such that $\mathrm{Ind}_{\alpha_n}(b) = \mathrm{Ind}_{\alpha}(b)$ for all $n \geq n_0$. If the α_n are sufficiently nice (for example if they are eventually simple) then we can find $\mathrm{Ind}_{\alpha_n}(b)$ and hence $\mathrm{Ind}_{\alpha}(b)$. The technique is illustrated in Figure 4.20.

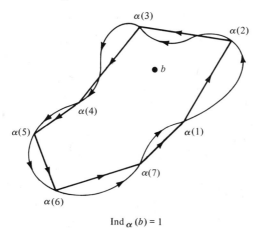

Ind $\alpha (b) = 1$

Figure 4.20

Summary. To say that a path is positively oriented if the interior of the path lies to the left as the path is traversed is a nice intuitive idea. If one attempts to make this idea a precise definition, several technical difficulties are encountered. We have thus defined positive orientation as in (3.17) and have shown that many paths which are positively oriented have their interiors lying to the left of the path. In Exercise 4, p. 109, it was indicated that there are paths which are neither positively nor negatively oriented. It is a theorem, the proof of which is difficult, that every simple closed curve is either positively or negatively oriented. Clearly the orientation of an oriented path can be reversed by a simple change of parameter.

We conclude the chapter with a rather technical theorem stating that $\text{Ind}_\alpha(z)$ is ± 1 for a large class of paths. This class will probably suffice for most readers' needs.

*4.33 THEOREM

Suppose that α is a simple closed path parametrized on $[0, 1]$ such that $\mathbb{C}\backslash\alpha^*$ has exactly two connected components, B the bounded component and U the unbounded component. In addition, suppose that the condition (i) below is satisfied. Then $\text{Ind}_\alpha(b)$ is either 1 or -1 for all $b \in B$.

(i) There is a $t_0 \in [0, 1]$ such that α' exists at t_0 and $\alpha'(t_0) \neq 0$; and, there is a segment $[z', z'']$ on the normal to α at $z = \alpha(t_0)$ such that $]z, z'] \subset U$ and $]z, z''] \subset B$.

Remark. If $\text{Ind}_\alpha = -1$, then $\text{Ind}_{-\alpha} = 1$, where $[-\alpha](t) = \alpha(1 - t)$.

PROOF. Since Ind_α is constant on B, it suffices to prove that $\text{Ind}_\alpha(b)$ is either 1 or -1 for one point $b \in B$. Take $b = z''$ and (by reparametrizing if necessary) suppose that $t_0 \neq 0$, $t_0 \neq 1$. Let

(ii) $\displaystyle\lim_{t \to t_0} \frac{\alpha(t) - \alpha(t_0)}{t - t_0} = |\alpha'(t_0)|e^{i\theta}$,

where $\theta \in]-\pi, \pi]$. The line $\{z + re^{i\theta} \mid r \in \mathbb{R}\}$ is tangent to α at $z = \alpha(t_0)$. Fix $\omega \in]0, \pi/4]$. For $\rho > 0$, let S_ρ denote the truncated cone $\{z + re^{i(\theta + v)} \mid |v| < \omega$ and $|r| < \rho\}$, and consider the truncated half-cones

$$X_\rho = \{w \in S_\rho \mid r > 0\}; \quad Y_\rho = \{w \in S_\rho \mid r < 0\}.$$

For each $t \in [0, 1]$, let

$$\alpha(t) = \alpha(t_0) + r(t)e^{i(\theta + v_t)},$$

where $r(t)$ is real and $-\pi/2 < v_t \leq \pi/2$. Since α is simple, $r(t)$ is nonzero if $t \neq t_0$. In this notation, we have

(iii) $\displaystyle\frac{\alpha(t) - \alpha(t_0)}{t - t_0} = \frac{r(t)}{t - t_0} e^{i(\theta + v_t)}$.

If there were a sequence (t_n) converging to t_0 for which (v_{t_n}) did not converge to zero, then by (iii) $[\alpha(t_n) - \alpha(t_0)]/(t_n - t_0)$ could not converge to $|\alpha'(t_0)|e^{i\theta}$. Hence there is an $\varepsilon > 0$ such that $\alpha(t) \in S_\rho$ if $|t - t_0| < \varepsilon$. Since $e^{i(\theta + v_t)}$ converges to $e^{i\theta}$ as t goes to t_0, equality (iii) shows that $r(t)/(t - t_0)$ converges to $|\alpha'(t_0)|$. In particular, there is an $\eta > 0$ such that $\eta < \varepsilon$ and $r(t)$ has the same sign as $t - t_0$ for $|t - t_0| < \eta$. Hence we have

$$\alpha\Big(]t_0 - \eta, t_0[\Big) \subset Y_\rho; \ \alpha\Big(]t_0, t_0 + \eta[\Big) \subset X_\rho.$$

For given ρ, let $y_\rho \in \ ]t_0 - \eta, t_0[$ satisfy $\alpha(y_\rho) \in Y_\rho$ and $x_\rho \in \ ]t_0, t_0 + \eta[$ satisfy $\alpha(x_\rho) \in X_\rho$. As ρ goes to zero, the corresponding ε and η must go to zero, so we have

(iv) $\lim\limits_{\rho \to 0} x_\rho = \lim\limits_{\rho \to 0} y_\rho = t_0.$

Fix s so that $D = D(s, b) \subset B$ and $D \cap S_\rho = \varnothing$ for all ρ. Let $C(s, b)$ intersect $[b, \alpha(x)]$ at a and $[b, \alpha(y)]$ at c. Let $\alpha_{\rho 1}$ be α restricted to $[0, y_\rho]$, let $\alpha_{\rho 2}$ be α restricted to $[x_\rho, 1]$, and let $\sigma_\rho, \beta_\rho, \tau_\rho$ be defined as in Figure 4.21.

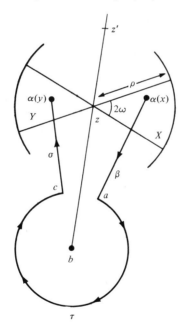

Figure 4.21

Let

(v) $\alpha_\rho = \alpha_{\rho 1} \smile \sigma_\rho \smile \tau_\rho \smile \beta_\rho \smile \alpha_{\rho 2}.$

The equality

$$\lim_{\rho \to 0} \left[\int_{\alpha_{\rho 1}} \frac{1}{z-b} \, dz + \int_{\alpha_{\rho 2}} \frac{1}{z-b} \, dz \right] = \int_{\alpha} \frac{1}{z-b} \, dz$$

is valid by (iv). The equality

$$\lim_{\rho \to 0} \left[\int_{\sigma_\rho} \frac{1}{z-b} \, dz + \int_{\beta_\rho} \frac{1}{z-b} \, dz \right] = 0$$

is valid because $(z-b)^{-1}$ is uniformly bounded on $\sigma_\rho^* \cup \beta_\rho^*$, σ_ρ^* and β_ρ^* become uniformly close as ρ goes to 0, and σ_ρ and β_ρ have opposite orientations. The equality

$$\lim_{\rho \to 0} \left| \frac{1}{2\pi i} \int_{\tau_\rho} \frac{1}{z-b} \, dz \right| = 1$$

is valid because for sufficiently small ρ, τ_ρ is arbitrarily close to either $\overrightarrow{C(s, b)}$ or $\overleftarrow{C(s, b)}$, so that the integral is close to either 1 or -1. Since $z \neq \alpha_\rho(t)$ for all t, the segment $[b, z']$ lies in the unbounded component of each α_ρ; thus $\mathrm{Ind}_{\alpha_\rho}(b) = 0$ for each ρ. It follows from (v) that $\mathrm{Ind}_\alpha(b)$ is ± 1.

EXERCISES

TYPE I

1. Which of the following open sets (shaded) are simply connected? Which connected?

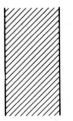
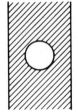

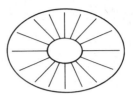

2. (a) Parametrize the curves α and β sketched below.
   (b) Find $\text{Ind}_\alpha\left(-\frac{1}{2}\right)$ and $\text{Ind}_\beta\left(-\frac{1}{2}\right)$.
   (c) Find the connected components of $\mathbf{C}\backslash\alpha^*$.

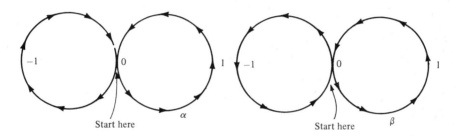

3. Find $\text{Ind}_\alpha\,(a)$ and $\text{Ind}_\alpha\,(b)$ for α, a, and b as sketched below. Do the same for γ.

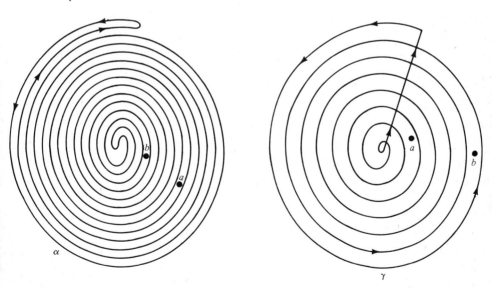

4. Find $\text{Ind}_\alpha\,(a)$ for α and a as sketched below.

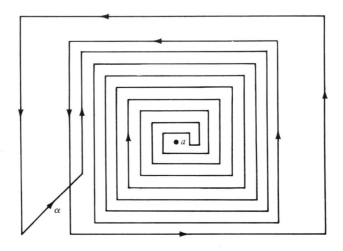

5. Find $\text{Ind}_\alpha\,(b)$ for α and b as sketched below.

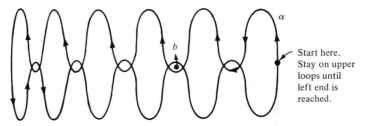

6. In the sketch of a closed path α (opposite page), the inner loops are meant to be infinite in number, and a_n is meant to be in the nth loop.
   (a) Find $\text{Ind}_\alpha\,(a_n)$ for each n.
   (b) Show that it is possible to define a path such as α.

7. Let U be a simply connected region not containing 0 and containing 1.
   (a) Show that $\int_\gamma (1/\zeta)\,d\zeta$ is independen tof the path γ in U connecting 1 and z.
   (b) Define
   $$L(z) = \int_\gamma \frac{1}{\zeta}\,d\zeta,$$
   γ any path in 0 connecting 1 to z. Show that $\exp\,(L(z)) = z$ for all $z \in U$.

[*Hint:* Compute the second derivative of exp $(L(z))$.]

Remark. $L(z)$ together with U is called a branch of log (z). Other branches of log z were discussed in (2.14). Additional branches are discussed in (7.7).

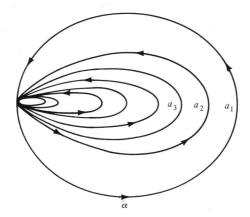

CHAPTER 5

Singularities and Residues

5.1 LAURENT SERIES

If h is analytic on a disk $D(R, 0)$, then h has a unique power series expansion

(i) $\quad h(z) = \sum_{n=0}^{\infty} a_n z^n$

centered at 0 and valid in $D(R, 0)$; [see (4.8)]. In this section we consider a function f which is analytic on an annulus

$$A_r^R = \{z \mid r < |z| < R\} \qquad (0 \le r < R \le \infty).$$

We will prove, in analogy with (i), that f has a series expansion of the form

(ii) $\quad f(z) = \sum_{n=-\infty}^{\infty} a_n z^n$

valid in A_r^R. The series in (ii) is called the *Laurent series* for f. We will obtain the Laurent series for f in two steps. First, we will use the Cauchy integral formula (4.4) to write f as a sum $h + g$, where h is analytic on $D(R, 0)$ and g is analytic outside of $\overline{D(r, 0)}$. Second, we will expand the integral formulas for h and g so obtained by using a geometric series and (3.14) on term-by-term integration of a uniformly convergent series. To this end, let v and μ satisfy $r < v < \mu < R$. (The cases $r = 0$ and $R = \infty$ are possible.) Consider a regular polygon[1] lying in the (open) annulus A_r^v together with those line segments from the vertices of the polygon to $C(\mu, 0)$ which bisect the angles of the polygon (see Figure 5.1). Each of the positively oriented closed paths γ determined by the portions of two adjacent segments between

---

[1] A polygon is regular if its sides are equal and its interior angles are equal.

174

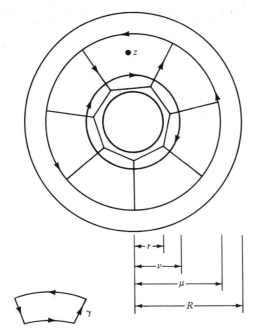

Figure 5.1

$C(v, 0)$ and $C(\mu, 0)$ and the portions of the circles between the segments lies in a corresponding convex region on which f is analytic. If z is interior to one of these curves, say β, then by the Cauchy integral representation (4.4), the equality

$$f(z) = \frac{1}{2\pi i} \int_\beta \frac{f(\zeta)}{\zeta - z} \, d\zeta$$

is valid ($\mathrm{Ind}_\beta (z) = 1$ by (4.3.ii)). For the other γ, $f(\zeta)/(\zeta - z)$ is analytic on a convex region containing γ, and so by (4.2) $\int_\gamma (f(\zeta)/(\zeta - z)) \, d\zeta = 0$ holds. Adding these integrals over all γ's and β, the integration over the line segments cancel and we obtain

$$f(z) = \frac{1}{2\pi i} \int_{\tau_\mu} \frac{f(\zeta)}{\zeta - z} \, d\zeta + \frac{1}{2\pi i} \int_{-\tau_v} \frac{f(\zeta)}{\zeta - z} \, d\zeta = h_\mu(z) + g_v(z),$$

where τ_μ and $-\tau_v$ are the obvious circles. The first integral defines $h_\mu(z)$ and the second integral defines $g_v(z)$. For given z, the polygon may be chosen so that the line segments miss z, and the final expressions for $g_\mu(z)$ and $h_v(z)$ do not depend on the polygon. Hence h_μ is analytic on $D(\mu, 0)$ and g_v is analytic on $\mathbf{C}\backslash\overline{D(v, 0)}$. Suppose that $\mu' < \mu$. Splitting the annulus $A^\mu_{\mu'}$ as above and applying Cauchy's theorem to $f(\zeta)/(\zeta - z)$ for $|z| < \mu'$, we see that $h_\mu(z) = h_{\mu'}(z)$. Hence we can define h on $D(R, 0)$ by letting $h(z) = h_\mu(z)$, where μ is any number satisfying $|z| < \mu < R$. Similarly, $g(z) = g_\nu(z)$ $(r < \nu < |z|)$ defines a function g analytic outside of $\overline{D(r, 0)}$, and the expression $f(z) = h(z) + g(z)$ is valid on A_r^R.

The Taylor expansion for h can be obtained as in (4.8); namely, taking $|z| < \mu < R$ we can use uniform convergence (1.5) and (3.14) to write

$$h(z) = \frac{1}{2\pi i} \int_{\tau_\mu} f(\zeta) \frac{1}{\zeta\left(1 - \dfrac{\zeta}{z}\right)} \, d\zeta$$

$$= \frac{1}{2\pi i} \int_{\tau_\mu} f(\zeta) \frac{1}{\zeta} \sum_{n=0}^{\infty} \left(\frac{z}{\zeta}\right)^n d\zeta = \sum_{n=0}^{\infty} \left[\frac{1}{2\pi i} \int_{\tau_\mu} \frac{f(\zeta)}{\zeta^{n+1}} \, d\zeta\right] z^n.$$

The final power series converges if $|z| < R$ and so has radius of convergence not less than R. Thus it converges uniformly on $\overline{D(R', 0)}$ if $R' < R$.

To obtain an expansion for g, let $r < \nu < |z|$ and write

$$g(z) = \frac{1}{2\pi i} \int_{-\tau_\nu} \frac{f(\zeta)}{\zeta - z} \, d\zeta = \frac{1}{2\pi i} \int_{-\tau_\nu} f(\zeta) \frac{1}{z\left(\dfrac{\zeta}{z} - 1\right)} \, d\zeta$$

$$= \frac{1}{2\pi i} \int_{-\tau_\nu} f(\zeta) \frac{1}{z} \left(- \sum_{n=0}^{\infty} \left(\frac{\zeta}{z}\right)^n\right) d\zeta$$

$$= \sum_{n=0}^{\infty} \left[\frac{1}{2\pi i} \int_{\tau_\nu} f(\zeta)\zeta^n \, d\zeta\right] z^{-n-1} = \sum_{n=1}^{\infty} \left[\frac{1}{2\pi i} \int_{\tau_\nu} \frac{f(\zeta)}{\zeta^{-n+1}} \, d\zeta\right] z^{-n}.$$

Again we have used uniform convergence of the integrand. The final series is a power series evaluated at $1/z$ and we have shown that it converges if $|1/z| < 1/r$; that is, the power series has radius of convergence at least $1/r$. Thus by (2.2) the final series converges uniformly on $C\backslash D(r', 0)$ if $r' > r$.

Next, by splitting A_ν^μ and applying Cauchy's theorem as above, the equality

$$\int_{\tau_\mu} \frac{f(\zeta)}{\zeta^{n+1}} \, d\zeta = \int_{\tau_\nu} \frac{f(\zeta)}{\zeta^{n+1}} \, d\zeta$$

is established for $r < \nu < \mu < R$ and any integer n.

Summarizing, we have the following theorem.

(iii) *Theorem.* If f is analytic on the annulus A_r^R $(0 \le r < R \le \infty)$, then there are complex numbers a_n, $n \in \mathbf{Z}$, such that $f(z) = \sum_{n=-\infty}^{\infty} a_n z^n$ for all

$z \in A_r^R$. (The definition of $\sum_{n=-\infty}^{\infty} a_n z^n$ is $\sum_{n=0}^{\infty} a_n z^n + \sum_{n=1}^{\infty} a_{-n} z^{-n}$.) The coefficients a_n are unique and the equality

$$a_n = \frac{1}{2\pi i} \int_{\tau_\mu} \frac{f(\zeta)}{\zeta^{n+1}} \, d\zeta \qquad (*)$$

is valid for all $n \in \mathbf{Z}$ and all $\mu \in \,]r, R[$. The series converges uniformly on $A_{r'}^{R'}$ if $r < r' < R' < R$.

The only statement in the theorem which we did not prove above is the uniqueness of the a_n. Suppose that

$$f(z) = \sum_{n=-\infty}^{\infty} a_n z^n, \qquad z \in A_r^R .$$

If $r < \mu < R$, then we have

$$\int_{\tau_\mu} \frac{f(\zeta)}{\zeta^{m+1}} \, d\zeta = \int_{\tau_\mu} \sum_{n=-\infty}^{\infty} a_n \zeta^{n-m-1} \, d\zeta$$

$$= \sum_{n=-\infty}^{\infty} a_n \int_{\tau_\mu} \zeta^{n-m-1} \, d\zeta = 2\pi i a_m.$$

The last equality holds since $\int_{\tau_\mu} \zeta^p \, d\zeta = 0$ if $p \neq -1$ and $\int_{\tau_\mu} \zeta^{-1} \, d\zeta = 2\pi i$. (We have also used uniform convergence of the integral.) Hence the a_n are unique and are given by $(*)$.

Thus we have obtained a two-sided series expansion for a function analytic on an annulus. We next go the other direction and define a function on an annulus from a two-sided infinite sequence of complex numbers.

(iv) *Radii of convergence of a Laurent series.* Let $(a_n)_{n=-\infty}^{\infty}$ be a two-sided sequence in $\mathbf{C}$ and define $\overline{\lim}_{n \to \infty} |a_{-n}|^{1/n} = r$ and $[\overline{\lim}_{n \to \infty} |a_n|^{1/n}]^{-1} = R$. If $r < R$, then $\sum_{n=-\infty}^{\infty} a_n z^n = f(z)$ converges on A_r^R and converges uniformly on A_ν^μ if $r < \nu < \mu < R$. Furthermore, f is analytic on A_r^R.

[We will not consider the case $R \leq r$. The same $0 - \infty$ conventions adopted in (2.2) apply to R.]

PROOF. By (2.2), the series $\sum_{n=1}^{\infty} a_{-n} \zeta^n$ converges on $D(r^{-1}, 0)$ and converges uniformly on $D(\nu^{-1}, 0)$ if $\nu > r$; hence $\sum_{n=1}^{\infty} a_{-n} z^{-n}$ converges uniformly on $\{z \mid |z| > \nu\}$. Also, $\sum_{n=0}^{\infty} a_n z^n$ converges uniformly on $D(\mu, 0)$ if $\mu < R$. That f is analytic is a consequence of the fact that a uniform limit of analytic functions is analytic (4.11). [One can also prove analyticity of f directly from (2.4).]

(v) *Different centers.* If f is analytic on an annulus $A_r^R(a) = \{z \mid r < |z - a| < R\}$, then $f(z + a)$ is analytic on $A_r^R(0)$ and the above result for $f(z + a)$ translates immediately to f. Thus we have

$$f(z) = \sum_{n=-\infty}^{\infty} a_n(z - a)^n$$

for $z \in A_r^R(a)$, where the a_n are uniquely given by

$$a_n = \frac{1}{2\pi i} \int_{\overrightarrow{C(\rho, a)}} \frac{f(z)}{(z - a)^{n+1}} \, dz, \qquad (r < \rho < R).$$

(vi) *Definitions and remarks.* Infinite series of the form $\sum_{n=-\infty}^{\infty} a_n(z - a)^n$ are called *Laurent series.* They converge if $\sum_{n=0}^{\infty} a_n(z - a)^n$ and $\sum_{n=1}^{\infty} a_{-n}(z - a)^{-n}$ converge, and by definition $\sum_{n=-\infty}^{\infty} a_n(z - a)^n$ is the sum of the two series. The *principal part* of the series is $\sum_{n=1}^{\infty} a_{-n}(z - a)^{-n}$, or in different notation $\sum_{n=-1}^{-\infty} a_n(z - a)^n$. We have seen that a Laurent series converges, if at all, on an annulus $A_r^R(a)$. If $f(z) = \sum_{n=-\infty}^{\infty} a_n(z - a)^n$, then the coefficient

$$a_{-1} = \frac{1}{2\pi i} \int_{\overrightarrow{C(\rho, a)}} f(z) \, dz$$

is particularly important because there are often methods of finding a_{-1} other than by direct calculation of the integral. In the case $r = 0$, a_{-1} is called the *residue of f at a*; we will denote it by $\text{Res}\,(f, a)$.

*5.2 A CONNECTION WITH FOURIER SERIES

Let f be analytic on $A_r^R(0)$, $r < R$, and suppose that $r < \rho < R$. Let $f_\rho(z) = f(\rho z)$, $z \in \mathbf{T} = \{\zeta \mid |\zeta| = 1\}$. The function f_ρ is analytic on an annulus containing $\mathbf{T}$. The Fourier coefficients of f_ρ are defined as usual by

$$\hat{f}_\rho(n) = \frac{1}{2\pi} \int_{[-\pi, \pi]} f(\rho e^{it}) e^{-int} \, dt \qquad (n \in Z);$$

see (3.18). The integral on the right is equal to

$$\frac{\rho^n}{2\pi i} \int_{\tau_\rho} \frac{f(\zeta)}{\zeta^{n+1}} \, d\zeta.$$

Hence the coefficients a_n in the Laurent expansion of f are related to the Fourier coefficients of f_ρ by the equality

$$\hat{f}_\rho(n) = a_n \rho^n.$$

In particular, if $r < 1 < R$ and $\rho = 1$, then $a_n = \hat{f}(n)$. Thus the corollary below follows from (5.1.iii).

(i) *Corollary.* Let f be analytic on an annulus centered at zero and containing $\mathbf{T} = \{z \mid |z| = 1\}$. The then Fourier series

$$S_n(e^{it}) = \sum_{k=-n}^{n} \hat{f}(k)e^{ikt}$$

converges to $f(e^{it})$ for all $t \in [-\pi, \pi]$.

We can also go in the other direction. That is, the Laurent expansion of an analytic function can be deduced from a result about Fourier series, together with Cauchy's theorem. Thus suppose that f is analytic on an annulus A_r^R and that $r < \rho < R$. Then $f\rho \in \mathbb{C}^2(\mathbf{T})$ [see (3.18) for definitions], and so the Fourier series for f_ρ converges uniformly on $\mathbf{T}$ by (3.19):

$$f(\rho e^{it}) = \sum_{n=-\infty}^{\infty} \hat{f}_\rho(n)e^{int}.$$

It is easy to prove that the series also converges absolutely, since $f \in \mathbb{C}^2(\mathbf{T})$. It follows from absolute convergence that $\sum_{n=-\infty}^{\infty} \hat{f}_\rho(n)e^{int} = \sum_{n=0}^{\infty} \hat{f}_\rho(n)e^{int} + \sum_{n=-\infty}^{-1} \hat{f}_\rho(n)e^{int}$. Substituting the expression for $\hat{f}_\rho$ obtained above, we have

(a) $\qquad f(\rho e^{it}) = \sum_{n=-\infty}^{\infty} a_n(\rho e^{it})^n \qquad \left(a_n = \frac{1}{2\pi i}\int_{\tau_\rho} \frac{f(\zeta)}{\zeta^{n+1}}\, d\zeta\right).$

The expansion in (a) is the Laurent expansion for f, except that the coefficients a_n appear to be dependent on ρ. Since f is analytic the integral defining a_n is the same for all $\rho \in\ ]r, R[$, as was shown in the proof of (5.1.iii).

5.3 EXAMPLES

(i) The function f defined by

$$f(z) = \frac{1 + 2z}{z^2(z + 1)} \qquad (z \neq 0,\ z \neq -1)$$

is analytic on $\mathbb{C}\backslash\{0, -1\}$ and so is analytic on each of the annuli $A_0^1(0)$, $A_0^1(-1)$, and $A_1^\infty(0)$. We will find the Laurent expansions of f in each of these annuli. We begin with the expansion centered at 0 and valid in $A_0^1(0)$. We have

$$f(z) = \frac{(1 + 2z)}{z^2} \cdot \frac{1}{1 - (-z)} = \frac{1 + 2z}{z^2} \sum_{n=0}^{\infty} (-1)^n z^n$$

$$= \frac{1}{z^2}\left[1 + \sum_{n=1}^{\infty}\left((-1)^n + 2(-1)^{n-1}\right)z^n\right]$$

$$= \frac{1}{z^2} + \frac{1}{z} - 1 + z - z^2 + \cdots$$

$$= \sum_{n=-2}^{\infty} a_n z^n \qquad (a_n = (-1)^{n+1} \text{ if } n \geq 0;\ a_{-1} = a_{-2} = 1).$$

Next consider the expansion about -1, valid on $A_0^1(-1)$. Write

$$f(z) = \frac{1}{z+1} \left[\frac{1+2z}{z^2} \right]. \tag{*}$$

The function in brackets is analytic on $D(1, -1)$, and so has a Taylor expansion about -1. The expansion can be found by finding all derivatives of $(1 + 2z)/z^2$ at -1. A simpler alternative is to write

$$\frac{1}{z} = \frac{1}{[(z+1)-1]} = \left[-\sum_{n=0}^{\infty} (z+1)^n \right], \quad ((z \in A_0^1(-1)).$$

Noting that $1/z^2 = -d\,(1/z)/dz$ and using (2.4), we obtain

$$\frac{1}{z^2} = \sum_{n=0}^{\infty} (n+1)(z+1)^n,$$

the Taylor expansion of $1/z^2$ about -1. Hence we have

$$\frac{1+2z}{z^2} = 2z \sum_{n=0}^{\infty} (n+1)(z+1)^n + \sum_{n=0}^{\infty} (n+1)(z+1)^n$$

$$= [2(z+1) - 2] \sum_{n=0}^{\infty} (n+1)(z+1)^n + \sum_{n=0}^{\infty} (n+1)(z+1)^n$$

$$= \sum_{n=0}^{\infty} (n-1)(z+1)^n.$$

Using (*), we obtain

$$f(z) = \frac{1}{z+1} \sum_{n=0}^{\infty} (n-1)(z+1)^n = \sum_{n=-1}^{\infty} n(z+1)^n.$$

Finally, to find the expansion centered at 0 and valid in $A_1^{\infty}(0)$, write

$$f(z) = \frac{1+2z}{z^2} \cdot \frac{1}{1+z} = \frac{1+2z}{z^2} \cdot \frac{1}{z(1+1/z)}$$

$$= \frac{1+2z}{z^3} \sum_{n=0}^{\infty} (-1)^n z^{-n}$$

$$= \sum_{n=0}^{\infty} (-1)^n z^{-n-3} + \sum_{n=0}^{\infty} 2(-1)^n z^{-n-2}$$

$$= \sum_{k=-\infty}^{-3} [(-1)^{k+1} + 2(-1)^k] z^k - \frac{2}{z^2}$$

$$= -\frac{2}{z^2} + \sum_{k=-\infty}^{-3} (-1)^k z^k.$$

In summary, the Laurent expansions we sought are as follows:

$$\frac{1 + 2z}{z^2(z + 1)} = \begin{cases} \dfrac{1}{z^2} + \dfrac{1}{z} + \displaystyle\sum_{n=0}^{\infty} (-1)^{n+1} z^n, \ z \in A_0^1(0) \\[2em] \displaystyle\sum_{n=-1}^{\infty} n(z + 1)^n, \ z \in A_0^1(-1) \\[2em] -\dfrac{2}{z^2} + \displaystyle\sum_{n=-\infty}^{-3} (-1)^n z^n, \ z \in A_1^{\infty}(0). \end{cases}$$

Notice that as a corollary we have (see (5.1.vi))

$$\int_{\overrightarrow{C(\rho, \, 0)}} \frac{1 + 2z}{z^2(1 + z)} \, dz = 2\pi i, \qquad \text{if } 0 < \rho < 1;$$

$$\int_{\overrightarrow{C(r, \, -1)}} \frac{1 + 2z}{z^2(1 + z)} \, dz = -2\pi i, \quad \text{if } 0 < r < 1;$$

$$\int_{\overrightarrow{C(R, \, 0)}} \frac{1 + 2z}{z^2(1 + z)} \, dz = 0, \qquad \text{if } 1 < R < \infty.$$

(ii) Consider the functions

$$f(z) = \exp\left(\frac{z}{1 - z}\right) \quad \text{and} \quad g(z) = \exp\left(\frac{1}{z - 1}\right),$$

which are analytic on $\mathbf{C}\backslash\{1\}$. Notice that $g(z) = f(1/z)$. Each function has Laurent expansions in $A_0^1(0)$, $A_0^{\infty}(1)$, and $A_1^{\infty}(0)$. First we find the Taylor expansions of f and g in $D(1, 0)$. For $|z| < 1$, we have

$$\exp\left(\frac{z}{1 - z}\right) = 1 + \sum_{j=1}^{\infty} \frac{1}{j!} \left(\frac{z}{1 - z}\right)^j$$

$$= 1 + \sum_{j=1}^{\infty} \frac{1}{j!} z^j \left(\sum_{k=0}^{\infty} z^k\right)^j$$

$$= 1 + \sum_{j=1}^{\infty} \frac{1}{j!} z^j \sum_{n=0}^{\infty} \binom{j + n - 1}{n} z^n.$$

The double series above converges absolutely and so equals

$$1 + \sum_{(j, \, n)} \binom{j + n - 1}{n} \frac{1}{j!} z^{j+n}.$$

Rearranging so as to factor out z^k, the series is

$$f(z) = 1 + \sum_{k=1}^{\infty} \left[\sum_{j=1}^{k} \frac{1}{j!} \binom{k - 1}{k - j}\right] z^k;$$

this is the power series expansion of f about zero. For g, write

$$\exp\left(\frac{1}{z-1}\right) = \sum_{n=0}^{\infty} \frac{1}{n!} \frac{1}{(z-1)^n}$$

$$= \sum_{n=0}^{\infty} \frac{(-1)^n}{n!} \left[\sum_{j=0}^{\infty} z^j\right]^n$$

$$= \sum_{n=0}^{\infty} \frac{(-1)^n}{n!} \sum_{m=0}^{\infty} \binom{n+m-1}{m} z^m$$

$$= \sum_{m=0}^{\infty} \left[\sum_{n=0}^{\infty} \binom{n+m-1}{m} \frac{(-1)^n}{n!}\right] z^m.$$

From these Taylor expansions for f and g, centered at 0, one can write the Laurent expansions for g and f, respectively, on $A_1^{\infty}(0)$.

Both f and g also have Laurent expansions about 1, valid on $A_0^{\infty}(1)$. For g, the expansion is simple:

$$g(z) = \sum_{n=0}^{\infty} \frac{1}{n!(z-1)^n} = \sum_{n=-\infty}^{0} \frac{(z-1)^n}{|n|!}.$$

For f, we have

$$f(z) = \sum_{n=0}^{\infty} \frac{1}{n!} \frac{z^n}{(1-z)^n}$$

$$= \sum_{n=0}^{\infty} \frac{(-1)^n}{n!} \left(\frac{1}{(z-1)^n} \cdot \sum_{j=0}^{n} \binom{n}{j}(z-1)^j\right)$$

$$= \sum_{n=0}^{\infty} \frac{(-1)^n}{n} \sum_{j=0}^{n} \binom{n}{j}(z-1)^{j-n}$$

$$= \sum_{m=0}^{\infty} \left[\sum_{k=0}^{\infty} \frac{(-1)^{m+k}}{(m+k)!} \binom{m+k}{k}\right](z-1)^{-m}.$$

Summarizing the Laurent expansions, we have

$$f(z) = \sum_{m=-\infty}^{0} \left[\sum_{n=0}^{\infty} \binom{n-m-1}{-m} \frac{(-1)^n}{n!}\right] z^m, \quad z \in A_1^{\infty}(0);$$

$$g(z) = 1 + \sum_{k=-\infty}^{-1} \left[\sum_{j=1}^{-k} \frac{1}{j!}\binom{-k-1}{-k-j}\right] z^k, \quad z \in A_1^{\infty}(0);$$

$$f(z) = \sum_{m=-\infty}^{0} \left[\sum_{k=0}^{\infty} \frac{(-1)^{m+k}}{(-m+k)!}\binom{-m+k}{k}\right](z-1)^m, \quad z \in A_0^{\infty}(1);$$

$$g(z) = \sum_{n=-\infty}^{0} \frac{(z-1)^n}{|n|!}, \quad z \in A_0^{\infty}(1).$$

(iii) Since $\sin (z) = 0$ if and only if $z = n\pi$ for an integer n, the function $1/(\sin z)$ has Laurent expansions in each of the annuli $A_{n\pi}^{(n+1)\pi}(0)$, n an integer, and also in each $A_0^\pi(n\pi)$. Since we have

$$\lim_{z \to 0} \frac{z}{\sin (z)} = 1,$$

the function $z/(\sin z)$ is analytic on $D(\pi, 0)$, with value 1 at 0 [by (4.6)]. If

$$\frac{z}{\sin z} = 1 + \sum_{n=1}^{\infty} a_n z^n$$

is the Taylor expansion of $z/(\sin z)$ centered at 0, then

$$\frac{1}{\sin z} = \frac{1}{z} + \sum_{n=0}^{\infty} a_{n+1} z^n$$

is the Laurent expansion of $1/(\sin z)$ on $A_0^\pi(0)$. Since $z/(\sin z)$ is an even function, $a_k = 0$ for odd k. We can find $(a_{2m})_{m=1}^\infty$ by using the equality

$$z = \left[z - \frac{z^3}{3!} + \frac{z^5}{5!} - \cdots \right] [1 + a_2 z^2 + a_4 z^4 + \cdots] \qquad (**)$$

and (2.9) to equate coefficients of corresponding powers of z on each side of equality $(**)$. We obtain

$$a_2 = \frac{1}{3} \; ; a_4 = \left(\frac{1}{3!}\right)^2 - \frac{1}{5!},$$

and, in general

$$a_{2m} = \sum_{j=0}^{m-1} a_{2j} \frac{-(-1)^{m-j}}{(2m - 2j + 1)!}.$$

EXERCISES

TYPE I

1. Find the Laurent expansions of the following functions about the points indicated, give the domain of validity of the expansion, identify the principal part of the expansion, and find the residues of the functions at the points
   (a) $1/[z(z - 1)]$ about 0 and 1
   (b) $1/[z^3(z - i)]$ about 0 and i
   (c) $\exp (1/z)$ about 0.

2. Calculate the first few terms of the Laurent expansion of $1/(\cos^2 z)$ on $A_0^\pi(\pi/2)$.

3. Use the result of Exercise 2 to calculate the first few terms of the Laurent expansion of tan (z) on $A_0^\pi(\pi/2)$.

4. (a) Find the Laurent expansion of $1/(\sin z)$ on $A_0^\pi(n\pi)$, n an integer [see Example (5.3.iii)].
   (b) Find the Laurent expansion of $1/(\sin z)$ on $A_\pi^{2\pi}(0)$. On $A_{n\pi}^{(n+1)\pi}(0)$.

5. Explain why $F_{1/2}(z)$ and Log (z) do not have Laurent expansions about 0.

6. Find the annulus of convergence $A_r^R(0)$ for the Laurent series $\sum_{n=-\infty}^{\infty} e^{-|n|}z^n$. Does the series converge on any part of the boundary of $A_r^R(0)$?

7. (a) Give an interpretation of (5.1.iv) where $R < r$.
   (b) Give some examples of what can happen if $R = r$.

TYPE II

1. In the proof of (5.1.iii), we considered a function $g(z)$ given by a certain integral and analytic in the exterior of a disk D. We used the integral expression for $g(z)$ in order to obtain an expansion for it of the form $\sum_{n=-\infty}^{-1} a_n z^n$. The expansion can also be obtained as outlined below.
   (a) Let g be analytic on $\mathbf{C}\backslash \overline{D(r, 0)}$ [that is, on $A_r^\infty(0)$] and suppose that $\lim_{z\to\infty} g(z) = a$. Show directly that

   $$g(z) = a + \sum_{n=-\infty}^{-1} a_n z^n \qquad (*)$$

   for suitable coefficients a_n.
   [*Hint:* The function $H(z) = g(1/z)$ is analytic in $D(r^{-1}, 0)\backslash\{0\}$.]
   Since $\lim_{z\to 0} H(z) = a$, H is analytic on $D(r^{-1}, 0)$. The power series expansion for H centered at 0 becomes $(*)$ for g.
   (b) Let g be analytic on $A_R^\infty(0)$, and let

   $$g(z) = \sum_{n=-\infty}^{\infty} a_n z^n$$

   be its Laurent expansion on $A_R^\infty(0)$. Show that $a_n = 0$ for $n > 0$ if and only if $\lim_{z\to\infty} g(z)$ exists.

2. Let f be analytic on $A_r^R(0)$ $(0 \le r < R \le \infty)$ with Laurent expansion

   $$f(z) = \sum_{n=-\infty}^{\infty} a_n z^n.$$

   Our definition of the series is simply $\sum_{n=-\infty}^{-1} a_n z^n + \sum_{n=0}^{\infty} a_n z^n$. Let μ and ν satisfy $r < \nu < \mu < R$.

(a) Prove that the double limit

$$\lim_{\substack{p \to \infty \\ m \to \infty}} \sum_{n=-m}^{p} a_n z^n$$

exists uniformly on $A_y^\mu(0)$.

(b) Show that the double sequence $(\sum_{n=-m}^{p} a_n z^n)_{m, p=1}^{\infty}$ diverges if $|z| < r$ or if $|z| > R$.

5.4 DEFINITIONS

The residue theorem is a generalization of Cauchy's theorem. Among other things, it gives a method of calculating $\int_\gamma f(z) \, dz$ for certain closed paths in sets where f is analytic except for a few isolated singularities. We first give several definitions, then state and prove two versions of the residue theorem. A somewhat more general version of the theorem is outlined in the exercises.

Let U be an open set and let S be a subset of U such that all cluster points of S, if any, lie on the boundary of U and not in U. That is, S is *isolated in U*. If f is analytic on $U \backslash S$, then the points of S are called *isolated singularities* of f. An isolated singularity s is *removable* if $f(s)$ can be defined so that f is analytic on an open disk containing s. If $s \in S$ and $D(r, s) \cap S = \{s\}$, then f has a Laurent expansion $f(z) = \sum_{n=-\infty}^{\infty} a_n(z - s)^n$ on the annulus $A_0^r(s) = \{z \mid 0 < |z - s| < r\}$. If there is an $m > 0$ such that $a_n = 0$ if $n < -m$ and $a_{-m} \neq 0$, then f is said to have a *pole of order m at s*. If the set $\{m \mid m > 0 \text{ and } a_{-m} \neq 0\}$ is infinite, then f is said to have an *essential singularity* at s. The *residue* of f at s is defined to be a_{-1} and denoted Res (f, s). Recall from (5.1.vi) that the equality

$$\frac{1}{2\pi i} \int_\tau f(z) \, dz = \text{Res } (f, s) \quad [\tau(t) = s + \rho e^{it}, (0 < \rho < r)]$$

is valid. It is useful to extend these definitions to the point ∞. If $\mathbf{C} \backslash \overline{D(n, 0)} \subset U$ for some n, then the function F given by $F(z) = f(1/z)$ is defined on $D(n^{-1}, 0) \backslash \{0\}$ and 0 is an isolated singularity of F. The behavior of f at ∞ is defined as the behavior of F at 0. Thus f has a removable singularity, a pole of order m, or an essential singularity at ∞ if F exhibits the corresponding phenomena at 0.

5.5 THE RESIDUE THEOREM FOR CONVEX SETS

Let U be an open convex set. Let S be a subset of U which is isolated as a subset of $\mathbf{C}$. Suppose that f is analytic on $U \backslash S$ and that every point of S is a singularity of f. Then the equality

(i) $\displaystyle\int_\gamma f(z) \, dz = 2\pi i \sum_{s \in S} \text{Res } (f, s) \text{ Ind}_\gamma (s)$

is valid for every closed path γ in U.

PROOF.   There is an R such that $\gamma^* \subset D(R, 0)$. The set $\mathbf{C} \backslash \overline{D(R, 0)}$ is connected and unbounded. Hence by (4.17) it is in the unbounded component of $\mathbf{C} \backslash \gamma^*$. By (4.22), we have $\operatorname{Ind}_\gamma (z) = 0$ for all $z \in \mathbf{C} \backslash \overline{D(R, 0)}$. The set $S \cap \overline{D(R, 0)}$ is finite by (1.11); denote it by X. If $x \in X$, then f has a Laurent expansion $f(z) = \sum_{n=-\infty}^{\infty} a_n (z - x)^n = P_x(z) + R_x(z)$. The function P_x is the principal part of f at x. Since P_x is analytic on $\mathbf{C} \backslash \{x\}$, the function $f - P_x$ is analytic on $U \backslash S$. The point x is a removable singularity of $f - P_x$, because it is defined near x by a power series. Thus in fact $f - P_x$ is analytic on $(U \backslash S) \cup \{x\}$. If y is a different point in X, then the principal part P_y of $f - P_x$ at y is precisely the principal part of f at y, because P_x is analytic near y. Subtracting P_y from $f - P_x$ we thus obtain a function analytic on $(U \backslash S) \cup \{x, y\}$. Exhausting the finite set X, we see that

$$ f - \sum_{x \in X} P_x \qquad (P_x \text{ principal part of } f \text{ at } x) $$

is analytic on $[U \backslash (S \backslash X)]$; that is, on $U \cap D(R, 0) = V$. By Cauchy's theorem (4.2) for the convex set V, we see that

$$ \int_\gamma f(z)\, dz = \sum_{x \in X} \int_\gamma P_x(z)\, dz = 2\pi i \sum_{x \in X} \operatorname{Res}(f, x)\, \operatorname{Ind}_y(x). $$

Remark.   In (5.5.i) one need consider only those singularities which are not removable, for as the reader can easily verify, the residue at a removable singularity is zero.

*5.6 THE RESIDUE THEOREM FOR SIMPLY CONNECTED SETS

Let U be a simply connected subset of $\mathbf{C}$. Let S be a finite subset of U. Suppose that f is analytic on $U \backslash S$ and that every point of S is a singularity of f. Then (5.5.i) holds for every closed path γ in U.

PROOF.   As in the proof of (5.5) we see that $f - \sum_{s \in S} P_s$ is analytic on U, where P_s is the principal part of f at s. The equality

$$ \int_\gamma \left[f(z) - \sum_S P_s(z) \right] dz = 0 $$

is valid by (4.28). This equality leads to (5.5.i) as before.

Remark.   The residue theorem for convex sets will suffice for most of our purposes. It is dependent on (4.2), while the simply connected version (5.6) is dependent on (4.28).

5.7 DERIVATIVE RULE FOR CALCULATING RESIDUES

Effective utilization of the residue theorem often depends on efficient methods of calculating a_{-1}. We have already given examples of such calculations in the section on Laurent series. Another method worth noting is the so-called derivative rule, frequently useful when the singularity under consideration is a pole.

To obtain it, note that if f has a pole of order m at a, say

$$f(z) = \frac{a_{-m}}{(z-a)^m} + \frac{a_{-m+1}}{(z-a)^{m-1}} + \cdots + a_0 + \sum_{n=1}^{\infty} a_n(z-a)^n,$$

then we have

$$(z-a)^m f(z) = a_{-m} + a_{-m+1}(z-a) + \cdots + a_{-1}(z-a)^{m-1} + \cdots,$$

so that the equality

$$\text{(i)} \quad a_{-1} = \frac{1}{(m-1)!} \lim_{z \to a} \frac{d^{(m-1)}}{dz^{(m-1)}} \left((z-a)^m f(z) \right)$$

is valid. This equality is the *derivative formula*. For an example, consider $f(z) = [\sin(z)]/(z-i)^3$. We have

$$(z-i)^3 f(z) = \sin(z),$$

and $[d^2 \sin(z)]/dz^2 = -\sin(z)$. Hence the equality

$$\text{Res}\left(\frac{\sin(z)}{(z-i)^3}, i \right) = -\frac{\sin(i)}{2}$$

is valid.

EXERCISES

TYPE I

1. Use the derivative rule to find Res (f, a) for the functions in (1.a) and (1.b) in the exercises on p. 183. Why does the rule not work for (1.c)?

2. Find Res (f, a) for the following f and a.
   (a) $f(z) = 1/(\sin z)$, $a = 0, \pm \pi, \ldots n\pi$, n an integer. (See Exercise 4, p. 184.)
   (b) $\exp[z/(1-z)]$, $a = 1$.
   (c) $\tan z$, $a = \pi/2$.

3. Use the residue theorem to calculate the following (contour) integrals.

   (a) $\displaystyle\int_{C(2,0)} \frac{1}{(z-i)^3(z+i)^2 z(z-3)} \, dz$

(b) $\displaystyle\int_{\overrightarrow{C(\pi/2,\,0)}} \frac{1}{\sin(z)}\,dz$

(c) $\displaystyle\int_{\overrightarrow{C([(2n+1)\pi]/2,\,0)}} \frac{1}{\sin(z)}\,dz$

(d) $\displaystyle\int_{\overrightarrow{C(\pi,\,0)}} \frac{1}{\cos^2(z)}\,dz$

(e) $\displaystyle\int_{\overrightarrow{C(1,\,0)}} \frac{1}{e^z - 1}\,dz$

(f) $\displaystyle\int_{\overrightarrow{C(8,\,0)}} \frac{1}{e^z - 1}\,dz.$

4. Show that the functions $[\sin(z)]/z$, $(e^z - 1)/z$, $[\mathrm{Log}\,(z + 1)]/z$ have removable singularities at 0.

TYPE II

1. In the residue theorem (5.5) we assumed that the set S is isolated as a subset of $\mathbf{C}$. The theorem is actually valid if we assume only that S is isolated as a subset of the convex set U. Prove this more general version.

5.8 APPLICATIONS OF THE RESIDUE THEOREM

The residue theorems (5.5) and (5.6) have many applications, both theoretical and applied. We will first apply (5.5) to prove the local, open, and inverse mapping theorems. This material, Sections (5.8) to (5.15), is not essential to the continuity of the book and may be omitted. We begin by computing Res $(f'/f, a)$.

Theorem If f is analytic on an open set containing a and has a zero of order m at a, then we have

$$\mathrm{Res}\left(\frac{f'}{f},\,a\right) = m.$$

If f is analytic on some $A_0^r(a)$ and f has a pole of order m at a, then we have

$$\mathrm{Res}\left(\frac{f'}{f},\,a\right) = -m.$$

PROOF. If f has a zero of order m and $\sum_{n=m}^{\infty} a_n(z - a)^n$ is the Taylor expansion of f, then we have

$$\frac{f'(z)}{f(z)} = \frac{ma_m(z - a)^{m-1} + (m + 1)a_{m+1}(z - a)^m + \cdots}{a_m(z - a)^m + a_{m+1}(z - a)^{m+1} + \cdots}$$

$$= \frac{m}{(z - a)} [1 + g(z)],$$

where g is analytic near a and $g(a) = 0$. Thus, the coefficient of $(z - a)^{-1}$ in the Laurent expansion of f'/f is m. If f has a pole of order m at a, then we have $f(z) = \sum_{n=-m}^{\infty} a_n(z - a)^n$ and

$$\frac{f'(z)}{f(z)} = \frac{-ma_{-m}(z - a)^{-m-1} + (-m + 1)(z - a)^{-m} + \cdots}{a_{-m}(z - a)^{-m} + a_{-m+1}(z - a)^{-m+1} + \cdots}$$

$$= \frac{-m}{(z - a)} [1 + h(z)],$$

where h is analytic near a and $h(a) = 0$. Thus, the coefficient of $(z - a)^{-1}$ in the Laurent expansion of f'/f is $-m$.

5.9 ARGUMENT PRINCIPLE

(i) Let U be an open convex set and let P be a finite subset of U. Let f be analytic on $U \backslash P$ and suppose that each point of P is a pole of f. Suppose also that the set $Z = \{a \in U \mid f(a) = 0\}$ is finite. Let γ be a closed path in $U \backslash (Z \cup P)$. Let $\sigma(a)$ and $\sigma(p)$ denote the *order* of the zero a and the pole p, respectively. We have

$$\frac{1}{2\pi i} \int_\gamma \frac{f'(z)}{f(z)} \, dz = \sum_{a \in Z} \sigma(a) \, \text{Ind}_\gamma (a) - \sum_{p \in P} \sigma(p) \, \text{Ind}_\gamma (p). \qquad (*)$$

(ii) If U is simply connected and other notation is as in (i), then $(*)$ also holds.

PROOF. The set $Z \cup P$ is the set of isolated singularities of f'/f in U. By (5.8), the residue at a pole p of f is $-\sigma(p)$ and the residue at a zero a of f is $\sigma(a)$. Thus $(*)$ for hypothesis (i) follows from (5.5) and for hypothesis (ii) from (5.6).

Remark. We distinguish between cases (i) and (ii) because the trail leading to the relevant Cauchy theorem is much shorter for (i). The "local" results which follow depend only (5.9.i).

5.10 DISCUSSION OF 5.9

If Log (f) were analytic in a region containing γ, then we could write

$$\frac{1}{2\pi i}\int_\gamma \frac{f'(z)}{f(z)}\,dz = \frac{1}{2\pi i}\int_\gamma [\text{Log }(f)]'(z)\,dz = 0.$$

But Log (f) can fail to be analytic in either of two ways:

(1) f can have a pole;

(2) f can have a zero in the interior of γ, so that Log (f) cannot be analytic in a region containing γ.

In case (2), $f(\zeta)$ travels around 0 m times (m the order of the zero a of f) each time that ζ travels around a on γ, which is $\text{Ind}_\gamma(a)$ times. Thus $f(\zeta)$ travels $m\text{Ind}_\gamma(a)$ times around 0, and Log $f(\zeta)$ picks up $2\pi i$ on each trip. Thus, in the supposed equality

$$\int_\gamma [\text{Log }f]'(z)\,dz = \text{Log }f(b) - \text{Log }f(b)$$

$$= \text{Log }|f(b)| - \text{Log }|f(b)| + i[\text{Arg }f(b) - \text{Arg }f(b)],$$

we get 0 for the real part but for the imaginary part we get a multiple of 2π.

5.11 COROLLARY OF ARGUMENT PRINCIPLE

Let f be analytic on an open set U. Suppose that $\overline{D(r, a)} \subset U$ and let $\tau(t) = a + re^{it}(t \in [-\pi, \pi])$. Let $T = f \circ \tau$. If $w \notin T^*$, then $\text{Ind}_T(w)$ is the number of zeros of $f - w$ in $D(r, a)$. (A zero of multiplicity m is counted m times.)

PROOF. There is a $\rho > r$ such that $D(\rho, a) \subset U$, and $D(\rho, a)$ is convex. By (5.9.i) the left side of the equalities below is the number of zeros of $f - w$ in $D(r, a)$.

$$\int_\tau \frac{(f-w)'(z)}{(f-w)(z)}\,dz = \int_{[-\pi,\pi]} \frac{(f'\circ\tau(t))\tau'(t)}{f\circ\tau(t) - w}\,dt = \int_T \frac{1}{\zeta - w}\,d\zeta.$$

Remark. The number of zeros of $f - w$ is, of course, the number of solutions to the equation $f(z) = w$. If $\text{Ind}_T(w) \neq 0$, that is, if w is inside T, then there is at least one $z \in D(r, a)$ such that $f(z) = w$. That is, every point interior to T comes from $D(r, a)$. From the above formula we also see that the converse is true; that is, each z in $D(r, a)$ is carried inside T.

5.12 LOCAL MAPPING THEOREM

Suppose that f is analytic and nonconstant on a connected open set U and that $z_0 \in U$ is a zero of $f - f(z_0)$ of multiplicity m. Then there is an open set V such that

(i) $z_0 \in V \subset U$

(ii) $f(V)$ is open

(iii) for each $w \in f(V) \backslash f(z_0)$ there are exactly m distinct points $z_1, z_2, \ldots,$ $z_m \in V$ such that $f(z_i) = w$.

PROOF. Let $w_0 = f(z_0)$. Let $r > 0$ be such that $\overline{D(r, z_0)} \backslash \{z_0\}$ contains no zeros of $f - w_0$ or of f'. Such an r exists by the identity theorem (4.18). For τ and T as in (5.11), let W be the component of $\mathbf{C} \backslash T^*$ containing w_0. By (5.11) and the constancy of Ind_T, m is the number of zeros (counting multiplicity) of $f - w$ in $D(r, z_0)$, for every $w \in W$. Since f' is not zero in $D(r, z_0) \backslash z_0$, the multiplicity of each of these m zeros is 1 if $w \neq w_0$. Letting

$$V = D(r, z_0) \cap f^{-1}(W),$$

we see that $f(V) = W$, and the proof is complete.

5.13 OPEN MAPPING THEOREM

If f is analytic and nonconstant on the connected open set U, then $f(U)$ is also open.

PROOF. Let $w_0 = f(z_0) \in f(U)$, and let V and W be as in (5.12); since W is open and $w_0 \in W \subset f(U)$, $f(U)$ is open.

5.14 INVERSE FUNCTION THEOREM

If M is one-to-one and analytic on an open set U (that is, M is *univalent*), then $M'(z) \neq 0$ for all $z \in U$, the inverse N of M is analytic on $V = M(U)$, and $N'(w) = [M'(N(w))]^{-1}$ for all $w \in V$.

PROOF. First we show that N is continuous. Let $\zeta \in M(U)$. If $\varepsilon > 0$ is given, then the set $W = M(D(\varepsilon, N(\zeta)))$ is open by (5.13), and it contains ζ. Hence, there is a $\delta > 0$ such that $D(\delta, \zeta) \subset W$. Thus we have $N(D(\delta, \zeta)) \subset D(\varepsilon, N(\zeta))$; that is, $|N(\xi) - N(\zeta)| < \varepsilon$ if $|\xi - \zeta| < \delta$. Hence N is continuous.

The set $V = M(U)$ is open by (5.13). If $M'(z_0) = 0$ for some $z_0 \in U$, then z_0 is a zero of multiplicity at least two of the functions $M - M(z_0)$. By (5.12), M cannot be one-to-one. Hence $M'(z_0) \neq 0$.

The difference quotient for N at z is

$$\frac{N(\zeta) - N(z)}{\zeta - z} = \left[\frac{M(N(\zeta)) - M(N(z))}{N(\zeta) - N(z)}\right]^{-1}.$$

Since $M'(N(z)) \neq 0$, the limit as ζ goes to z exists and equals $[M'(N(z))]^{-1}$ by the continuity of N.

5.15 LOCAL CONVERSE TO 5.14

Let f be analytic on an open set U and suppose that $f'(z_0) \neq 0$, z_0 a fixed point in U. Then there is an $r > 0$ such that $D(r, z_0) \subset U$ and f is one-to-one on $D(r, z_0)$.

PROOF.   The point z_0 is a zero of multiplicity 1 of $f - f(z_0)$. Apply (5.12).

Remark.   Locally univalent functions are called *conformal*. We will discuss them in more detail in Chapter 7.

EXERCISES

TYPE II

1. Let $M(z) = z^2$ on **C**.
   (a) Show that M is one-to-one (and hence univalent) on each half-space $H_\theta = \{z \neq 0 : \theta < \text{Arg}(z) < \theta + \pi\}$.
   (b) Find the function S_θ, which is inverse to M on H_θ, for each θ.
   (c) Since 0 is a root of multiplicity 2 of m, by (5.12) there is a V such that $M(z) = w$ has two solutions in z for such $w \in M(V)\backslash\{0\}$. How large can V be in this case? What are the solutions?

2. Show that exp is one-to-one on each strip $R_k = \{z : k < \text{Im}(z) < k + 2\pi\}$. Find the inverse of exp on each of these strips. Use the argument principle to evaluate the integral

$$\int_{\overrightarrow{C_m}} \frac{e^z}{e^z - 1}\, dz$$

when $\overrightarrow{C_m}$ is the circle centered at 0 of radius $(2m + 1)\pi$, m is an integer.

3. Let f be analytic and one-to-one on an open set U. Suppose that $\overline{D(r, z_0)} \subset U$ and let $\tau = \overrightarrow{C(r, z_0)}$, as in (5.11). Let $T = f \circ \tau$.
   (a) Show that $\text{Ind}_T (f(z)) = \text{Ind}_\tau (z)$ for all z in $U\backslash\tau^*$.
   (b) Define F on $f(U)\backslash T^*$ by

$$F(w) = \frac{1}{2\pi i}\int_\tau \frac{f'(\zeta)}{f(\zeta) - w}\, d\zeta.$$

Show that

$$F(f(z)) = \frac{1}{2\pi i} \int_T \frac{f^{-1}(w)}{w - f(z)} \, dw$$

for $z \in U \backslash \tau^*$.

(c) If $z \in D(r, z_0)$, use Cauchy's formula to show that $F(f(z)) = z$
That is, F is the inverse of f on $f(D(r, z_0))$.

4. Let U, f, τ, and T be as (5.11). Show that $C \backslash T^*$ has exactly one bounded component, namely, $f(D(r, a))$.

5.16 EVALUATION OF INTEGRALS ON A FINITE INTERVAL

If f is a rational function of x and y, then the integral $\int_{[-\pi, \pi]} f(\cos t, \sin t) \, dt$ is equal to

$$\int_{[-\pi, \pi]} f\left(\frac{e^{it} + e^{-it}}{2}, \frac{e^{it} - e^{-it}}{2i}\right) dt = \frac{1}{i} \int_{\vec{T}} f\left(\frac{z + z^{-1}}{2}, \frac{z - z^{-1}}{2i}\right) \frac{1}{z} \, dz.$$

Hence by (5.5) the equality

$$\int_{[-\pi, \pi]} f(\cos t, \sin t) \, dt = 2\pi \sum_{a \in S} \text{Res}\left[f\left(\frac{z + z^{-1}}{2}, \frac{z - z^{-1}}{2i}\right) \frac{1}{z}, a \right]$$

is valid, where S is the set of singularities in $D = D(1, 0)$ of the function

$$g(z) = f\left(\frac{z + z^{-1}}{2}, \frac{z - z^{-1}}{2i}\right) \frac{1}{z}.$$

As an example, we evaluate

$$\int_{[0, \pi]} \frac{\cos 2\theta}{1 + k^2 - 2k \cos \theta} \, d\theta, \ 0 < |k| < 1.$$

In the above form, the integral is equal to

$$\frac{1}{2} \int_{[-\pi, \pi]} \frac{\cos^2 t - \sin^2 t}{1 + k^2 - 2k \cos t} \, dt,$$

and so the function f of x and y is

$$f(x, y) = \frac{[x^2 - y^2]}{2(1 + k^2 - 2kx)}.$$

Thus we have

$$g(z) = \frac{1}{z} f\left(\frac{z + z^{-1}}{2}, \frac{z - z^{-1}}{2}\right) = -\frac{1}{4} \frac{z^4 + 1}{kz^2(z - k^{-1})(z - k)}.$$

The poles of this function in D are at 0 and k. The pole at 0 is of order 2. To find Res $(g, 0)$, write the Taylor expansion

$$-\frac{1}{4}\frac{z^4 + 1}{k(z - k^{-1})(z - k)} = -\frac{1}{4k} + \frac{-k^2 + 1}{4k^2} z + \cdots.$$

Thus the Laurent expansion at 0 begins

$$-\frac{1}{4}\frac{z^4 + 1}{kz^2(z - k^{-1})(z - k)} = -\frac{1}{4k}\frac{1}{z^2} + \frac{(-k^2 + 1)}{4k^2}\frac{1}{z} + \cdots.$$

The pole at k is simple, and so Res (g,k) is simply the constant term in the Taylor expansion centered at k of the function with values

$$-\frac{1}{4}\frac{z^4 + 1}{kz^2(z - k^{-1})}.$$

Thus we have

$$\text{Res } (g,0) = \frac{-k^2 + 1}{4k^2} \; ; \qquad \text{Res } (g,k) = -\frac{1}{4k^2}\frac{k^4 + 1}{k^2 - 1}$$

We thus obtain the equality

$$\int_{[0, \pi]} \frac{\cos 2\theta}{1 + k^2 - 2k\cos\theta}\, d\theta = 2\pi\left(\frac{-1}{2k^2}\right)\left(\frac{k^4 - k^2 + 1}{k^2 - 1}\right).$$

The "derivative rule" (5.7) could have been used to calculate the residues.

5.17 CALCULATION OF FOURIER TRANSFORMS

Let f be a continuous complex-valued function on $\mathbf{R}$ for which $\lim_{N \to \infty} \int_{[-N, N]} |f(x)|\, dx$ is finite. We call such a function integrable. It is easy to show that $\lim_{N \to \infty} \int_{[-N, N]} f(x)\, dx$ also exists. We write this limit as $\int_{\mathbf{R}} f(x)\, dx$. The *Fourier transform* of f is the function $\hat{f}$ defined on $\mathbf{R}$ by

$$\hat{f}(y) = \frac{1}{\sqrt{2\pi}} \int_{\mathbf{R}} f(x)e^{-ixy}\, dx.$$

Notice that $\hat{f}(0)$ is simply the integral of f over $\mathbf{R}$ divided by $\sqrt{2\pi}$. (The Fourier transform is the analogue for $\mathbf{R}$ of the Fourier coefficients of a function on $\mathbf{T}$, or on $[-\pi, \pi]$.) The residue theorem is useful in the evaluation of certain Fourier transforms and, in particular, of integrals $\int_{\mathbf{R}} f(x)\, dx$. The method is as follows. If f can be extended to a function which is analytic except for isolated singularities on an open set containing **R**, then the poles of $\{f(z)e^{-izy}\}$ are the same as those of f. It is often possible to evaluate

$$\lim_{M, N \to \infty} \int_{[-M, N]} f(x)e^{-ixy}\,dx$$

by including the oriented segment $\overrightarrow{[-M, N]}$ in a closed path containing all or some of the isolated singularities of f, and using the residue theorem (5.5) or (5.6). In (5.18–5.22) we present several situations in which the technique applies, and give examples.

5.18 STRONG GROWTH CONDITION

Suppose that f is analytic on **C**, except for finitely many isolated singularities, none of which lie on **R**. Suppose also that there are constants $M > 0$, $R > 0$, and $\eta > 1$ such that

(i) $|z^\eta f(z)| < M$ if $|z| > R$.

Under these conditions, we have

(ii) $\hat{f}(y) = \begin{cases} \sqrt{2\pi}\, i \displaystyle\sum_{a \in U} \operatorname{Res}\left(\{f(z)e^{-izy}\}, a\right) & \text{if } y < 0 \\[2ex] -\sqrt{2\pi}\, i \displaystyle\sum_{a \in L} \operatorname{Res}\left(\{f(z)e^{-izy}\}, a\right) & \text{if } y \geq 0, \end{cases}$

where U and L denote the set of singularities of f in the upper and lower open half-planes, respectively.

    First notice that condition (i) for $x \in \mathbf{R}$ implies that f is integrable on **R**. For $N > R$, let ω_N and v_N be the oriented semicircles of radius N from N to $-N$ in the upper and lower half-planes, respectively. Let $\sigma_N = \omega_N \smallsmile \overrightarrow{[-N, N]}$ and $\delta_N = v_N \smallsmile \overrightarrow{[-N, N]}$. (See Figure 5.2.) If $U_N = U \cap D(N, 0)$ and $L_N = L \cap D(N, 0)$, then by (5.5) we have

$$\int_{\sigma_N} f(z)e^{-izy}\,dz = 2\pi i \sum_{a \in U_N} \operatorname{Res}\left(\{f(z)e^{-izy}\}, a\right);$$

$$\int_{\delta_N} f(z)e^{-izy}\,dz = -2\pi i \sum_{a \in L_N} \operatorname{Res}\left(\{f(z)e^{-izy}\}, a\right).$$

The estimate

$$\left| \int_{\omega_N} f(z)e^{-izy}\,dz \right| = \left| iN \int_{[0, \pi]} f(Ne^{it}) \exp\left(-iyN[\cos t + i \sin t]\right) dt \right|$$

$$\leq \frac{M\pi}{N^{\eta-1}} \operatorname{lub}\left\{ \exp\left(yN \sin t\right) \mid t \in [0, \pi] \right\}$$

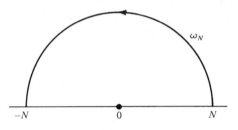

Figure 5.2

shows that the integral over ω_N goes to zero as $N \to \infty$, provided that $y \le 0$. Similarly, the integral over v_N goes to zero if $y \ge 0$. The equalities in (ii) follow immediately.

Notice that (i) is satisfied for all rational functions whose denominators have no zeros on the x-axis and for which the degree of the denominator exceeds the degree of the numerator by at least two.

Example A.   As a first example let

$$f(x) = \frac{1}{1 + x^2} \cdot$$

We have

$$\left| z^2 \frac{1}{1 + z^2} \right| < \frac{4}{3} \quad \text{if } |z| > 2,$$

so that (i) is satisfied with $R = 2$, $M = 4/3$, $\eta = 2$.   The poles of $[e^{-izy}/(1 + z^2)]$ are simple poles at $\pm i$.   To calculate the residues at $+i$ and $-i$, write

$$\frac{e^{-izy}}{1 + z^2} = \frac{1}{(z - i)} \cdot \frac{1}{(z + i)} \cdot e^{-izy}.$$

The residue at i is thus $e^y/2i$ and at $-i$ it is $-e^{-y}/2i$.   Hence by (ii) we have

$$\hat{f}(y) = \frac{1}{\sqrt{2\pi}} \int_{\mathbf{R}} \frac{e^{-ixy}}{1 + x^2}\, dx = \sqrt{\frac{\pi}{2}}\, e^{-|y|}.$$

In particular, we have

$$\int_{\mathbf{R}} \frac{1}{1 + x^2}\, dx = \pi.$$

Example B.   As a second example, let $f(x) = x^2/(1 + x^4)$. The singularities of f are the roots of $1 + x^4 = 0$, so that the set of poles of $\{f(z)e^{-izy}\}$ is

$\{e^{\pm i\pi/4}, e^{\pm i3\pi/4}\}$, and each of these poles is simple. To calculate the residues, let a be a pole and write

$$\frac{z^2}{1 + z^4} = \frac{a^2 + 2a(z - a) + (z - a)^2}{(1 + a^4) + 4a^3(z - a) + 6a^2(z - a)^2 + 4a(z - a)^3 + (z - a)^4}$$

$$= \frac{a^2}{4a^3} \cdot \frac{1}{(z - a)} + \frac{4}{3} + \cdots. \qquad (\textit{Note: } a^4 = -1.)$$

Thus we have

$$\text{Res}\,(\{f(x)e^{-ixy}\}, a) = e^{-iay}/4a.$$

The poles in the upper half-plane are $e^{i\pi/4}$ and $e^{i3\pi/4}$, so

$$\hat{f}(y) = \frac{\sqrt{2\pi i}}{4}\exp\left(\frac{y}{\sqrt{2}}\right)\left[\exp\left(-i\left(\frac{y}{\sqrt{2}} + \frac{\pi}{4}\right)\right) + \exp\left(i\left(\frac{y}{\sqrt{2}} - \frac{3\pi}{4}\right)\right)\right]$$

$$(y < 0).$$

Since f is symmetric, we have

$$\hat{f}(y) = \frac{1}{\sqrt{2\pi}}\int_{\mathbf{R}} f(x)e^{-ixy}\,dx = \frac{1}{\sqrt{2\pi}}\int_{\mathbf{R}} f(-x)e^{-ix(-y)}\,dx = \hat{f}(-y);$$

that is, $\hat{f}(y) = \hat{f}(-y)$, if $y > 0$. Notice that $\hat{f}$ is again an integrable function on $\mathbf{R}$. Evaluation at $y = 0$ yields the equality

$$\int_{\mathbf{R}} \frac{x^2}{1 + x^4}\,dx = 2\pi i\left[\frac{1}{4}e^{-i\pi/4} + \frac{1}{4}e^{-i3\pi/4}\right] = \frac{\pi}{\sqrt{2}}.$$

5.19 WEAKENED GROWTH CONDITION

Suppose that f is analytic on $\mathbf{C}$ except for a finite set S of singularities such that $S \cap \mathbf{R} = \phi$. If we weaken the growth condition (5.18.i) to
(i) there are constants M and R such that $|zf(z)| < M$ if $|z| \geq R$, then the double limit

(ii) $\quad \hat{f}(y) = \lim\limits_{N, L \to \infty} \dfrac{1}{\sqrt{2\pi}}\displaystyle\int_{[-L, N]} f(x)e^{-ixy}\,dx \qquad (y \neq 0)$

exists and (5.18.ii) holds if $y \neq 0$. [We take (ii) as the definition of $\hat{f}(y)$ under the hypothesis (i).]

In this case, the "residue method" is used to prove the *existence* of the limit as well as to prove (5.18.ii). The condition (5.19.i) does not imply that f is integrable on $\mathbf{R}$.

If $y < 0$, let

$$\delta = \overrightarrow{[N, N + iK]} \smile \overrightarrow{[N + iK, -L + iK]} \smile \overrightarrow{[-L + iK, -L]}$$

and let $\sigma = \overrightarrow{[-L, N]} \smile \delta$, for N, L, K greater than R (see Figure 5.3).
Suppose also that R is large enough so that $S \subset D(R, 0)$. By (5.5), we have

$$\int_\sigma f(z)e^{-izy}\, dz = 2\pi i \sum_{a \in U} \text{Res}\,(a),$$

where Res (a) is the residue of $f(z)e^{-izy}$ at a. The integral over δ is dominated
by

(iii) $$\frac{M(L + N)e^{yK}}{K} + \frac{M}{N} \int_0^K e^{ty}\, dt + \frac{M}{L} \int_0^K e^{ty}\, dt.$$

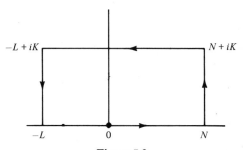

Figure 5.3

The first term comes from the integral over $\{\text{Img}\,(z) = K\}$ and the other two
from the integrals over the vertical sides. Letting $K \to \infty$ in (iii) gives an
estimate

(iv) $$\left| \int_{[-L, N]} f(x)e^{-ixy}\, dx - 2\pi i \sum_{a \in U} \text{Res}\,(a) \right| \le M \int_{[0, \infty[} e^{ty}\, dt \left[\frac{1}{N} + \frac{1}{L} \right].$$

Letting $(N, L) \to (\infty, \infty)$ in (iv) gives $\hat{f}(y) = \sqrt{2\pi} i \sum_{a \in U} \text{Res}\,(a)$.   A
similar calculation holds for a rectangle in the lower half-plane, if $y > 0$.

The equality (5.18.ii) does not extend (in general) to $y = 0$ in this case.

Example C. As an example, consider the function $f(x) = x/(x^2 + 1)$. The
condition (5.19.i) is satisfied for f. The isolated singularities in $\mathbf{C}$ are simple
poles at $\pm i$. Writing

$$\frac{z}{z^2 + 1} = \frac{1}{(z - i)} \cdot \frac{z}{(z + i)}\,;\qquad \frac{z}{z^2 + 1} = \frac{1}{(z - (-i))}\frac{z}{z - i}$$

we see that the residues of $z/(z^2 + 1)e^{-izy}$ at i and $-i$ are $\frac{1}{2}e^y$ and $\frac{1}{2}e^{-y}$, respectively. Hence, the equality (5.18.ii) for the function $\{x/(1 + x^2)\}$ is valid:

$$\hat{f}(y) = \left\{\frac{x}{1 + x^2}\right\}^{\wedge} (y) = \frac{-\operatorname{sgn}(y)}{2} e^{-|y|}, \quad y \in \mathbf{R}\backslash\{0\},$$

Notice the discontinuity at 0 of the function $\{x/(1 + x^2)\}^{\wedge}$. Notice also that the function $x/(1 + x^2)e^{-ixy}$ is not absolutely integrable on $\mathbf{R}$; that is, $\int_{\mathbf{R}} |x/(1 + x^2)|\, dx = \infty$. The convergence in (ii) for this function f is dependent on the presence of the exponential. The limit

$$\lim_{L \to \infty, N \to \infty} \int_{-L}^{N} [x/(1 + x^2)]e^{-ixy}\, dx$$

is an improper integral, the existence of which we have established by the residue theorem in the case $y \neq 0$.

5.20 POLES ON THE *x*-AXIS

Suppose that f is analytic except for a finite number of singularities on $\mathbf{C}$, has a pole of order 1 at 0, has no other singularities on the *x*-axis, and that (5.19.i) holds. We have

(i) $\displaystyle\lim_{\substack{L, N \to \infty \\ \delta \to 0}} \left[\int_{[-L, -\delta]} f(x)e^{-ixy}\, dx + \int_{[\delta, N]} f(x)e^{-ixy}\, dx \right]$

$$= -\operatorname{sgn}(y)\pi i \operatorname{Res}(f, 0) + \begin{cases} 2\pi i \displaystyle\sum_{s \in U} \operatorname{Res}(\{f(z)e^{-izy}\}, s) & \text{if } y < 0 \\ -2\pi i \displaystyle\sum_{s \in L} \operatorname{Res}(\{f(z)e^{-izy}\}, s) & \text{if } y > 0. \end{cases}$$

To prove (i), let l_N, l_K, l_L, and σ be as in Figure 5.4. Letting $K \to \infty$ gives the estimate

(ii) $\displaystyle\left| \int_{\overrightarrow{[-L, -\delta]}} + \int_{\sigma} + \int_{\overrightarrow{[\delta, N]}} f(z)e^{-izy}\, dz - 2\pi i \sum_{\substack{\operatorname{Img}(a) \geq 0 \\ a \in S}} \operatorname{Res}(a) \right|$

$$\leq M\left(\frac{1}{N} + \frac{1}{L}\right)\int_{[0, \infty[} e^{ty}\, dt$$

for $y < 0$, as in (5.19). To find the integral over σ, write

$$f(z) = \frac{a_{-1}}{z} + a_0 + a_1 z + \cdots = \frac{a_{-1}}{z} + g(z),$$

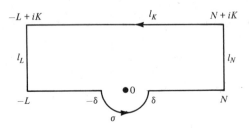

Figure 5.4

g is analytic near zero. Then $f(z)e^{-izy} = a_{-1}/z + h(z)$, h is analytic near zero. By direct calculation, we have

$$\int_\sigma f(z)e^{-izy}\, dz = \pi i a_{-1} + \int_\sigma h(z)\, dz.$$

The last integral goes to zero as $\delta \to 0$, independent of N and L, because h is bounded. Hence (i) is established for $y < 0$. Applying the same technique for $y > 0$ and a rectangle in the lower half-plane, we obtain all of (i).

As a special case, we have

$$\lim_{\substack{L,N \to \infty \\ \delta \to 0}} \int_J \frac{e^{-ixy}}{x}\, dx = -\pi i \operatorname{sgn}(y),$$

$$\lim_{\substack{L,N \to \infty \\ \delta \to 0}} \int_J \frac{\sin(xy)}{x}\, dx = \pi \operatorname{sgn}(y),$$

where $J = [-L, -\delta] \cup [\delta, N]$.

These equalities are of considerable importance in pure and applied Fourier analysis.

A function f with a simple pole at 0 (as above) cannot be integrable because $\int_{[-\delta, \delta]} (1/|x|)\, dx = \infty$ for any $\delta > 0$. Nevertheless, we found that $\lim_{\delta \to 0} \int_{|x| > \delta} f(x)\, dx$ does exist. Another method of proof of the existence of the limit, perhaps offering a different insight, is obtained as follows. For simplicity, suppose that $|f|$ is integrable at ∞. We have

$$\int_{|x| > \delta} f(x)\, dx = \int_{]-\infty, -\delta]} f(x)\, dx + \int_{[\delta, \infty[} f(x)\, dx$$

$$= \int_{[\delta, \infty[} f(-x)\, dx + \int_{[\delta, \infty[} f(x)\, dx$$

$$= \int_{[\delta, \infty[} [f(-x) + f(x)]\, dx.$$

The function f is approximately odd near zero, because it has a simple pole there and near 0 the term $a_{-1}z^{-1}$ dominates. Hence near zero $f(-x)$ tends to cancel $f(x)$; as δ gets small, this cancellation gives us a limit. To be explicit, we have

$$f(-x) + f(x) = \left[\frac{a_{-1}}{-x} + a_0 - a_1 x + \cdots\right] + \left[\frac{a_{-1}}{x} + a_0 + a_1 x + \cdots\right]$$

$$= 2a_0 + 2\sum_{k=1}^{\infty} a_{2k}x^{2k}.$$

The last displayed function is integrable at 0.

The method of this section applies if the pole is elsewhere on the x-axis, and also for a finite number of poles on the x-axis.

5.21 A SPECIAL INTEGRAL

Calculations involving the formula

(i) $$\int_{[0, \infty[} e^{-x^2}\, dx = \sqrt{\pi}/2$$

occur frequently. For a proof of (i), let the integral be α. Then we have

$$\alpha^2 = \lim_{R\to\infty} \left[\left[\int_{[0, R]} e^{-t^2}\, dt\right]\left[\int_{[0, R]} e^{-x^2}\, dx\right]\right]$$

$$= \lim_{R\to\infty} \int_{[0, R]}\int_{[0, R]} e^{-(t^2+x^2)}\, dx\, dt.$$

The *iterated integral* is equal to the *double integral*

$$\int_{[0, R]\times[0, R]} e^{-(t^2+x^2)}\, d(x, t).$$

Transforming to polar coordinates by $T(r, \theta) = (r\cos\theta, r\sin\theta)$, the integral dominates

$$\int_{[0, \pi/2]\times[0, R]} e^{-r^2}\, r\, d(r, \theta)$$

and is dominated by the same integral taken over $[0, \pi/2] \times [0, \sqrt{2}R]$. The limit as R goes to ∞ of these two integrals is clearly the same. To find this limit, write the double integral as

$$\int_{[0, R]}\int_{[0, \pi/2]} e^{-r^2}r\, d\theta\, dr = \frac{\pi}{2}\int_{[0, R]} e^{-r^2}r\, dr.$$

The limit as $R \to \infty$ of the integral on the right is $\pi/4$.

5.22 THE FOURIER TRANSFORM $\{e^{-x^2}\}^\wedge$

The function $\{e^{-x^2}e^{-ixy}\}$ (y fixed in $\mathbf{R}$) on $\mathbf{R}$ extends to an analytic function on $\mathbf{C}$. Cauchy's theorem applies to it for the closed curve in Figure 5.5. The integrals over γ_2 and γ_3 are constants multiplied by $e^{-\rho^2}$, and so go to zero as ρ goes to ∞. Hence by Cauchy's theorem (4.2) we have

(i) $\quad\displaystyle\lim_{\rho\to\infty}\int_{[-\rho,\,\rho]} e^{-z^2}e^{-izy}\,dz = -\lim_{\rho\to\infty}\int_{\gamma_1} e^{-z^2}e^{-izy}\,dz.$

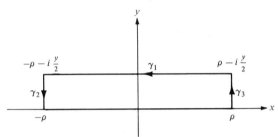

Figure 5.5

Parametrizing γ_1 on $[-\rho,\rho]$ we have

(ii) $\quad\displaystyle\int_{\gamma_1} e^{-z^2}e^{-izy}\,dz$

$= -\displaystyle\int_{[-\rho,\,\rho]} \exp\left[-\left(t-i\frac{1}{2}y\right)^2\right]\exp\left[-i\left(t-i\frac{1}{2}y\right)y\right]dt = e^{-y^2/4}\int_{[-\rho,\,\rho]} e^{-t^2}\,dt.$

By (i), (ii), and (5.21.i), the equality

$$\{\exp(-x^2)\}^\wedge(y) = \frac{1}{\sqrt{2}}\exp\left(-\frac{1}{4}y^2\right)$$

follows. Using a change of variables, we obtain the important equality

$$\left\{\exp\left(-\frac{1}{2}x^2\right)\right\}^\wedge(y) = \exp\left(-\frac{1}{2}y^2\right);$$

that is, the function $f(x) = \exp(-\frac{1}{2}x^2)$ is its own Fourier transform.

EXERCISES

TYPE I

1. Find the Fourier transforms of the following functions. Explain carefully why the evaluation works. Separate your equality into real and

imaginary parts so that you obtain equalities for sine and cosine transforms.

(a) $\dfrac{x^2}{x^4 + 6x^2 + 13}$

(c) $\dfrac{1}{x^2 + x + 1}$

(b) $\dfrac{1}{x^4 + 1}$

(d) $\dfrac{1}{x^2 + 1}$.

2. Find

$$\lim_{\substack{R, N \to \infty \\ \delta \to 0}} \int_{[-R, -\delta] \cup [\delta, N]} f(x) e^{-ixy}\, dx$$

for the following functions. State why the limit exists.

(a) $f(x) = \dfrac{2x^2 + 1}{x^3 + x}$

(b) $f(x) = \dfrac{1}{x(x^2 + 1)}$.

3. Evaluate

$$\frac{1}{2\pi} \int_{[-\pi, \pi]} \frac{1 - r^2}{1 - 2r \cos t + r^2}\, dt$$

for $0 \le r < 1$ by the method of (5.16). (The integral is the Poisson kernel. It is discussed in Chapter 6.)

*5.23 CLASSIFICATION OF ISOLATED SINGULARITIES BY GROWTH CONDITIONS

See (5.4) for the relevant definitions.

Theorem I. Suppose that $a \in \mathbf{C}$ and that $r > 0$. Suppose that f is analytic on an open set containing $E = \overline{D(r, a)}\backslash\{a\}$. The following assertions describe the nature of the singularity a.

(i) a is a removable singularity of f if and only if f is bounded on E.

(ii) a is a pole of order $m > 0$ of f if and only if $\lim_{z \to a} |z - a|^{m-1}|f(z)| = \infty$ and $(z - a)^m f(z)$ is bounded on E.

(iii) a is an essential singularity of f if and only if $\overline{f(D(\varepsilon, a)\backslash\{a\})} = \mathbf{C}$ for every $\varepsilon > 0$.

PROOF. If a is removable, then f is clearly bounded on $D(r, a)$. In the Laurent expansion of f about a, we have

$$a_{-n} = \frac{1}{2\pi i} \int_{\overrightarrow{C(\rho, a)}} f(\zeta)(\zeta - a)^{n-1}\, d\zeta \qquad (0 < \rho < r).$$

If f is bounded, by letting $\rho \to 0$ we see that $a_{-n} = 0$ if $n > 0$. Hence we have $f(z) = \sum_{n=0}^{\infty} a_n (z - a)^n$. Letting $f(a) = a_0$, we obtain an extension of f to a. Thus (i) is established.

If $(z - a)^m f(z)$ is bounded, then (i) applies to it and it has a power series expansion

$$(z - a)^m f(z) = a_{-m} + a_{-m+1}(z - a) + \cdots.$$

If, in addition, $\lim_{z \to a} |(z - a)^{m-1} f(z)| = \infty$, then $a_{-m} = 0$ is not possible. Hence f has a pole of order m at a. If f has a pole of order m at a, then we have

$$|z - a|^{m-1} |f(z)| \geq \left| \frac{a_{-m}}{z - a} \right| - |a_{-m+1} + (z - a)[a_{-m+2} + \cdots]|$$

and

$$|z - a|^m |f(z)| \leq |a_{-m}| + |a_{-m+1}(z - a) + a_{-m+2}(z - a)^2 + \cdots|.$$

In the first inequality the term on the right goes to ∞ as $z \to a$. In the second inequality, the term on the right goes to $|a_{-m}|$ as z goes to a. This proves (ii).

If a is a pole, then by (ii) for any $R > 0$ there is an $\varepsilon > 0$ such that

$$f(D(\varepsilon, a) \backslash a) \subset \{z \mid |z| > R\}.$$

Hence if the equality in (iii) holds, a is not a pole. Clearly f is not bounded if the equality holds, so a cannot be removable. Thus a is an essential singularity if the equality holds. If the equality fails, then there is a triple (ε, δ, w) $[\varepsilon > 0, \delta > 0, w \in \mathbf{C}]$ such that

$$|f(z) - w| > \delta \qquad \text{if } |z - a| < \varepsilon.$$

Therefore the function $(f - w)^{-1}$ has a removable singularity at a [by (i)], and so a power series expansion

$$[f(z) - w]^{-1} = a_m (z - a)^m + a_{m+1}(z - a)^{m+1} + \cdots (m \geq 0, a_m \neq 0)$$

is valid in $D(\varepsilon, a)$. It follows that

$$f(z) - w = \frac{1}{(z - a)^m} \left[\frac{1}{a_m + a_{m+1}(z - a) + \cdots} \right]$$

$$= \frac{1}{(z - a)^m} \left[\frac{1}{a_m} + g(z) \right],$$

where g is analytic on $D(\varepsilon, a)$. Hence a is a pole of f of order m ($m = 0$ is possible, in which case f has a removable singularity).

Theorem II. Suppose that f is analytic on an open set containing $\{|z| > R\}$ for some $R > 0$. Thus ∞ is an isolated singularity of f. We have

(i) ∞ is removable if and only if f is bounded;

(ii) ∞ is a pole of order m if and only if $f(z)/z^m$ is bounded and $\lim_{|z| \to \infty} |f(z)/z^{m-1}| = \infty$;

(iii) ∞ is an essential singularity of f if and only if $\overline{f\{z \mid |z| > M\}} = \mathbf{C}$, every $M \geq R$.

A proof based on Theorem I and the definitions in (5.4) is left to the reader.

*5.24 ENTIRE FUNCTIONS

A function f analytic on $\mathbf{C}$ is called an *entire function*. By (4.8), f has a power series expansion about 0, say $f(z) = \sum_{n=0}^{\infty} a_n z^n$, valid for all $z \in \mathbf{C}$. From Theorem II of (5.23) we see that f has a pole of order $m > 0$ at ∞ if and only if $a_n = 0$ for $n > m$ and $a_m \neq 0$; that is, if and only if f is a polynomial of order m. If f is bounded, then ∞ is not an essential singularity of f and not a pole; hence $a_n = 0$ for $n > 0$ and f is constant. (The fact that a bounded entire function is constant is *Liouville's theorem*.) Again by Theorem II of (5.23) we see that if f is not a polynomial then $f(z \mid |z| > R)$ is dense in $\mathbf{C}$ for every $R > 0$. Entire functions which are not polynomials are called *transcendental*. *Picard's theorem* states that an entire transcendental function assumes as a value *every* complex number, except perhaps one value. Picard's theorem is very hard to prove. Elementary entire transcendental functions are exp, sin, cos, sinh, and cosh.

TYPE I

1. Suppose that f is analytic on $A_0^1(0) = \{z \mid 0 < |z| < 1\}$ and that $f(1/n) = e^{1/n}$ for all positive integers n. Prove that $f(z) = e^z$.
2. Let P and Q be polynomials. Describe the behavior of $P(z)/Q(z)$ at ∞.
3. Show directly that for any $w \in \mathbf{C}\setminus\{0\}$ and any $R > 0$, there is a $z \in \mathbf{C}$ with $|z| > R$ such that $\exp(z) = w$. Compare (5.23.II). Does this property hold for exp replaced by a polynomial P?

CHAPTER 6

Harmonic Functions

6.1 DEFINITION

Let U be an open set in **C**. If $u \in \mathbb{C}^2(U)$ (u has continuous first and second partials at all points of U) and

$$\frac{\partial^2 u}{\partial x^2} + \frac{\partial^2 u}{\partial y^2} = 0$$

then u is said to be *harmonic* on U. This famous differential equation is called Laplace's equation.

We have already seen one important physical example of harmonic functions. In the study of fluid flow (see exercises on p. 103), we encountered harmonic potential functions for velocity. At this point the reader may find it instructive to return to these exercises. In Section (6.8) we review the discussion of fluid flow problems and give two other physical examples.

Notice that a complex-valued function $u + iv$ is harmonic if and only if the real and imaginary parts u and v are harmonic.

6.2 EXAMPLES

We mentioned harmonic functions briefly in (1.22), pointing out that it is an immediate result of the Cauchy-Riemann equations that the real and imaginary parts of an analytic function are harmonic. The continuity condition on u and v in (1.22) can of course be dropped, in light of (4.5). We list below some examples of harmonic functions by simply writing the real or imaginary parts of various familiar analytic functions.

(a) The function Arg (z) is harmonic on $\{z \mid -\pi < \text{Arg } (z) \le \pi, z \ne 0\}$.

(b) Fix $\psi \, \varepsilon \,]-\pi, \pi[$. The function Arg $(ze^{i\psi})$ is harmonic on $\{z \mid \psi + \text{Arg } z \ne \pm \pi, z \ne 0\}$.

(c) The function $\log |z|$ is harmonic on $\mathbf{C}\backslash\{0\}$.

(d) The functions $e^x \cos (y)$ and $e^x \sin (y)$ are the real and imaginary parts of e^{x+iy}, and so are harmonic on $\mathbf{C}$.

(e) The functions $x^2 - y^2$ and $2xy$ are the real and imaginary parts of z^2, and so are harmonic on $\mathbf{C}$.

(f) The function $\sin (x)(e^{-y} + e^y)/2 = \sin (x) \cosh (y)$ is harmonic on $\mathbf{C}$.

(g) Since $\sum_{n=1}^{\infty} n^{-x} [\cos (y \operatorname{Log} n)]$ and $\sum_{n=1}^{\infty} n^{-x}[\sin (y \operatorname{Log} n)]$ are the real and imaginary parts, respectively, of the Riemann-zeta function $\zeta(z)$, they are harmonic on $\{(x, y) \mid x > 1\}$. (See Exercise 4, p. 77.)

The following definition and theorem show that *every* real harmonic function is locally the real part of an analytic function.

6.3 DEFINITION

Let u be a real-valued harmonic function on an open set U. If v is a real-valued function on U such that $u + iv$ is analytic, then v is called a *harmonic conjugate* of u on U.

6.4 THEOREM

Let U be simply connected, u harmonic on U, and fix $z_0 \in U$. Define v on U by

(i) $$v(z) = \operatorname{Im}\left[\int_{\gamma}\left[\frac{\partial u}{\partial x}(\zeta) - i\frac{\partial u}{\partial y}(\zeta)\right] d\zeta\right],$$

where γ is a path in U from z_0 to z. The integral is independent of path and v is a harmonic conjugate of u on U. Every other harmonic conjugate of u has the form $v + k$ for a constant k.

Compare (i) with Exercises 7 and 8 on p. 102.

PROOF. The Cauchy-Riemann equations are satisfied by the function $\partial u/\partial x - i(\partial u/\partial y)$, so that the integral in (i) is independent of the path by (4.28). The integral defines an analytic function $f = \phi + iv$ for which it is straightforward to verify that

$$f'(z) = \frac{\partial u}{\partial x}(z) - i\frac{\partial u}{\partial y}(z). \tag{1}$$

Computing the partial derivatives of $f = \phi + iv$ and using (1), we obtain

$$\frac{\partial v}{\partial x} = -\frac{\partial u}{\partial y} \quad \text{and} \quad \frac{\partial v}{\partial y} = \frac{\partial u}{\partial x}; \tag{2}$$

that is, $u + iv$ satisfies the Cauchy-Riemann equations. Hence it is analytic on U.

If $u + iw$ is also analytic on U, then $i[v - w]$ is analytic on U. Hence $v - w$ is real and analytic on the connected set U. By (4.20) $v - w$ is constant on U.

Remark. If we assume that U is convex, then (4.2) can be used above instead of (4.28). This makes a weaker theorem with a simpler proof, but it is still very useful.

6.5 EXAMPLE

We use (6.4.i) to construct a harmonic conjugate v of the harmonic function $u(x, y) = x^3 - 3xy^2$ on $\mathbb{C}$. At the point $x + iy$, we have

$$\frac{\partial u}{\partial x} = 3x^2 - 3y^2; \frac{\partial u}{\partial y} = -6xy.$$

Selecting $z_0 = 0$ in (6.4.i) and letting γ be a line segment, we have

(i) $v(z) = \text{Im} \displaystyle\int_{\overrightarrow{[0, z]}} [3x^2 - 3y^2 + i6xy] \, d\zeta \qquad (\zeta = x + iy).$

If we notice that the integrand is $3\zeta^2$, we find that

(ii) $v(z) = \text{Im} \displaystyle\int_{\overrightarrow{[0, z]}} 3\zeta^2 \, d\zeta = \text{Im} \, z^3 = 3x^2y - y^3.$

Had we failed to notice that the integrand in (i) was $3\zeta^2$, we could have let $\gamma(t) = tz = tx + ity$ on $[0, 1]$. We would then find that

(iii) $v(z) = \text{Im} \displaystyle\int_{[0, 1]}^{t} ([3(tx)^2 - 3(ty)^2] + i6[txty])(x + iy) \, dt.$

Calculation of (iii) yields, of course, the same result as (ii).

6.6 THEOREM (LAPLACE'S EQUATION IN POLAR COORDINATES)

Suppose that U is an open set, $0 \notin U$, and $u \in \mathbb{C}^2(U)$. Then u is harmonic on U if and only if

(i) $r^2 \dfrac{\partial^2 u}{\partial r^2} + r \dfrac{\partial u}{\partial r} + \dfrac{\partial^2 u}{\partial \theta^2} = 0$

on U.

PROOF.   Suppose first that u is harmonic. For a point $z \in U$ and an $\varepsilon > 0$ such that $D(\varepsilon, z) \subset U$, there is a function v such that $u + iv$ is analytic in $D(\varepsilon, z)$. If we use the Cauchy-Riemann equations in the equalities

(ii)
$$\frac{\partial u}{\partial \theta} = - \frac{\partial u}{\partial x} r \sin (\theta) + \frac{\partial u}{\partial y} r \cos (\theta)$$

$$r \frac{\partial u}{\partial r} = \frac{\partial u}{\partial x} r \cos (\theta) + \frac{\partial u}{\partial y} r \sin (\theta)$$

we obtain

$$r \frac{\partial u}{\partial r} = \frac{\partial v}{\partial \theta} \quad \text{and} \quad r \frac{\partial v}{\partial r} = - \frac{\partial u}{\partial \theta} .$$

From these last equalities, (i) follows.

If (i) holds, simply use the equalities (ii) in (i) to prove that $\partial^2 u/\partial x^2 + \partial^2 u/\partial y^2 = 0$. The reader should carry out the calculations.

6.7 ALGEBRAIC PROPERTIES

The class of function $\mathfrak{H}(U)$ consisting of all complex-valued functions harmonic on the open set U is a complex vector space; that is, $f + g$ is harmonic if f and g are harmonic and αf is harmonic if $\alpha \in \mathbf{C}$ and f is harmonic. Similarly, real-valued harmonic functions form a real vector space. Products of harmonic functions need not be harmonic; for example, x and $x^2 - y^2$ are harmonic on $\mathbf{C}$, but $x(x^2 - y^2) = x^3 - xy^2$ is not.

*6.8 PHYSICAL INTEPRETATIONS

Harmonic functions occur naturally in many areas of the physical sciences. We list a few examples.

(i) *Fluid flow.*   We discussed fluid flow problems in some detail in Exercises 9–20 on p. 102. We summarize here the results of those exercises. Suppose that a layer of incompressible fluid is flowing in a plane. Let $v = v_x + iv_y$ be the velocity of the fluid, written as a complex number. Suppose that the flow is such that v is independent of depth, and suppose also that $\partial v_y/\partial x = \partial v_x/\partial y$ (the flow is irrotational).   Then there is an analytic function $V = \Phi + i\Psi$ on the region of the flow which satisfies

$$\frac{\partial \Phi}{\partial x} = v_x, \quad \frac{\partial \Phi}{\partial y} = v_y.$$

The function Φ is of course harmonic; it is the *velocity potential* for v. The function V is called the complex potential function for v. The function $\bar{v}$ is analytic.

(ii) *Heat transfer.* Suppose that a solid body occupies a set U of three-space. At each point $p = (x, y, z)$ in U and at each time t the temperature is $T(p, t)$. If the temperature is not constant, then heat will flow from the warmer parts of U to the colder parts. The direction of this flow will be $-\text{grad } T(p, t)$, the direction of the maximal temperature decrease. (Heat presumably flows in the direction opposite to that of increasing temperature.) If $\mathbf{u}$ is a unit vector, then the *flux* $\Phi_{\mathbf{u}}(p, t)$ in the direction $\mathbf{u}$ is defined as the number of calories per second per unit area crossing a plane with normal $\mathbf{u}$ at point p and at time t. It has been observed that the flux is a multiple of the rate of change of T in the direction $\mathbf{u}$; that is, the flux is a multiple of the directional derivative $D_{\mathbf{u}}T$. That multiple is called the *thermal conductivity of the body*. We assume that it is constant in both p and t. Hence we have

$$\Phi_{\mathbf{u}} = -kD_{\mathbf{u}}T = -k \,(\text{grad } T) \cdot \mathbf{u}$$

for a constant k. Since heat energy (measured in calories) will move from warm regions to cold regions, k is positive.

Consider a point $P_0 \in U$ and center a cube of side length h with faces parallel to the coordinate planes at P_0. Let us number the coordinate axes with indices 1, 2, 3. Let $\bar{\mathbf{u}}_1$, $\bar{\mathbf{u}}_2$, and $\bar{\mathbf{u}}_3$ be unit vectors along them, and let $P_0 = (p_1, p_2, p_3)$. We will calculate the net change of calories in the cube in a time interval Δt in two different ways. Consider first the two faces of the cube perpendicular to the first axis. Since $\Phi_{\mathbf{u}_1}(x_1, x_2, x_3, t)$ is continuous as a function of (x_1, x_2, x_3), by the mean value theorem for integrals there is a pair $(\hat{x}_2, \hat{x}_3)$ such that

$$\left[-\Phi_{\mathbf{u}_1}\left(p_1 + \frac{h}{2}, \hat{x}_2, \hat{x}_3, t \right) + \Phi_{\mathbf{u}_1}\left(p_1 - \frac{h}{2}, \hat{x}_2, \hat{x}_3, t \right) \right] h^2$$

$$= \iint \left[-\Phi_{\mathbf{u}_1}\left(p_1 + \frac{h}{2}, x_2, x_3, t \right) + \Phi_{\mathbf{u}_1}\left(p_1 - \frac{h}{2}, x_2, x_3, t \right) \right] dx_2 \, dx_3 \quad (1)$$

where integration is over the square

$$\left\{ (x_2, x_3) \,\middle|\, p_2 - \frac{h}{2} \leq x_2 \leq p_2 + \frac{h}{2}, \, p_3 - \frac{h}{2} \leq x_3 \leq p_3 + \frac{h}{2} \right\}.$$

Suppressing the variables $\hat{x}_2, \hat{x}_3$ and using the definition of $\Phi_{\mathbf{u}}$ we can write the left side of (1) in the form

$$k \left[\frac{\partial T}{\partial x_1}\left(p_1 + \frac{h}{2}, t \right) - \frac{\partial T}{\partial x_1}\left(p_1 - \frac{h}{2}, t \right) \right] h^2.$$

Making a similar calculation for the other two pairs of faces and summing, we determine that the change per second in calories in the cube at time t is equal to

$$kh^2 \sum_{j=1}^{3} \left[\frac{\partial T}{\partial x_j} \left(p_j + \frac{h}{2}, t \right) - \frac{\partial T}{\partial x_j} \left(p_j - \frac{h}{2}, t \right) \right].$$

The net change in calories in a time interval from t to $t + \Delta t$ is found by integrating the last expression from t to $t + \Delta t$. Applying the mean value theorem for integrals again, we find that the net change in calories in the time interval from t to $t + \Delta t$ is equal to

$$kh^2 \, \Delta t \sum_{j=1}^{3} \left(\frac{\partial T}{\partial x_j} \left(p_j + \frac{h}{2}, t' \right) - \frac{\partial T}{\partial x_j} \left(p_j - \frac{h}{2}, t' \right) \right) \qquad (2)$$

for some t' satisfying $t \leq t' \leq t + \Delta t$.

Let c denote the heat capacity of the body; that is, the calories needed to raise a unit volume of the body a unit temperature. The total number of calories in the cube at time t is given by

$$\iiint\limits_{\text{cube}} cT(p, t) \, dp.$$

Hence the net change of calories in the cube in the time interval from t to $t + \Delta t$ is given by

$$\iiint\limits_{\text{cube}} c[T(p, t + \Delta t) - T(p, t)] \, dp.$$

By the mean value theorem for integrals there is a point p' in the cube such that

$$c[T(p', t + \Delta t) - T(p', t)]h^3 = \iiint\limits_{\text{cube}} c[T(p, t + \Delta t) - T(p, t)] \, dp.$$

$$(3)$$

We equate the two expressions for the change in calories to obtain

$$ch^3[T(p', t + \Delta t) - T(p', t)]$$

$$= kh^2 \, \Delta t \sum_{j=1}^{3} \left[\frac{\partial T}{\partial x_j} \left(p_j + \frac{h}{2}, t' \right) - \frac{\partial T}{\partial x_j} \left(p_j - \frac{h}{2}, t' \right) \right].$$

If one divides both sides of the above equality by $h^3 \, \Delta t$ and computes the limit as $h \to 0$ and $t \to 0$, then one obtains, with some effort, the equality

$$c \frac{\partial T}{\partial t} = k \left[\sum_{j=1}^{3} \frac{\partial^2 T}{\partial x_j \, \partial x_j} \right]. \qquad (4)$$

This equality holds at each point of U.

In summary, the temperature function $T(x_1, x_2, x_3, t)$ of a body satisfies the partial differential equation (4) provided that the body has constant specific heat and thermoconductivity. If T is not dependent on time, then the temperature is said to have reached a steady state and (3) reduces to Laplace's equation in three variables. Finally, if there is no change in temperature in the x_3-direction, then we obtain

$$\frac{\partial^2 T}{\partial x_1^2} + \frac{\partial^2 T}{\partial x_2^2} = 0.$$

In this case T is an harmonic function of two variables. This latter case is the two-dimensional steady-state heat transfer problem. Elementary complex analysis is useful in solving this problem as we will see in Chapter 7.

(iii) *An electric field potential function.* Consider a wire of infinite length perpendicular to the complex plane at 0. Suppose that a uniform positive charge density of $1/2\pi$ units per unit length is distributed over the wire. The vector force **F** of the wire at a point in space is the force exerted by the charged wire on a negative unit charge placed at the point. We calculate the force **F**(z) at a point $z \in \mathbf{C}$. By symmetry the vertical components of **F**(z) will be zero so that **F** is directed from z to 0. Hence we can write

$$\mathbf{F}(z) = |\mathbf{F}(z)| \frac{-z}{|z|}.$$

It is an experimental fact that the magnitude of the force exerted by one charged particle on another is directly proportional to the product of the charges and inversely proportional to the square of the distance between them. In the case of the wire, consider a segment of the wire of length Δu. The force exerted by this charged segment on a unit negative charge at z is approximately equal to $\Delta u / [2\pi(x^2 + y^2 + s^2)]$ where s is the midpoint of the segment. Summing over all such segments and taking the limit as Δu goes to zero, we obtain the equality

$$|\mathbf{F}(z)| = \frac{1}{2\pi} \int_{]-\infty, \, \infty[} \frac{1}{x^2 + y^2 + u^2} \, du$$

where $z = x + iy$.

Referring to Figure 6.1, we see that the relations

$$\rho = r \cos(\theta), \quad u = \rho \tan(\theta), \quad \frac{1}{r^2} = \frac{\cos^2(\theta)}{\rho^2}$$

hold. Using these to change variables in the integral, we obtain

$$|\mathbf{F}(z)| = \frac{1}{2\pi} \int_{[-\pi, \pi]} \frac{\cos^2(\theta)\rho\sec^2(\theta)}{\rho^2} \, d\theta = \frac{1}{\rho}.$$

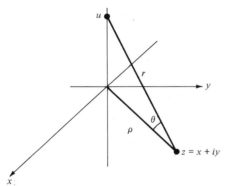

Figure 6.1

Hence we have

$$\mathbf{F}(z) = -\frac{z}{|z|^2}.$$

(The force acts as though the wire were concentrated at 0 and obeyed a $1/r$ law instead of a $1/r^2$ law.) In complex notation, we have

$$\text{grad}\left(\log\frac{1}{\rho}\right) = \left(\frac{\partial}{\partial x}\log\left(\frac{1}{\rho}\right)\right) + i\left(\frac{\partial}{\partial y}\log\left(\frac{1}{\rho}\right)\right) = -\frac{x}{\rho^2} - i\frac{y}{\rho^2}.$$

Thus we have

$$\mathbf{F}(z) = \text{grad}\left(\log\frac{1}{|z|}\right).$$

Whenever a vector function $\mathbf{E}$ is the gradient of a real-valued function ϕ, ϕ is said to be a *potential function for* $\mathbf{E}$. In this example, $\log(1/|z|)$ is the potential function for the electric field $\mathbf{F}$. As we have seen before, $\log(1/|z|)$ $= -\log|z|$ is harmonic on $\mathbf{C}\backslash\{0\}$.

Even mathematicians have no access to infinitely long wires. However, the above ideal problem is a close approximation to reality if the distance from the end points of the wire is great in comparison with the distance from the point to the wire.

EXERCISES

TYPE I

1. Find the real and imaginary parts of the following analytic functions, and thereby find pairs of conjugate harmonic functions. (Notice that Laplace's equation would be cumbersome to verify directly in some of these cases.)

   (a) $\tan(z)$ on $S = \{z \mid \cos(z) \neq 0\}$ (d) $\sinh(z)$ on $\mathbf{C}$
   (b) z^4 on $\mathbf{C}$ (e) $1/z^2$ $(z \neq 0)$
   (c) $\cosh(z)$ on $\mathbf{C}$ (f) $1/z^4$ $(z \neq 0)$.

2. Use (6.4.i) to find a harmonic conjugate on $\mathbf{C}$ for the function with value $x^2 - y^2$.

3. Show that $x/(x^2 + y^2)$ is harmonic on $\mathbf{C}\backslash\{0\}$.

4. Determine which of the following functions are harmonic and find a harmonic conjugate for the harmonic ones.

   (a) $u(x, y) = x$ on $\mathbf{C}$
   (b) $u(x, y) = y$ on $\mathbf{C}$
   (c) $u(x, y) = x^2$ on $\mathbf{C}$
   (d) $u(x, y) = x^m$ on $\mathbf{C}$ (m a positive integer)
   (e) $u(x, y) = \cos(x) - \cos(y)$
   (f) $u(x, y) = e^{xy}$
   (g) $u(x, y) = x + y$
   (h) $u(x, y) = \cos(xy)$.

5. The analogue of Laplace's equation for a twice differentiable function f of one real variable defined on $[a, b]$ is simply

   $$\frac{d^2 f}{dx^2}(x) = 0, \qquad x \in \,]a, b[\qquad\qquad (*)$$

   Hence the analogue of harmonicity for the function f is that $(*)$ hold. Find all such "one-dimensional" harmonic functions.

6. For what nonnegative integral values of n and m is $x^n y^m$ harmonic on $\mathbf{C}$? Find harmonic conjugates in each harmonic case.

7. For what values of the positive integers n and m is $1/(x^n y^m)$ harmonic on some open subset of $\mathbf{C}$?

8. Show by example that $|f(z)|$ need not be harmonic for an analytic function f.

9. (a) If f is analytic on an open set U and ϕ is harmonic on $f(U)$, show that $\phi \circ f$ is harmonic on U.
   [*Hint:* Consider first the case in which ϕ is real and use a local harmonic conjugate.]

(b) If f is analytic and nonzero on an open set U, show that $\log |f(z)|$ is harmonic on U.

10. Show that $r^n \cos(n\theta)$ and $r^n \sin(n\theta)$ ($z = re^{i\theta}$) are harmonic on **C** for all positive integers n.

TYPE II

1. Let $P(x, y)$ be a polynomial in x and y with real coefficients. Is it true that P is harmonic if and only if it is the real part of a *complex* polynomial having complex coefficients? Prove your answer. Give an example of a real harmonic polynomial which is not the real part of any complex polynomial with real coefficients.

2. Suppose that u is a real-valued harmonic function on a simply connected open subset U of **C**. Suppose also that u is not identically zero. Show that there is an open subset of U on which $1/u$ is harmonic if and only if u is constant on U.

3. Let ϕ be a harmonic function on **C**. If ϕ is bounded, show that ϕ is constant.

[*Hint:* Exponentiate and use Liouville's theorem.]

6.9 THE POISSON KERNEL

For every real number r such that $0 \le r < 1$, the two-sided infinite series

$$\text{(i)} \quad P(r, t) = \sum_{n=-\infty}^{\infty} r^{|n|} e^{int}$$

converges uniformly for $t \in \mathbf{R}$. That is, the sequence of functions S_N given on **R** by

$$S_N(t) = \sum_{n=-N}^{N} r^{|n|} e^{int}$$

converges uniformly on **R**. To see this, write

$$\text{(ii)} \quad S_N(t) = 1 + \sum_{n=1}^{N} r^n (e^{int} + e^{-int}).$$

Evidently, $|r^n(e^{int} + e^{-int})| \le 2r^n$. Hence the Weierstrass M-test (1.5) gives uniform convergence of S_N (compare with the geometric series $2 \sum_{n=1}^{\infty} r^n$).

The equality (i) defines the *Poisson kernel* $P(r, t)$. It is defined for $0 \le r < 1$ and all $t \in \mathbf{R}$. From (ii), the equality

$$\text{(iii)} \quad P(r, t) = 1 + 2 \sum_{n=1}^{\infty} r^n \cos(nt)$$

follows immediately. The following equalities are important.

(iv) $P(r, \theta - t) = \mathrm{Re}\left[\dfrac{e^{it} + re^{i\theta}}{e^{it} - re^{i\theta}}\right] = \dfrac{1 - r^2}{1 - 2r\cos(\theta - t) + r^2}$.

PROOF OF (iv)   If $z = re^{i\theta}$, then we have

$$\sum_{n=-\infty}^{\infty} r^{|n|}e^{in(\theta-t)} = \mathrm{Re}\left[1 + 2\sum_{n=1}^{\infty} r^n e^{in(\theta-t)}\right]$$

$$= \mathrm{Re}\left[1 + 2\sum_{n=1}^{\infty} (ze^{-it})^n\right] = \mathrm{Re}\left[1 + 2\cdot\frac{ze^{-it}}{1 - ze^{-it}}\right]$$

$$= \mathrm{Re}\left[\frac{1 + ze^{-it}}{1 - ze^{-it}}\right] = \mathrm{Re}\left[\frac{e^{it} + re^{i\theta}}{e^{it} - re^{i\theta}}\right].$$

Further reduction to the right side of (iv) is straightforward and should be carried out by the reader.

It is also convenient to consider the kernel $Q(z, t)$, defined for z in the open disk D and real t, by

(v) $Q(z, t) = P(r, \theta - t) = \displaystyle\sum_{n=-\infty}^{\infty} r^{|n|}e^{in(\theta-t)}$ $(z = re^{i\theta})$.

In this notation, the first equality in (iv) becomes

(vi) $Q(z, t) = \mathrm{Re}\left[\dfrac{e^{it} + z}{e^{it} - z}\right]$.

From (vi) it is clear that $Q(z, t)$ is the real part of an analytic function for each fixed t, and so it is a harmonic function of z on D for each t.

6.10 FURTHER PROPERTIES OF THE POISSON KERNEL

We have

(i) $P(r, t) > 0$ for $0 \le r < 1$ and $-\pi < t < \pi$,

(ii) $(1/2\pi)\int_{[-\pi, \pi]} P(r, t)\, dt = 1$ for all $r \in [0, 1[$,

(iii) for fixed $\delta > 0$, $\lim_{r\to 1} P(r, \delta) = 0$,

(iv) if $r < 1$ is fixed, then $P(r, t)$ is decreasing in t on $[0, \pi]$,

(v) $P(r, t) = P(r, -t)$.

PROOF.   The inequality (i) is clear from (6.9.iv) (take $\theta = 0$). Equality (ii) is established using the definition (6.9.i) and integrating the series term-by-term (by uniform convergence). Assertion (iii) follows from (6.9.iv), as does assertion (iv). Equality (v) is clear from either (6.9.iii) or (6.9.iv).

6.11 THE POISSON INTEGRAL

If f is a piecewise continuous function on $\mathbf{T} = \{z \mid |z| = 1\}$, let

(i) $Pf(z) = \dfrac{1}{2\pi} \displaystyle\int_{[-\pi, \pi]} Q(z, t)f(e^{it})\, dt \qquad (z \in D)$.

In different notation, we have

(ii) $Pf(z) = \dfrac{1}{2\pi} \displaystyle\int_{[-\pi, \pi]} P(r, \theta - t)f(e^{it})\, dt \qquad (z = re^{it})$.

The function Pf is called the *Poisson integral* of f.

Remark. The Poisson integral can be defined for more general functions f. Essentially all that is required is that the integral in (i) exist in some sense. We have restricted ourselves to piecewise continuous functions in order to avoid integration problems. (Recall that the definition of piecewise continuity of f is that there be a partition of $[-\pi, \pi]$, say $\{t_0 < t_1 < \cdots < t_n\}$, such that f has a continuous extension from each $]t_{k-1}, t_k[$ to $[t_{k-1}, t_k]$.)

6.12 THEOREM

If f is piecewise continuous on $\mathbf{T}$, then Pf is harmonic on $D(= \{z \mid |z| < 1\})$.

PROOF.   If f is real-valued, then by (6.9.vi) we have

$$Pf(z) = \operatorname{Re}\left[\frac{1}{2\pi} \int_{[-\pi, \pi]} \frac{e^{it} + z}{e^{it} - z}\, f(e^{it})\, dt\right].$$

The integral is analytic on D, so Pf is harmonic on D. If we have $f = u + iv$, u and v real-valued, then harmonicity of Pf follows from the equality $Pf = Pu + iPv$.

We come now to the principal theorem about the Poisson integral.

6.13 THEOREM

Suppose that f is piecewise continuous on $\mathbf{T}$ and that S is the set of points in $\mathbf{T}$ at which f is not continuous. Define F on $\bar{D}\backslash S$ by

$$F(z) = \begin{cases} Pf(z) & \text{if } |z| < 1 \\ f(z) & \text{if } |z| = 1 \text{ and } z \notin S. \end{cases}$$

The function F is continuous on $\bar{D}\backslash S$.

PROOF. The proof is dependent on the properties listed in (6.10). Let $\zeta = e^{i\omega} \in \mathbf{T} \backslash S$ and write $z \in D$ as $z = re^{i\theta}$, ω and θ in $]-\pi, \pi]$. By (6.10.ii), we have

$$Pf(z) - f(\zeta) = \frac{1}{2\pi} \int_{[-\pi, \pi]} P(r, \theta - t) f(e^{it}) \, dt - f(\zeta)$$

$$= \frac{1}{2\pi} \int_{[-\pi, \pi]} P(r, \theta - t)[f(e^{it}) - f(\zeta)] \, dt. \qquad (1)$$

Suppose first that $\omega < \pi$. For every $\delta > 0$ satisfying $[\omega - \delta, \omega + \delta] \subset \,]-\pi, \pi[$, the second integral in (1) is equal to

$$\frac{1}{2\pi} \int_{|\omega - t| > \delta} P(r, \theta - t)[f(e^{it}) - f(\zeta)] \, dt$$

$$+ \frac{1}{2\pi} \int_{[\omega - \delta, \omega + \delta]} P(r, \theta - t)[f(e^{it}) - f(\zeta)] \, dt. \qquad (2)$$

Let $\varepsilon > 0$. There is a $\delta' > 0$ such that

$$|f(e^{it}) - f(\zeta)| < \tfrac{1}{2}\varepsilon \quad \text{if } |\omega - t| < \delta'.$$

It follows from (6.10.i) and (6.10.ii) that the integral over $[\omega - \delta, \omega + \delta]$ in (2) is less than $\tfrac{1}{2}\varepsilon$ for all θ if $\delta < \delta'$ and $[\omega - \delta, \omega + \delta] \subset [-\pi, \pi]$. For such a δ and all θ satisfying $|\theta - \omega| < \tfrac{1}{2}\delta$, we have

$$|t - \theta| > \tfrac{1}{2}\delta \quad \text{if } |t - \omega| > \delta. \qquad (3)$$

Hence by (6.10.iv) the first integral in (2) is dominated by

$$\frac{1}{2\pi} P(r, \tfrac{1}{2}\delta) \left[\int_{[-\pi, \pi]} |f(e^{it})| \, dt + |f(\zeta)| \right], \qquad (4)$$

provided that $|\theta - \omega| < \tfrac{1}{2}\delta$. By (6.10.iii), the expression in (4) goes to zero as r goes to 1. Hence there is an $\eta > 0$ such that (4) is less than $\tfrac{1}{2}\varepsilon$ if $1 - r < \eta$. In conclusion, for given $\varepsilon > 0$ there is a $\delta > 0$ and an $\eta > 0$ such that

$$|Pf(z) - f(\zeta)| < \varepsilon \qquad (5)$$

holds if

$$1 - |z| < \eta \quad \text{and} \quad |\theta - \omega| = |\operatorname{Arg}(z) - \operatorname{Arg}(\zeta)| < \tfrac{1}{2}\delta. \qquad (6)$$

Suppose now that $(z_n)_{n=1}^{\infty}$ is a sequence in $\bar{D} \backslash S$ converging to ζ. If all but finitely many z_n are in D, then there is an n_0 such that (6) holds for z_n for all $n \geq n_0$; thus, it follows from (5) that $\lim_{n \to \infty} Pf(z_n) = f(\zeta)$. If $|z_n| = 1$ for all but finitely many z_n, then there is an n_0 such that $F(z_n) = f(z_n)$ for $n \geq n_0$; thus, the equality $\lim_{n \to \infty} F(z_n) = f(\zeta)$ is valid by the continuity of f at ζ. If both sets $\{n \mid z_n \in D\}$ and $\{n \mid |z_n| = 1\}$ are infinite, then the subsequences of (z_n) in D and $\mathbf{T}$ each converge to ζ; the corresponding values of F converge to $F(\zeta)$. It follows that $(F(z_n))_{n=1}^{\infty}$ converges to $F(\zeta)$, and thus F is continuous at ζ.

If $\omega = \pi$, then the second integral in (1) can be written as an integral over $[0, 2\pi]$, by the periodicity of the integrand. One then selects δ satisfying $[\omega - \delta, \omega + \delta] \subset [0, 2\pi]$, and essentially the same proof is valid.

6.14 DIRICHLET PROBLEM FOR THE DISK

A Dirichlet problem is a boundary value problem P of the form:

Given a function f on all or part of the boundary of a set U, find a function F defined on $\bar{U}$ (except perhaps for a few exceptional points in the boundary) which is harmonic on the interior of U and equal to f on the boundary.

If f is piecewise continuous on $\mathbf{T}$, then (6.12) and (6.13) show that the Poisson integral Pf solves the Dirichlet problem for $D(1, 0)$. The situation has more generality than it appears, for any simply connected region can be mapped to the disk by a bijective analytic function, and these maps preserve harmonicity. We will elaborate on this point in Chapter 7 [see (7.16)]. The Dirichlet problem is important for applied work because many types of flows are known to satisfy Laplace's equation throughout the region in which the flow is taking place [see (6.8)]. If the behavior of the flow on the boundary can be determined (the boundary function is known), the solution of the resulting Dirichlet problem yields a description of the flow in the entire region.

The Poisson integral solution is also unique. This fact is an immediate corollary of the maximum principle for harmonic functions, which in turn follows from the mean value property. Before proving these facts, we first point out how the Poisson integral can be modified for an arbitrary disk and also give some examples.

6.15 THE POISSON INTEGRAL FOR ARBITRARY DISKS

If f is piecewise continuous on $C(R, a)$, then the function

$$g(e^{it}) = f(Re^{it} + a)$$

is piecewise continuous on $C(1, 0)$ and $Pg(z)$ is well-defined for $z \in D(1, 0)$. The Poisson integral $P_R f$ of f in $D(R, a)$ is defined by

$$P_R f(z) = Pg\left(\frac{z - a}{R}\right) = \frac{1}{2\pi} \int_{[-\pi, \pi]} \text{Re}\left[\frac{Re^{it} + (z - a)}{Re^{it} - (z - a)}\right] f(Re^{it} + a) \, dt.$$

The function $P_R f$ is harmonic on $D(R, a)$. Setting F equal to $P_R f$ on $D(R, a)$ and equal to f on $C(R, a)$, we obtain a function continuous on $\overline{D(R, a)}$, except at the discontinuities of f.

6.16 EXAMPLES

The Poisson integral $Pf(z)$ is difficult to calculate directly for most functions f. However, it can always be obtained as an infinite series. Let f be a piecewise continuous function on $\mathbf{T} = \{z \mid |z| = 1\}$. Using the infinite series in (6.9.i) for the Poisson kernel, for $z = re^{i\theta}(r < 1)$, we have

$$Pf(z) = \frac{1}{2\pi} \int_{[-\pi, \pi]} \left[\sum_{n=-\infty}^{\infty} r^{|n|} e^{in\theta} e^{-int} \right] f(e^{it}) \, dt$$

$$= \sum_{n=-\infty}^{\infty} \left[\frac{1}{2\pi} \int_{[-\pi, \pi]} f(e^{it}) e^{-int} \, dt \right] r^{|n|} e^{in\theta}$$

$$= \sum_{n=-\infty}^{\infty} \hat{f}(n) r^{|n|} e^{in\theta}$$

where $\hat{f}(n)$ is the nth Fourier coefficient of f, defined by the bracketed expression. The interchange of summation and integration is valid because the series $\sum_{n=-\infty}^{\infty} r^{|n|} e^{-int}$ converges uniformly for fixed $r < 1$, and (3.14) applies. One can thus calculate $Pf(z)$ as an infinite series by finding the Fourier coefficients $\hat{f}(n)$. In the following examples the reader may want to keep a physical example in mind. For example, in each case the given function f could be the steady-state temperature (which depends only on x and y) on the surface of an infinite cylinder. The solution then gives the temperature in the interior of the cylinder.

Example (i).   Let f be defined on $\mathbf{T}$ by $f(e^{it}) = t$, $t \in \,]-\pi, \pi]$; that is, $f(z) = \text{Arg}\,(z)$. Then f is continuous at every point of $\mathbf{T}$, except -1. We have $\hat{f}(0) = 0$. For $n \neq 0$, we use integration by parts to obtain

$$\hat{f}(n) = \frac{1}{2\pi} \int_{[-\pi, \pi]} te^{-int} \, dt = \frac{i \cos(n\pi)}{n} = \frac{i(-1)^n}{n}.$$

Thus we have

$$Pf(z) = \sum_{\substack{n=-\infty \\ n \neq 0}}^{\infty} \frac{i(-1)^n}{n} r^{|n|} e^{in\theta} \qquad (z = re^{i\theta})$$

$$= \sum_{n=1}^{\infty} \frac{2(-1)^{n+1} r^n \sin(n\theta)}{n}$$

$$= 2 \, \text{Im} \left(\sum_{n=1}^{\infty} \frac{(-1)^{n+1} z^n}{n} \right).$$

By (6.13), the function F defined by

$$F(z) = \begin{cases} 2 \text{ Im} \displaystyle\sum_{n=1}^{\infty} \frac{(-1)^{n+1}z^n}{n}, & |z| < 1 \\[2mm] \text{Arg}\,(z) & |z| = 1 \end{cases}$$

is continuous on $\bar{D}\backslash\{-1\}$. It is interesting to note that the series diverges if $z = -1$. We proved in (2.18) that the series actually converges uniformly on every set $\{z \mid |z| \le 1 \text{ and } |z + 1| > \eta\}$, $\eta > 0$. Hence the sum is a continuous function on $\bar{D}\backslash\{-1\}$. Combining this continuity with the continuity of F, we obtain

$$t = 2 \text{ Im} \sum_{n=1}^{\infty} \frac{(-1)^{n+1}e^{int}}{n} \qquad (-\pi < t < \pi).$$

Just for fun, we take $t = \pi/2$ and obtain

$$\frac{\pi}{2} = 2 \text{ Im} \sum_{n=1}^{\infty} \frac{(-1)^{n+1}i^n}{n} \qquad (-\pi < t < \pi).$$

The summands of this series are real for even n and we have $i^{2k-1} = (-1)^{k+1}i$ for odd integers $2k - 1$. Hence we obtain

$$\pi = 4 \sum_{k=1}^{\infty} \frac{(-1)^{k+1}}{2k - 1}.$$

Example (ii).   Let

$$f(e^{it}) = \begin{cases} 0 & -\pi < t \le 0 \\ 1 & 0 < t \le \pi. \end{cases}$$

We have $\hat{f}(0) = \frac{1}{2}$ and

$$\hat{f}(n) = \frac{1}{2\pi} \int_{[0,\,\pi]} e^{-int}\, dt = \frac{1 - (-1)^n}{2\pi in} \qquad (n \ne 0).$$

Thus for $z = re^{i\theta}$ we obtain the equality

$$Pf(z) = \frac{1}{2} + \sum_{k=-\infty}^{\infty} \frac{r^{|2k-1|}(e^{i\theta})^{2k-1}}{\pi i(2k - 1)}$$

$$= \frac{1}{2} + \sum_{k=1}^{\infty} \frac{2r^{2k-1}\sin((2k-1)\theta)}{\pi(2k - 1)}$$

$$= \frac{1}{2} + \text{Im}\left(\frac{2}{\pi}\sum_{k=1}^{\infty}\frac{(re^{i\theta})^{2k-1}}{(2k - 1)}\right).$$

If $r = 1$, the final series converges unless $\theta = 0$. The function f is discontinuous at 1 and at -1 and continuous on $\mathbf{T}\backslash\{1, -1\}$. Since the function F defined as Pf on $D(1, 0)$ and f on $\mathbf{T}$ is continuous on $\bar{D}\backslash\{-1, 1\}$, it follows that the equality

$$\frac{1}{2} + \frac{2}{\pi}\sum_{k=1}^{\infty} \frac{\sin((2k-1)\theta)}{2k-1} = \begin{cases} 0 & \text{if } -\pi < \theta < 0 \\ 1 & \text{if } 0 < \theta < \pi. \end{cases}$$

holds. At $\theta = \pi$ and $\theta = 0$, the series converges to $\frac{1}{2}$. Hence, in this case, the Poisson integral is fair; it splits the values at the discontinuities equally between the values on each side of the discontinuity.

There is another way to evaluate $Pf(z)$ in this case. Write

(a) $$Pf(z) = \frac{1}{2\pi}\int_0^\pi \frac{1-r^2}{1+r^2-2r\cos(\theta - t)}\,dt \qquad (z = re^{i\theta}).$$

It is easy to verify that

$$\frac{d}{dt}\left[2\,\text{Arctan}\left(\frac{1+r}{1-r}\tan\frac{t}{2}\right)\right] = \frac{1-r^2}{1+r^2-2r\cos(t)}.$$

Hence a primitive for the integrand in (a) is

$$-2\,\text{Arctan}\left[\left(\frac{1+r}{1-r}\right)\tan\left(\frac{\theta - t}{2}\right)\right]$$

and we obtain

$$Pf(z) = \frac{1}{\pi}\left[\text{Arctan}\left(\frac{1+r}{1-r}\tan\frac{\theta}{2}\right) - \text{Arctan}\left(\frac{1+r}{1-r}\tan\left(\frac{\theta-\pi}{2}\right)\right)\right].$$

One can simplify this expression by simplifying $\tan(\pi P f(z))$. The result is the equality

$$Pf(z) = \frac{1}{\pi}\text{Arctan}\left(\frac{1-r^2}{-2r\sin\theta}\right) \qquad (z = re^{i\theta}, r < 1).$$

Example (iii).   Let $f(e^{it}) = \sin(t)$. Since $f(z) = \text{Im}(z)$, it is easy to read off a solution to the Dirichlet problem for f. The function $\text{Im}(z)$ is harmonic on $D(1, 0)$ and agrees with f on $\mathbf{T}$. To compute $Pf(z)$, note that

$$\int_{-\pi}^{\pi} e^{-int}\sin(t)\,dt = \int_{-\pi}^{\pi} e^{-int}\frac{e^{it} - e^{-it}}{2i}\,dt$$

$$= \int_{[-\pi, \pi]} \frac{e^{-i(n-1)t} - e^{-i(n+1)t}}{2i}\,dt.$$

The final integral is 0 if $|n| \neq 1$, is $-\pi i$ if $n = 1$, and is πi if $n = -1$. Hence for $z = re^{i\theta}$ we have

$$Pf(z) = \sum_{n=-\infty}^{\infty} r^{|n|} e^{in\theta} \frac{1}{2\pi} \int_{[-\pi, \pi]} e^{-int} \sin(t)\, dt$$

$$= \frac{ir[-e^{i\theta} + e^{i\theta}]}{2} = \frac{re^{i\theta} - re^{-i\theta}}{2i} = \operatorname{Im}(z).$$

Hence, as expected, $Pf(z) = \operatorname{Im}(z)$.

We have not yet proved the uniqueness of the Poisson integral solution of the Dirichlet problem for $D(1, 0)$; we will do so in (6.20).

Example (iv). Let $f(e^{it}) = e^t \ (-\pi < t \leq \pi)$. For each n, we have

$$\frac{1}{2\pi} \int_{[-\pi, \pi]} e^{-int} e^t\, dt = \frac{1}{2\pi} \int_{[-\pi, \pi]} e^{t(1-in)}\, dt = \frac{(e^\pi - e^{-\pi})(1 + in)(-1)^n}{2\pi(1 + n^2)}.$$

Hence for $z = re^{i\theta} \ (r < 1)$, we obtain

(a) $\quad Pf(z) = \dfrac{e^\pi - e^{-\pi}}{2\pi} \displaystyle\sum_{n=-\infty}^{\infty} \dfrac{(-1)^n(1 + in)}{1 + n^2} r^{|n|} e^{in\theta}$

$$= \frac{e^\pi - e^{-\pi}}{\pi} \left[\frac{1}{2} + \sum_{n=1}^{\infty} \frac{r^n(-1)^n[\cos(n\theta) - n\sin(n\theta)]}{1 + n^2} \right].$$

The series on the right converges for all θ (by partial summation). Hence e^θ is given by the expression in (a) provided that $-\pi < \theta < \pi$. Taking $\theta = 0$, we obtain as an interesting application the equality

$$\frac{1}{2} + \sum_{n=1}^{\infty} \frac{(-1)^n}{n^2 + 1} = \frac{\pi}{e^\pi - e^{-\pi}} = \frac{1}{2} \frac{\pi}{\sinh(\pi)}.$$

*6.17 REMARKS ON THE POISSON AND CAUCHY INTEGRALS

In our discussion of the Cauchy integral (4.14), we posed the problem of finding an analytic function on $D(1, 0) = D$ which agrees with a given continuous function f on the boundary $\mathbf{T}$ of D. The Cauchy integral solves this problem provided that $\hat{f}(n) = 0$ for $n < 0$. The reason for this restriction on f is that the absence of terms for negative integers in the Cauchy kernel

$$K(z, t) = \frac{e^{it}}{e^{it} - z} = \sum_{n=0}^{\infty} z^n e^{-int} \quad (t \in [-\pi, \pi],\ z \in D)$$

annihilates $\hat{f}(n)$ for $n < 0$. To be explicit, suppose that f is continuous on **T** and that its Fourier series converges absolutely to f. We have

$$Cf(z) = \frac{1}{2\pi i} \int_{\vec{T}} \frac{f(\zeta)}{\zeta - z} \, d\zeta = \frac{1}{2\pi} \int_{[-\pi, \pi]} f(e^{it}) \left[\sum_{n=0}^{\infty} e^{-int} z^n \right] dt$$

$$= \frac{1}{2\pi} \int_{[-\pi, \pi]} \left[\sum_{k=-\infty}^{\infty} \hat{f}(k) e^{ikt} \right] \left[\sum_{n=0}^{\infty} e^{-int} z^n \right] dt$$

$$= \frac{1}{2\pi} \sum_{n=0}^{\infty} \left[\sum_{k=-\infty}^{\infty} \hat{f}(k) \int_{[-\pi, \pi]} e^{i(k-n)t} \, dt' \right] z^n$$

$$= \sum_{n=0}^{\infty} \hat{f}(n) z^n.$$

Thus the series expansion $f(e^{it}) = \sum_{n=-\infty}^{\infty} \hat{f}(n)(e^{it})^n$ is valid, but the power series expansion for $Cf(z)$ is $Cf(z) = \sum_{n=0}^{\infty} \hat{f}(n) z^n$. The negative coefficients have vanished. The equality

$$\lim_{\substack{z \to e^{it} \\ |z| < 1}} Cf(z) = f(e^{it})$$

holds if and only if $\hat{f}(n) = 0$ for $n < 0$.

In the definition of the *Poisson kernel*, a two-sided infinite series is used. The result is that analyticity of the integral transform is lost, harmonicity is retained, and the appropriate boundary values are obtained. Notice that $Pf(z) = Cf(z)$ if f is analytic, for then $\hat{f}(n) = 0$ for $n < 0$.

EXERCISES

TYPE I

1. Use the example in (6.16.i) to show that

$$\pi = 8 \sum_{n=1}^{\infty} \frac{\sin \left[(\pi/4)n\right](-1)^{n+1}}{n}.$$

2. Compute $Pf(z)$ as an infinite series for the function f defined on **T** by
$$f(e^{it}) = t^2 \qquad (-\pi < t \leq \pi).$$

3. Solve the Dirichlet problem for $D(1, 0)$ for the function f defined on **T** by
$$f(e^{it}) = \cos(t).$$

4. Solve the Dirichlet problem for $D(1, 0)$ for the function given by $f(e^{it}) = e^{t^2}$.

5. Generalize example (6.16.ii) to the case where the function f is given by

$$f(e^{it}) = \begin{cases} 1 & \text{if } -\pi < t < t_0 \\ 0 & \text{if } t_0 \le t \le \pi \end{cases}$$

where t_0 is fixed and $-\pi < t_0 < 0$.

6. Find $Pf(z)$ if $f(e^{it}) = t + \pi$.

TYPE II

1. Let K_N be the Féjer kernel [see (3.21)]. Let f be a continuous function on $\{z \mid |z| = 1\}$ such that $\hat{f}(n) = 0$ if $n < 0$. Define

$$F_N(e^{it}) = \frac{1}{2\pi} \int_{[-\pi, \pi]} K_N(\theta - t) f(e^{it}) \, dt.$$

For $|z| < 1$, let $F_N(z)$ be the polynomial in z obtained by replacing e^{it} by z in the polynomial $F_N(e^{it})$. Show that $\lim_{N \to \infty} F_N(z) = Pf(z)$ for all z such that $|z| \le 1$ and that the convergence is uniform on any compact subset of $\{z \mid |z| < 1\}$.

6.18 THE MEAN VALUE PROPERTY

Suppose that f is analytic on an open set containing the closed disk $\overline{D(\rho, a)}$, $\rho > 0$. The derivative of the mean value

(i) $$M(r) = \frac{1}{2\pi} \int_{[-\pi, \pi]} f(a + re^{it}) \, dt$$

with respect to $r (r \in \,]0, \rho[)$ is equal to $(1/2\pi) \int_{[-\pi, \pi]} f'(a + re^{it}) e^{it} \, dt$, and so is zero by Cauchy's theorem [only a weak form is needed; see (3.6)]. The mean value (i) is thus a constant function of r on $[0, \rho]$. The constant is $f(a)$, since $\lim_{r \to 0} M(r) = f(a)$; we have thus proved the *mean value property*

(ii) $$f(a) = \frac{1}{2\pi} \int_{[-\pi, \pi]} f(a + re^{it}) \, dt, \text{ all } r \in \,]0, \rho]$$

for analytic functions. If u is real-valued and harmonic on an open set containing $\overline{D(\rho, 0)}$, and v is a harmonic conjugate for u on an open set containing $\overline{D(\rho, 0)}$, then (ii) is valid for $f = u + iv$; equating real and imaginary parts, we see that u also has the mean value property. An arbitrary complex-valued function u is harmonic if and only if its real and imaginary parts are harmonic. Hence we proved the following theorem.

Theorem. If u is harmonic on an open set containing $\overline{D(\rho, a)}$, then we have

(iii) $u(a) = \dfrac{1}{2\pi} \displaystyle\int_{[-\pi, \pi]} u(a + re^{it})\, dt \quad (r \in\,]0, \rho])$.

6.19 MAXIMUM PRINCIPLE FOR HARMONIC FUNCTIONS

If u is a real-valued nonconstant harmonic function on an open connected set U, then u has no local maximum and no local minimum on U.

PROOF.   We first show that u cannot be constant on any disk contained in U. Let $E = \{z \in U \mid u$ is constant on a disk centered at $z\}$. It is clear that E is open. We will show that $U \backslash E$ is also open. If $z_1 \in U \backslash E$, let $D(r, z_1) \subset U$. Let v be a harmonic conjugate for u on $D(r, z_1)$. If $D(r, z_1)$ contains points of E, then it contains disks on which u is constant. By the Cauchy-Riemann equations, v is also constant on these disks, and hence the analytic function $u + iv$ is constant on these disks. By (4.18), $u + iv$ is constant on $D(r, z_1)$. Hence u is constant on $D(r, z_1)$. Thus $D(r, z_1)$ contains no points of E and so $U \backslash E$ is open. Since U is connected, either E or $U \backslash E$ is void. If $E = U$, it is easy to see that u is in fact constant on all of U. Hence $E = \varnothing$.

Assume that u has a maximum at $z_0 \in U$. Since u is not constant on any disk centered at z_0, there is an $r > 0$ such that $u(z_1) < u(z_0)$ for some z_1 such that $|z_1 - z_0| = r$ and such that $u(z) \le u(z_0)$ if $|z - z_0| = r$. The inequality $u(z) < u(z_0)$ must hold on an arc of $\{z_0 + re^{it} \mid t \in\,]-\pi, \pi]\}$ containing z_1 by the continuity of u. It follows easily that (6.18.iii) is contradicted.

Similarly, u cannot have a minimum.

6.20 UNIQUENESS FOR THE DIRICHLET PROBLEM ON $D = D(1, 0)$

If two functions u and v are harmonic on D, continuous on $\overline{D}$, and equal on $\mathbf{T} = \{|z| = 1\}$, then $u - v$ is continuous on $\overline{D}$ and zero on $\mathbf{T}$. By (6.19), we have $u - v = 0$ on $\overline{D}$ and hence $u = v$.

6.21 DEFINITION

A continuous function u defined on an open set U containing a has the *mean value property* at a if (6.18.iii) holds for a positive sequence $\{r_n\}_{n=1}^{\infty}$ such that $\lim_{n \to \infty} r_n = 0$ and $\overline{D(r_n, a)} \subset U$.

The following theorem is the converse of the theorem in (6.18). These theorems together characterize harmonic functions in a manner totally different from the definition.

6.22 THEOREM

If u is continuous and has the mean value property at every point of an open set U, then u is harmonic on U.

PROOF. Suppose first that u is real-valued. Let $\overline{D(\rho, a)} \subset U$, and put $h = P_\rho u - u$ [see (6.15)] on $\overline{D(\rho, a)}$. We have $h = 0$ on $\{z \mid |z - a| = \rho\}$. Since u and $P_\rho u$ both have the mean value property on $D(\rho, a)$, so does h. If there are points in $D(\rho, a)$ where h is positive, let M be the maximal value of h. There are points $z \in D(\rho, a)$ such that $h(z) = M$. Among these z, let z_1 be such that $R = |z_1 - a|$ is maximal. Let $D(r, z_1) \subset D(\rho, a)$. The inequalities

$$h(z) \leq M \text{ on } \{z \mid |z - z_1| = r \text{ and } |z - a| \leq R\}$$

$$h(z) < M \text{ on } \{z \mid |z - z_1| = r \text{ and } |z - a| > R\}$$

show that (6.18.iii) cannot hold for h on $\{z \mid |z - z_1| = r\}$. Since (6.18.iii) fails for every r such that $D(r, z_1) \subset D(\rho, a)$, h cannot have the mean value property.

Similarly, $h(z) < 0$ for points in $D(\rho, a)$ is impossible. Hence we have $u = P_\rho u$ on $D(\rho, a)$; and so u is harmonic on $D(\rho, a)$. Since a is arbitrary, u is harmonic on U.

If u is complex-valued and has the mean value property, then so do its real and imaginary parts. The real and imaginary parts are therefore harmonic, and so u is harmonic.

6.23 THEOREM (SCHWARZ'S REFLECTION PRINCIPLE)

Let U be an open simply connected subset of the upper half-plane $\{z \mid \text{Im} (z) > 0\}$ which has the property that $\overline{U} \cap \mathbf{R}$ is a nonvoid closed interval $[a, b]$. Let $U^* = \{\bar{z} \mid z \in U\}$ and let $V = U \cup \,]a, b[\,\cup U^*$. Let $f = u + iv$ be analytic on U and suppose that the extended function v obtained by setting $v = 0$ on $]a, b[$ is continuous on $U \cup \,]a, b[$. There is an analytic function F on V such that $F = f$ on U. The equality $\overline{F(\bar{z})} = F(z)$ holds for all $z \in V$.

PROOF. Extend v to V by letting $v(z) = -v(\bar{z})$ for $z \in U^*$. It is not difficult to show that v is then continuous on V. We know that v is harmonic on U. We will show that v is harmonic on V by verifying that it satisfies the mean value property at all points of $]a, b[\,\cup U^*$. If $c \in U^*$ and $D(r, c) \subset U^*$, then we have

$$\frac{1}{2\pi} \int_{[-\pi, \pi]} v(c + re^{it}) \, dt = -\frac{1}{2\pi} \int_{[-\pi, \pi]} v(\bar{c} + re^{-it}) \, dt = -v(\bar{c}) = v(c),$$

so that v does have the mean value property at c. If $c \in \,]a, b[$ and $D(r, c) \subset V$, then we have

$$\frac{1}{2\pi} \int_{[-\pi, \pi]} v(c + re^{it})\, dt = -\frac{1}{2\pi} \int_{[-\pi, 0]} v(c + re^{-it})\, dt$$

$$+ \frac{1}{2\pi} \int_{[0, \pi]} v(c + re^{it})\, dt = 0 = v(c).$$

The function v is thus harmonic on V by (6.22).

The set V is easily seen to be simply connected. By (6.4), there is a real harmonic function W on V such that $W + iv$ is analytic on V. Again by (6.4), there is a constant A such that $W - A = u$ on U. Let $w = W - A$ on V. Then $w + iv$ is analytic on V and equal to $u + iv$ on U.

It is easy to see by direct consideration of difference quotients that the function g given by $g(z) = \overline{F(\bar{z})}$ is analytic on V. Since $v = 0$ on $]a, b[$, the equality $\overline{F(\bar{z})} = F(z)$ holds on $]a, b[$. By (4.18), the equality holds throughout V.

6.24 REMARKS

The hypothesis that U be simply connected can easily be weakened; connectedness is all that is needed. We give the result as an exercise.

The theorem has important applications in constructing conformal mappings. We will treat some of these in Chapter 7.

EXERCISES

TYPE I

1. Refer to Exercise 5, p. 214, and give an interpretation of the mean value property for "one-dimensional" harmonic functions.
   [*Hint:* What is a "one-dimensional circle"?]

2. Let u be a nonnegative function on an open set U which is twice continuously differentiable. Suppose also that there is a nonnegative continuous function f on U for which

$$\Delta u = \rho f$$

   on U. Prove that u has no local maximum on U.

   [*Hint:* Use Green's theorem.]

TYPE II

1. Show that the conclusion of (6.23) holds if U is a connected open set.

2. Give a different proof of (6.23) by simply defining F on U^* by $F(z) = \overline{f(\bar{z})}$ and using Morera's theorem (4.10) to show that F is analytic on V.

Conformal and Univalent Mappings

In (5.12) we proved that a function with a zero of multiplicity 1 at a point is one-to-one in a neighborhood of the point. This fact led to the inverse function theorem (5.14) and its local converse (5.15). In this chapter we make a more detailed study of analytic functions which are either locally or globally one-to-one. We begin with some definitions.

7.1 DEFINITIONS

An analytic function f defined on an open set U containing z_0 is *conformal* at z_0 if $f'(z_0) \neq 0$. It is *univalent* on U if it is analytic and one-to-one on U. If f is univalent on U, then by theorem (5.14) it is conformal at every point of U.

One important application of univalent functions is in the solution of Dirichlet problems. In outline, this application is as follows. Suppose that U is an open set with boundary B and that f is a continuous function on B. We want to solve the Dirichlet problem for the function f on the set $\overline{U} = U \cup B$. That is, we want to find a continuous function on $\overline{U}$ which is equal to f on B and is harmonic on $\overline{U}$. Suppose that M is a univalent mapping from U to the unit disk D which has a continuous extension to $\overline{U} = U \cup B$, and that the extended mapping takes B into $\mathbf{T} = \{z \mid |z| = 1\}$. By Theorem (5.14), the inverse M^{-1} of M is also univalent. By Theorem (6.13), there is a continuous function F on $\overline{D}$ which is harmonic on D and equal to $f \circ M^{-1}$ on $\mathbf{T}$. The function $F \circ M$ is equal to f on B. It is a crucial property (proved below) of analyticity that the function $F \circ M$ is harmonic on U. Hence $F \circ M$ solves the Dirichlet problem for $\overline{U}$. To complete the argument, we prove that $F \circ M$ is harmonic on U. If F is real-valued, let G be a harmonic conjugate for F on D (see 6.4). The function $(F + iG) \circ M$ is analytic on U

and has real part $F \circ M$. Thus $F \circ M$ is harmonic by (1.22). It follows easily that $F \circ M$ is also harmonic for complex F.

Of course the technique described above does not require that we map U to D. The idea is to find an M on U which transfers the given Dirichlet problem to one that can be solved. We will give several examples, beginning with (7.16). First we develop general properties of conformal and univalent mappings and then give several examples. We begin with the following geometrical interpretation of conformality.

7.2 THEOREM

Suppose that f is analytic on an open set U, that $z_0 \in U$, and that $f - f(z_0)$ has a zero of multiplicity n at z_0. If $S \subset U$ is a set such that the limit

(i) $\quad \lim\limits_{\substack{z \to z_0 \\ z \in S}} \dfrac{z - z_0}{|z - z_0|} = e^{i\theta}$

exists, then we have

(ii) $\quad \lim\limits_{\substack{z \to z_0 \\ z \in S}} \dfrac{f(z) - f(z_0)}{|f(z) - f(z_0)|} = \dfrac{f^{(n)}(z_0)}{|f^{(n)}(z_0)|} e^{i\theta n}.$

PROOF.   Let

$$f(z) = f(z_0) + \sum_{k=1}^{\infty} a_k(z - z_0)^k$$

be the Taylor expansion of f centered at z_0. Since $f - f(z_0)$ has a zero of multiplicity n at z_0, we have $a_1 = a_2 = \cdots = a_{n-1} = 0$ and $a_n = f^{(n)}(z_0)/n! \neq 0$. Hence the equality

$$\frac{f(z) - f(z_0)}{|f(z) - f(z_0)|} = \frac{a_n(z - z_0)^n + a_{n+1}(z - z_0)^{n+1} + \cdots}{|a_n(z - z_0)^n + a_{n+1}(z - z_0)^{n+1} + \cdots|}$$

holds for all z sufficiently close to z_0. Factoring out $a_n(z - z_0)^n$, we obtain

$$\frac{f(z) - f(z_0)}{|f(z) - f(z_0)|} = \frac{f^{(n)}(z_0)(z - z_0)^n}{|f^{(n)}(z_0)(z - z_0)^n|} \left[\frac{1 + \dfrac{a_{n+1}}{a_n}(z - z_0) + \cdots}{\left|1 + \dfrac{a_{n+1}}{a_n}(z - z_0) + \cdots\right|} \right].$$

The expression in brackets goes to 1 as z goes to z_0. Hence (ii) follows from (i).

7.3 COROLLARY

Let f be as in (7.2). If γ is a curve in U parametrized on $[0, 1]$ with initial point z_0 such that $\gamma(t) \neq z_0$ if $t > 0$, then

(i) $\displaystyle \lim_{t \to 0} \frac{\gamma(t) - z_0}{|\gamma(t) - z_0|} = e^{i\theta}$

$$\text{implies that} \quad \lim_{t \to 0} \frac{f(\gamma(t)) - f(z_0)}{|f(\gamma(t)) - f(z_0)|} = \frac{f^{(n)}(z_0)}{|f^{(n)}(z_0)|} e^{in\theta}.$$

Statement (i) can be interpreted geometrically to mean that the tangent ray to γ at z_0 is in the direction θ and is mapped by f into a tangent ray of $f \circ \gamma$ at $f(z_0)$ in the direction $n\theta + \omega$, where $\omega = \text{Arg}\,[f^{(n)}(z_0)/|f^{(n)}(z_0)|]$. If $n = 1$ ($f'(z_0) \neq 0$), then the tangent ray is moved by the constant ω. Since ω is constant, the angle between the tangent rays to two curves intersecting at z_0 is preserved. In Figure 7.1, the situation is depicted for the segment $\gamma(t) = z_0 + te^{i\theta}$ and the function $f(z) = z^2$ at a point $z_0 \neq 0$.

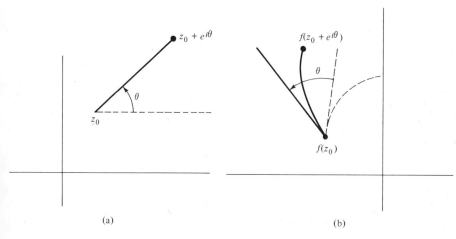

(a)   (b)

Figure 7.1

7.4 DISCUSSION OF CONFORMALITY

For a function f defined on a neighborhood of z_0 and satisfying the restriction $f(z) \neq f(z_0)$ if $z \neq z_0$, write the difference quotient for f as

(i) $\displaystyle \frac{f(z) - f(z_0)}{z - z_0} = \frac{f(z) - f(z_0)}{|f(z) - f(z_0)|} \cdot \left| \frac{f(z) - f(z_0)}{z - z_0} \right| \cdot \frac{|z - z_0|}{z - z_0},$

$$= A(z)M(z)Q(z)$$

where A, M, and Q are defined (in order) by the expressions on the right of the first equality in (i). Suppose that $f'(z_0)$ exists and is not zero. Then we have $\lim_{z \to z_0} M(z) = |f'(z_0)|$.   If S is a set such that $\lim_{z \to z_0, \, z \in S} Q(z) = e^{i\theta}$ exists, then it follows that $\lim_{z \to z_0, \, z \in S} A(z) = e^{i\omega}$ must exist. Since $f'(z_0) = e^{i\psi}|f'(z_0)|$ holds for a fixed ψ, we have $e^{i\psi} = e^{i[\theta + \omega]}$. This is an alternate proof that an analytic function f preserves angles at z_0 if $f'(z_0) \neq 0$. It is a little surprising that this statement has a converse.   That is, under the hypothesis that $\lim_{z \to z_0, \, z \in S} Q(z) = e^{i\theta}$ implies that $\lim_{z \to z_0, \, z \in S} A(z)$ exists and is equal to $e^{i(\theta + \psi)}$ for a constant ψ independent of S (f preserves angles at z_0), we can prove that $f'(z_0)$ exists. Furthermore, in the hypothesis, S need range only over line segments. A precise result follows.

7.5 THEOREM (CONVERSE TO 7.3)

Suppose that f is defined and differentiable (see (1.20)) on a neighborhood of z and that $f(\zeta) \neq f(z)$ if $\zeta \neq z$. If there is a constant ψ such that

(i) $\lim\limits_{t \to 0} \dfrac{f(z + te^{i\theta}) - f(z)}{|f(z + te^{i\theta}) - f(z)|} = e^{i(\theta + \psi)}$

for all θ, then $f'(z_0)$ exists.

PROOF.   By the definition of a differential and some algebraic manipulation (as in (1.21)) we see that there are constants α and β such that

(ii) $\dfrac{f(z + \xi) - f(z)}{\xi} = \alpha + \beta \dfrac{\bar{\xi}}{\xi} + \dfrac{R(\xi)}{\xi}$

and $\lim_{\xi \to 0} R(\xi)/\xi = 0$ [see (1.20)]. Taking $\xi = te^{i\theta}$, we obtain the equality

(iii) $\dfrac{|f(z + te^{i\theta}) - f(z)|}{t}$

$$= e^{i\theta} \left[\alpha + \beta e^{-2i\theta} + \dfrac{R(te^{i\theta})}{te^{i\theta}} \right] \dfrac{|f(z + te^{i\theta}) - f(z)|}{f(z + te^{i\theta}) - f(z)}$$

for all θ.   The limit as $t \to 0$ of the right side in (iii) exists and equals $[\alpha + \beta e^{-2i\theta}]e^{-i\psi}$ by (i). Since the left side is real-valued for all θ, β must be zero. The limit as $\xi \to 0$ in (ii) thus exists, and so $f'(z)$ exists.

Remark.   In the above proof, select $\theta = 0$ and $\theta = \pi/2$ and subtract the corresponding values of $[\alpha + \beta e^{-2i\theta}]e^{-i\psi}$; this gives $\beta = re^{i\psi}$ for a real number r. Select $\theta = \pi/4$ and $\theta = -\pi/4$ and subtract to obtain $\beta = i\rho e^{i\psi}$ for a second real number ρ proving that $\beta = 0$. This last proof uses only four values for θ: 0, $\pi/4$, $\pi/2$, and $-\pi/4$. The hypothesis for (7.5) can correspondingly be weakened. The preservation of only four particular angles and the existence of a differential imply analyticity.

The following theorem will be useful in several of our examples.

7.6 THEOREM

Let f be a univalent function from an open set U onto V. Let $\mathbf{a}$ be the collection of connected components of U. Then V is open and the family of sets $\{f(A) \mid A \in \mathbf{a}\}$ is precisely the family of connected components of V.

PROOF. Since f is one-to-one on U, the sets $\{f(A) \mid A \in \mathbf{a}\}$ are pairwise disjoint and we have

$$f(U) = f\left(\bigcup_{\mathbf{a}} A\right) = \bigcup_{\mathbf{a}} f(A) = V.$$

The components of U are open (4.16.iii) and so each $f(A)$ is open by the open mapping theorem (5.13). Also V is open. Let B be a connected component of V. We have

$$\bigcup_{\mathbf{a}} f(A) \supset B.$$

Hence B is contained in a union of pairwise disjoint open sets. It follows from the connectedness of B that $f(A) = B$ for some $A \in \mathbf{a}$. Hence every component of V is $f(A)$ for some $A \in \mathbf{a}$.

The function f^{-1} is univalent from V onto U. By the above argument, for each $A \in \mathbf{a}$ there is a component B of V such that $A = f^{-1}(B)$. We have $f(A) = B$, and so $f(A)$ is a component of V.

7.7 EXAMPLES OF UNIVALENT FUNCTIONS

(i) Let $f(z) = z^2$. Since $f'(z) = 2z$, f is conformal on $\mathbf{C}\backslash\{0\}$. If l is a line through the origin, then f is one-to-one, and hence univalent, on each of the open half-spaces determined by l (the reader should check this) and maps each closed half-space onto $\mathbf{C}$. Let x_0 and y_0 be positive constants. We saw in (1.1) that f maps the lines $\{\operatorname{Im}(z) = y_0\}$ and $\{\operatorname{Re}(z) = x_0\}$ onto parabolas (see Figure 7.2). Let H be the open half-space $\{z \mid \operatorname{Im}(z) > 0\}$, let

$$X = \{x + iy \mid x = x_0 \quad \text{and} \quad y > 0\} \cup \{x + iy \mid y = y_0\},$$

and let $H_1 = H\backslash X$. The components $\{A_j\}_{j=1}^4$ of H_1 are shown in Figure 7.2. The image of H under f is $\mathbf{C}\backslash\{x \mid x \geq 0\}$. The image of X is all of the parabola $\{(x + iy)^2 \mid y = y_0\}$ together with that portion of $\{(x + iy)^2 \mid x = x_0\}$ for which $y > 0$ (see Figure 7.2). The components $\{B_j\}_{j=1}^4$ of $f(H_1)$ are also shown in Figure 7.2. By Theorem (7.6), $f(A_j)$ is some B_k for each j; which one? We need only pick a point z_j in each component A_j and see in which

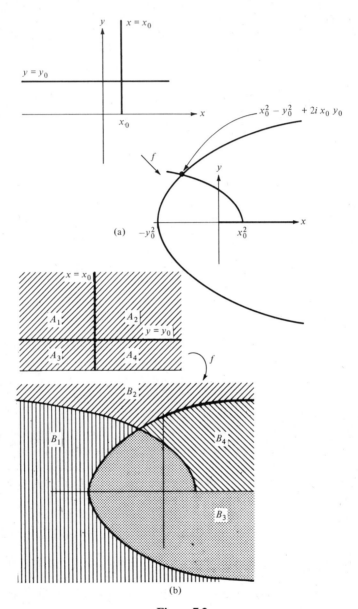

(a)

(b)

Figure 7.2

$B_i f(z_j)$ falls. Doing this (or using other evident arguments) we see that $f(A_j) = B_j$ for each j.

Example (ii). The function exp is one-to-one and hence univalent on the strip $S = \{z \mid a < \text{Im }(z) < a + 2\pi\}$, where a is any fixed real number. It maps S onto $\mathbf{C}\backslash\{re^{ia} \mid r \geq 0\}$ and it maps $\bar{S}$ onto $\mathbf{C}\backslash\{0\}$. The inverse of exp on exp (S) is $\text{Log}_{a + \pi}$ [see Section (2.14)]. On the semistrip $S_1 = \{z \in S \mid 0 < \text{Re }(z)\}$, we have

$$\exp (S_1) = \{z \mid |z| > 1 \quad \text{and} \quad z \neq re^{ia} \text{ for all } r > 0\}.$$

See Figure 7.3.

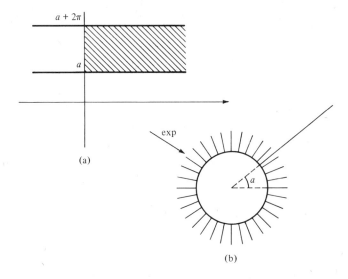

(a)

exp

(b)

Figure 7.3

(iii) *Logarithm.* The function Log is one-to-one and analytic on the set V consisting of the complement of the nonpositive part of the x-axis. It thus provides a univalent mapping of V onto the strip $S = \{z \mid -\pi < \text{Im }(z) < \pi\}$. If we restrict Arg (z) to $]0, \pi[$, then we obtain a univalent mapping from the upper half-plane to the open strip $\{0 < \text{Im }(z) < \pi\}$. If we let $B = \{z \mid |z| < 1 \text{ and } \text{Im }(z) > 0\}$, then Log $(B) = \{z \mid 0 < \text{Im }(z) < \pi$ and Re $(z) < 0\}$.

(iv) *Logarithm in a simply connected region.* It is easy to verify that Log_θ is univalent on $A = \mathbf{C}\backslash\{re^{i(\theta + \pi)} \mid 0 \leq r < \infty\}$ [see (2.14)]. Notice that A is simply connected. A univalent logarithm can be defined on any simply connected region not containing 0. Let U be such a region and fix $z_0 \in U$. By (4.28) the integral

$$L(z) = L(z, z_0) = \int_\gamma \frac{1}{\zeta} \, d\zeta$$

is independent of the path γ in U from z_0 to z. Since

$$L'(z) = \frac{1}{z},$$

L is locally one-to-one by (5.15). Let us prove the equality

$$\exp\left[L(z, z_0)\right] = \frac{z}{z_0}. \tag{1}$$

PROOF OF (1). The function $\phi(z) = \exp\left[L(z, z_0)\right]$ is analytic on U and the relation $\phi'(z) = \phi(z)/z$ holds. From the resulting equality

$$\frac{d}{dz}\left(\frac{\phi(z)}{z}\right) = 0,$$

we have $\phi(z) = cz$ for a constant c. Since $\phi(z_0) = 1$, we have $c = z_0^{-1}$ and the proof of (1) is complete.

If $L(z, z_0) = u + iv$, then the relation $e^{u+iv} = z/z_0 = \exp\left(\text{Log } |z/z_0| + i \text{ Arg } (z/z_0)\right)$ shows that the equalities

$$\text{Re}\left[L(z, z_0)\right] = \log |z/z_0|; \quad \text{Im}\left[L(z, z_0)\right] = \text{Arg}\left(\frac{z}{z_0}\right) + 2\pi n$$

hold for some integer n. Thus for each z the equality

$$L(z, z_0) = \text{Log}\left(\frac{z}{z_0}\right) + 2\pi i n \tag{2}$$

is valid for some integer n, which may depend on z.

Next we show that

if $L(z, z_0)$ attains a value w, it omits $\{w + 2\pi i k \mid k \in \mathbf{Z}\}$.   (3)

To prove (3), suppose that $L(z, z_0) = w$. If $L(z_1, z_0) = w + 2\pi i n$ for an integer n, then we have

$$\frac{z}{z_0} = e^w, \frac{z_1}{z_0} = e^{w+2\pi i n} = e^w,$$

and hence $z = z_1$. Since L is single-valued, n is 0.

In conclusion, we prove that $L(z, z_0)$ is one-to-one and hence univalent on U. If $L(z, z_0) = L(z_1, z_0)$, then by (2) we have $|z| = |z_1|$ and Arg

$(z/z_0) = \text{Arg } (z_1/z_0)$. Hence we have $\text{Arg } (z) = \text{Arg } (z_1) + 2\pi\delta$, where $\delta \in \{-1, 0, 1\}$. By (3), δ must be 0.

(v) *Square root.* For a simply connected region U such that $0 \notin U$, fix $z_0 \in U$ and a value $\sqrt{z_0}$ of $z_0^{1/2}$. Let

$$S(z, z_0) = \sqrt{z_0}\, e^{(1/2)L(z, z_0)},$$

where $L(z, z_0)$ is as in (iv). It is clear that

$$[S(z, z_0)]^2 = z$$

holds. If $S(z_1, z_0) = S(z, z_0)$, then we obtain $L(z, z_0) = L(z_1, z_0)$, implying that $z = z_1$. Hence $S(z, z_0)$ is one-to-one (hence univalent) on U. The function $S(z, z_0)$ is another branch of the square root function. Special cases were discussed in (2.16).

EXERCISES

TYPE I

1. Show that the given function f is univalent on the given set S, find the image of S under the mapping, and identify the image of the boundary of S under f in each case.
   (a) $f(z) = z^m$, $S = \{z \mid 0 < \text{Arg } (z) < \pi/m\}$, m a positive integer.
   (b) $f(z) = \sin (z)$, $S = \{z \mid \text{Im } (z) > 0 \text{ and } |\text{Re } (z)| < \pi/2\}$.
   (c) $f(z) = \cot (z)$, $S = \{z \mid 0 < \text{Re } (z) < \pi\}$.

2. Let $f(z) = z + 1/z$.
   (a) Show that f is univalent on $S = \{z \mid \text{Im } (z) > 0 \text{ and } |z| > 1\}$.
   (b) Show that $\text{Im } [z + 1/z] = y - y/(x^2 + y^2)$ and $\text{Re } [z + 1/z] = x + x/(x^2 + y^2)$.
   (c) Sketch the curves $\text{Im } [z + 1/z] = c$ where c is a constant (examine the cases $c > 0$, $c = 0$, and $c < 0$).
   (d) Refer to the comments just prior to the Exercises on p. 105 and show that $z + 1/z$ solves the problem of fluid flow around an obstacle whose boundary is $x^2 + y^2 = 1$. Compute the formula for the velocity at each point.

3. What is the image of the set $\{z \mid |z| < 1\}$ under the mapping $z \to z + 1/z$?

4. (a) Show that $az + 1/(az)$ $(a > 0)$ is univalent on $\{z \mid |z| > 1\}$.
   (b) Show that the image of $\{z \mid |z| = 1\}$ under the mapping $w = az + 1/(az)$ is the ellipse having equation

$$\frac{a^2 v^2}{(1 - a^2)^2} + \frac{a^2 u^2}{(1 + a^2)^2} = 1,$$

   where $w = u + iv$.

5. Find the image of the annulus $A = \{z \mid a < |z| < b\}$ $(0 < a < b)$ under Log.

6. Show that the image under z^2 of the circle $C(a, a)$, a positive, is the cardioid having polar equation $\rho = 2a^2(1 + \cos \theta)$.

7.8 LINEAR FRACTIONAL TRANSFORMATIONS

Let a, b, c, and d be complex numbers satisfying the relation $ad - bc \neq 0$. Define a function T by

(i) $\quad T(z) = \dfrac{az + b}{cz + d}$.

If $c = 0$, then d cannot be zero and in this case T is a very simple analytic function on $\mathbf{C}$. If $c \neq 0$, then T is defined and analytic on $\mathbf{C} \backslash \{-d/c\}$. If $z = -d/c$, then we have

$$az + b = -\left(\frac{ad - bc}{c}\right) \neq 0,$$

and so the equality

$$\lim_{z \to -d/c} |T(z)| = \infty$$

holds. For this reason we set $T(-d/c) = \infty$. Since

$$\lim_{|z| \to \infty} T(z) = \frac{a}{c},$$

we also set $T(\infty) = a/c$. In the case $c = 0$ we have $\lim_{|z| \to \infty} |T(z)| = \infty$ and so we let $T(\infty) = \infty$. With these conventions, T is defined from $\mathbf{C} \cup \{\infty\}$ to $\mathbf{C} \cup \{\infty\}$. It is convenient to write $\overline{\mathbf{C}}$ for the set $\mathbf{C} \cup \{\infty\}$.

If T is a function from $\overline{\mathbf{C}}$ to $\overline{\mathbf{C}}$ for which there are complex numbers a, b, c, and d such that $ad - bc \neq 0$ and (i) holds, then T is called a *linear fractional transformation*. Notice that two distinct 4-tuples (a, b, c, d) and (a', b', c', d') can define the same linear fractional transformation.

By (i), we have

$$T'(z) = \frac{ad - bc}{(cz + d)^2} \quad \left(z \neq -\frac{d}{c}\right).$$

Hence $T'(z) \neq 0$ if $z \neq -d/c$. (We use the convention $d/c = \infty$ if $c = 0$.) Thus every linear fractional transformation is conformal on $\mathbf{C} \backslash \{-d/c\}$. The reader can easily check (and should check) to see that the equation

$$w = \frac{az + b}{cz + d}$$

for given $w \in \overline{C}$ has the unique solution

$$z = \frac{-dw + b}{cw - a}.$$

Hence a linear fractional transformation T is one-to-one on $\overline{C}$.   On $C \setminus \{-d/c\}$, T is thus univalent. If T is given by (i), then T^{-1} is given by

$$T^{-1}(z) = \frac{-dz + b}{cz - a}.$$

We have $(-d)(-a) - bc = ad - bc \neq 0$, so that T^{-1} is again a linear fractional transformation.

7.9 MATRIX INTERPRETATION

A 2×2 complex-matrix $\begin{pmatrix} a & b \\ c & d \end{pmatrix}$ defines a linear fractional transformation provided that its determinant $ad - bc$ is different from zero. Suppose that $\begin{pmatrix} a & b \\ c & d \end{pmatrix}$ defines T and that $\begin{pmatrix} a_1 & b_1 \\ c_1 & d_1 \end{pmatrix}$ defines a second linear fractional transformation S. We calculate the composition map $S \circ T$ as follows:

$$(S \circ T)(z) = \frac{a_1 T(z) + b_1}{c_1 T(z) + d_1} = \frac{a_1 \dfrac{az + b}{cz + d} + b_1}{c_1 \dfrac{az + b}{cz + d} + d_1}$$

$$= \frac{(aa_1 + b_1 c)z + (a_1 b + b_1 d)}{(ac_1 + d_1 c)z + (c_1 b + d_1 d)}.$$

Thus it appears that $S \circ T$ is again a linear fractional transformation defined by coefficients which are the terms of the matrix product $\begin{pmatrix} a_1 & b_1 \\ c_1 & d_1 \end{pmatrix}\begin{pmatrix} a & b \\ c & d \end{pmatrix}$. Since the determinant of this product matrix is not zero, $S \circ T$ is indeed a linear fractional transformation.   Hence the composition of two linear fractional transformations is again a linear fractional transformation whose coefficients are computed from those of the original transformation by matrix multiplication.

Let G denote the set of all 2×2 complex matrices with nonzero determinant, and let $\mathfrak{L}$ denote the set of all linear fractional transformations. We have shown that the mapping ϕ from G onto $\mathfrak{L}$ given by

$$\phi \begin{pmatrix} a & b \\ c & d \end{pmatrix} = T,$$

(*T* as in 7.8.i), satisfies

$$\phi(AB) = \phi(A) \circ \phi(B).$$

Clearly $\phi(I)$ is the identity transformation. The mapping ϕ is thus a homo-morphism from the group G (with matrix multiplication) onto the group $\mathfrak{L}$ (with composition). (The reader unfamiliar with groups and homomorphisms can safely ignore this last paragraph.)

7.10 DEFINITIONS AND REMARKS

If $a \in \mathbf{C}$, then the linear fractional transformation T_a defined by $T_a(z) = z + a$ is called a *translation*. If $a \neq 0$, then the linear fractional trans-formation M_a defined by $M_a(z) = az$ is called a *multiplicative transformation* or simply a *multiplier*. If $|a| = 1$, then M_a is called a *rotation* and if $a > 0$ it is called a *dilation* (or *stretching*). A multiplier is thus a composition of a dilation and a rotation. The mapping $J(z) = 1/z$ is called an *inversion*. We denote the identity transformation by I; $I(z) = z$. Let $T \in \mathfrak{L}$ be given by $\begin{pmatrix} a & b \\ c & d \end{pmatrix}$. The diagram

$$z \to cz \to cz + d \to \frac{1}{cz + d} \to \frac{\dfrac{bc - ad}{c}}{cz + d} \to \frac{\dfrac{bc - ad}{c}}{cz + d} + \frac{a}{c} = \frac{az + b}{cz + d} = T(z)$$

shows that T can be written as a composition of translations, multipliers, and inversion.

7.11 IMAGES OF LINES AND CIRCLES

If T_a is a translation (in $\mathfrak{L}$), then the image of the circle $C(r, b)$ under T_a is the circle $C(r, b + a)$. The image of the line $\{b + re^{i\theta} \mid r \in \mathbf{R}\} = l$ under T_a is the line $l + a = \{(a + b) + re^{i\theta} \mid r \in \mathbf{R}\}$. If $a \neq 0$, then the multi-plier M_a takes $C(r, b)$ to the circle $C(r|a|, ab)$ and takes l to the line

$$\{ab + re^{i(\theta + \mathrm{Arg}(a))} \mid r \in \mathbf{R}\}.$$

Thus translations and multipliers each carry lines to lines and circles to circles. We showed in Example 3 of Section (1.1) that the inversion J carries a line to either a line or a circle and carries a circle to either a line or a circle. The reader may want to return to Section (1.1) at this point. From the decomposition in (7.10), it follows that every linear fractional transformation carries the collection of lines and circles to the same collection. This mapping property is extremely useful. We expand on it in the next theorem.

7.12 THEOREM

Let l be a line and let C be a circle. There is a $T \in \mathfrak{L}$ which carries C to l.

FIRST PROOF.   There is a translation which carries C to a circle C' through the origin. The inversion J carries C' to a line l' [see (1.1), Example 3]. A translation will carry l' to a line l'' through the origin, a rotation of l'' will give an l''' parallel to l, and finally a translation will carry l''' to l. The composition T of these maps is in $\mathfrak{L}$ and carries C to l.

SECOND PROOF.   Let (a, b, c) be three distinct points on C.   Let

(i) $\quad T(z) = \dfrac{b - c}{b - a} \dfrac{z - a}{z - c}.$

We have

(ii) $\quad T(a) = 0,\ T(b) = 1,\ T(c) = \infty.$

Since the image of C under T must be a line or a circle, the image is $\mathbf{R}$, by (ii). Let a', b', c' be three distinct points on l. As above, there is an $S \in \mathfrak{L}$ taking (a', b', c') to $(0, 1, \infty)$. Hence S takes l to $\mathbf{R}$. The composition $S^{-1} \circ T$ takes C to l. In fact, it takes the ordered triple (a, b, c) to the ordered triple (a', b', c').

7.13 EXAMPLES

(i) Suppose we want to take $\{z \mid |z - 1| = 1\}$ onto the line $\{z \mid \text{Re } (z) = \text{Im } (z)\}$ by $T \in \mathfrak{L}$. We could use either of the techniques used to prove (7.12), but it is actually just as easy to proceed directly. Thus suppose that $T(z) = (az + b)/(cz + d)$. The problem is to find $\begin{pmatrix} a & b \\ c & d \end{pmatrix}$. Since a circle is to go to a line, we have $c \neq 0$. Normalize by letting $c = 1$. If we demand that $T(0) = 0$ and $T(2) = \infty$, we obtain

$$\frac{b}{d} = 0 \quad \text{and} \quad 2 + d = 0,$$

so that $b = 0$ and $d = -2$. Finally, we demand that $T(1 + i) = 1 + i$, so that

$$\frac{a(1 + i)}{(1 + i) - 2} = 1 + i$$

holds. Solving for a, we obtain $a = -1 + i$. Hence a suitable T is given by

$$T(z) = \frac{(-1 + i)z}{z - 2}.$$

We have, of course, used the fact that the three points $(0, 2, 1 + i)$ determine the circle and that $(0, 1 + i, \infty)$ determine the line. The mapping is illustrated in Figure 7.4. In Section (7.14) we will see that the shaded areas correspond under the mapping T.

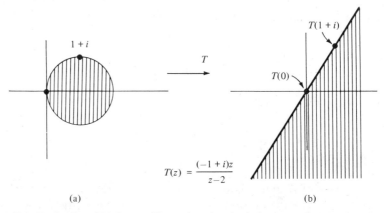

(a)   (b)

Figure 7.4   In Section (7.14) we will see that the shaded areas correspond under the mapping T.

(ii) In Example (i) we found a transformation which does a particular job. Suppose now that we are given a T and a circle (or line) C and we ask what happens to C under T. For example, to find the image of $C = \{z \mid |z| = 1\}$ under

$$T(z) = \frac{z + 1}{z - 1}$$

we need only find the image of three distinct points on C under T. We have $T(-1) = 0$, $T(1) = \infty$, and $T(i) = -i$. Hence T carries C to the y-axis.

Remarks.   If $T(z) = (az + b)/(cz + d)$, then at least one of a or c is not zero. If $a \neq 0$, then we can write $T(z) = (z + (b/a))/[(c/a)z + (d/a)]$. If $c \neq 0$, then we have

$$T(z) = \frac{\left(\dfrac{a}{c}\right) z + \dfrac{b}{c}}{z + \dfrac{d}{c}}.$$

Thus T is actually determined by only three constants. There is nothing magical about three points determining T if one looks at the problem from the proper geometric viewpoint. Consider a straight line to be a circle through the point ∞. From this viewpoint the collection of circles and lines referred to in Section (7.11) becomes simply the collection of circles. From elementary geometry it is well known that three points determine a circle. Since T maps any given circle, C, onto another circle, C', we need only to compute the value of T at three points of C to determine C'. Hence it is reasonable that the value of T at three points suffices to determine T uniquely.

7.14 ORIENTATION

If C is a circle, an ordered triple (a, b, c) on C induces an orientation for C in the following manner. One starts at a and goes along the arc which does not contain c to get to b. If C is a line, an ordered triple (a, b, c) on C also induces an orientation for C. Again one goes first from a to b along the segment (or ray) which does not include c. Any of a, b, c can be ∞. If T is a linear fractional transformation, then T induces an orientation of C onto $T(C)$. Thus if C is oriented by (a, b, c), then $T(C)$ is oriented by $(T(a), T(b), T(c))$. With respect to this induced orientation, T maps the left-connected component of $\mathbf{C} \backslash C$ onto the left-connected component of $\mathbf{C} \backslash T(C)$. This fact is nearly obvious for translations, is easily verified for multipliers, and it can be verified by consideration of cases (see Exercise 9, p. 245) for inversion. By (7.10), it then follows for all T.

In mapping a circle to a line or a line to a circle, one often wants to specify where the components of the complement are mapped as well as the curves. For example, in taking $\{z \mid |z - 1| = 1\}$ to $\{z \mid \text{Re}(z) = \text{Im}(z)\}$ in (7.13.i) we mapped $(0, 1 + i, 2) \rightarrow (0, 1 + i, \infty)$, thereby taking the clockwise orientation of the circle to the "upward" orientation of the line and the component $D(1, 1)$ to $\{z \mid \text{Re}(z) > \text{Im}(z)\}$. By requiring that $(0, 1 + i, 2) \rightarrow (0, -1 - i, \infty)$, we would find a T mapping $D(1, 1)$ to $\{z \mid \text{Re}(z) < \text{Im}(z)\}$.

As another example, we find a T which maps the upper half-plane $H = \{x + iy \mid y > 0\}$ onto $D(1, 0)$. Normalizing by setting $c = 1$ so that $T(z) = (az + b)/(z + d)$ and requiring that

(i) $T: (-1, 0, \infty) \rightarrow (-1, 1, i)$

we obtain the equations

$$\frac{b}{d} = 1; \quad \frac{-a + b}{d - 1} = -1; \quad a = i.$$

These give us the equality

$$T(z) = \frac{2iz + (1 + i)}{2z + (1 + i)}.$$

If instead of (i), we require that

(ii) $S: (-1, 0, \infty) \rightarrow (-1, 1, -i)$,

then we obtain

$$S(z) = \frac{2iz + (1 - i)}{2z + (1 - i)}$$

and S maps the lower half-plane to $D(1, 0)$. If we require that

$$R: (0, i, \infty) \rightarrow (1, i, -1),$$

then R will take the left half-plane to $D(1, 0)$. Solution of the necessary equations yields

$$R(z) = \frac{z + 1}{-z + 1}.$$

7.15 MORE EXAMPLES OF UNIVALENT FUNCTIONS

Since the composition of two univalent functions is univalent, we can obtain some examples by composing linear fractional transformations with previously discussed examples. Let D denote $D(1, 0)$, U the open upper half-plane $\{z \mid \text{Im}(z) > 0\}$ and S the infinite horizontal strip $\{z \mid 0 < \text{Re}(z) < \pi\}$.

(i) *The disk to the strip.* If T is a linear fractional transformation from D to U, then Log (T) takes D to S. As an example, we mention

$$f(z) = \text{Log}\left(i \, \frac{-z + 1}{z + 1}\right).$$

(ii) *Semidisk to disk.* The unit semidisk $B = D \cap U$ can be mapped to D as follows. First map B to the first quadrant Q_1 by $(z + 1)/(-z + 1)$. (This mapping takes the unit circle to the imaginary axis and the real axis to itself.) Take Q_1 to U by z^2 and finally take U to D by $(-z + i)/(z + i)$. The function

$$f(z) = \frac{-(z + 1)^2 + i(-z + 1)^2}{(z + 1)^2 + i(-z + 1)^2}$$

thus carries B univalently to D. The inverse of f is easily found to be

$$f^{-1}(z) = \frac{F_{1/2}\left(\dfrac{-iz + i}{z + 1}\right) - 1}{F_{1/2}\left(\dfrac{-iz + i}{z + 1}\right) + 1}.$$

See (2.16) for the definition of $F_{1/2}$.

EXERCISES

TYPE I

1. Which of the following matrices define a linear fractional transformation?
   Write a formula for the transformation in each case.

   (a) $\begin{pmatrix} 1 & 0 \\ 0 & 1 \end{pmatrix}$ (b) $\begin{pmatrix} 1 & 1 \\ 1 & 1 \end{pmatrix}$ (c) $\begin{pmatrix} 1 & 1 \\ 1 & -1 \end{pmatrix}$ (d) $\begin{pmatrix} 1 & 0 \\ 1 & 0 \end{pmatrix}$.

2. In (a–d), find a linear fractional transformation T with the specified
   mapping property.
   (a) $T: C(1, 0) \to C(8, 5)$
   (b) $T: C(1, 0) \to \{z \mid \text{Im } (z) = 2\}$
   (c) $T: \{z \mid \text{Re } (z) = 0\} \to C(1, i)$
   (d) $T: \{z = x + iy \mid x + 2y = 1\} \to C(1, 0)$.

3. Find the image of $C(1, i)$ under each of the transformations found in
   Exercise 2. For (a) and (b), sketch $C(1, 0)$ and $C(1, i)$ on one coordinate system and their images on another coordinate system. Observe
   the preservation of angles.

4. Map the half-plane $H = \{x + iy \mid y > 0\}$ onto $D(1, 0)$ by a T satisfying
   $T(0) = i$, $T(1) = -1$, $T(\infty) = -i$.

5. Map the half-plane $H = \{x + iy \mid x > 1\}$ onto $D(1, 0)$ by a linear
   fractional transformation.

6. Let $0 < \theta < \pi/2$, $y_0 > 0$, and $L = \{iy_0 + te^{i\theta} \mid t \in R\}$. Find a T such
   that T maps the upper half-plane of L onto the unit disk.

7. Map the unit disk $D(1, 0)$ onto itself by T's satisfying the following
   conditions.
   (a) $T(0) = i$, $T(i) = -1$, $T(-1) = -i$
   (b) $T(0) = -i$, $T(i) = -i$, $T(-1) = (1 - i)/\sqrt{2}$.

8. Construct two linear fractional transformations that map the unit circle
   to itself but that switch the connected components of the complement of
   the circle. (Notice the preservation of "left side.")

9. Let $J(z) = 1/z$. Let C be an oriented circle or line. Verify that J maps
   the left component of $\mathbf{C}\backslash C$ onto the left component of $\mathbf{C}\backslash J(C)$ by considering the following cases.
   (a) C is a line through the origin
   (b) C is a line which does not go through the origin
   (c) C is a circle containing the origin
   (d) C is a circle which does not contain the origin.
   Sketch an example of C and $J(C)$ for each of the cases (a–d). [Example
   3 (1.1) may be useful.]

10. Inversion need not preserve counterclockwise orientation of a circle. To see this, verify the following assertions. "Positive" orientation for circles means counterclockwise and for lines means "up" if possible, otherwise "right."

   (i) Inversion reverses a positively oriented circle with 0 in its interior, preserves one with 0 in its exterior.

   (ii) Inversion preserves positive orientation of any line through 0.

   (iii) Inversion preserves positive orientation of a circle containing 0 if the center of the circle is in the left half-plane and reverses orientation if the center is in the right half-plane; it preserves orientation if the center is on the negative y-axis and reverses orientation if the center is on the positive y-axis.

TYPE II

1. With reference to (7.10), let

$$\mathfrak{T} = \{T_a \mid a \in \mathbf{C}\}$$

$$\mathfrak{M} = \{M_a \mid a \in \mathbf{C}\backslash\{0\}\}$$

$$\mathfrak{J} = \{I, J\}.$$

   Show that each of $\mathfrak{T}$, $\mathfrak{M}$, $\mathfrak{J}$ is a subgroup of $\mathfrak{L}$. Let ϕ be the homomorphism mentioned in (7.9). Find the matrix groups $\phi^{-1}(I)$, $\phi^{-1}(\mathfrak{J})$, $\phi^{-1}(\mathfrak{T})$, and $\phi^{-1}(\mathfrak{M})$. (That is, find the matrices which correspond to I, to inversion, to translations, and to multipliers.)

2. See Exercise 1 for notation. The diagram in (7.10) shows that each element S of the group $\mathfrak{L}$ can be written in the form

$$S = TMJ\tilde{T}\tilde{M}$$

   where T and $\tilde{T}$ are translations, M and $\tilde{M}$ multipliers, and J inversion. Show that every $S \in \mathfrak{L}$ can be written as a product of four elements chosen from the three groups $\mathfrak{T}$, $\mathfrak{M}$, and $\mathfrak{S}$. Show also that there are elements $S \in \mathfrak{L}$ which cannot be written as the product of three elements from the three groups $\mathfrak{T}$, $\mathfrak{M}$, and $\mathfrak{S}$.

7.16 THE DIRICHLET PROBLEM FOR A HALF-PLANE

To find the general solution for the Dirichlet problem for a half-plane we examine the linear fractional transformation

$$z = \frac{w - i}{w + i}$$

carrying the half-plane Im $(w) > 0$ onto the disk $|z| < 1$. Since the real axis Im $(w) = 0$ is mapped to the circle $|z| = 1$ under this transformation, for each real t there is a unique s (a real number or ∞) such that

$$e^{it} = \frac{s - i}{s + i}. \tag{1}$$

If t is restricted to $]-\pi, \pi]$, then t is uniquely determined by s. To "lift" the Poisson formula from the disk to the half-plane we simply substitute $(s - i)/(s + i)$ for e^{it} and $(w - i)/(w + i)$ for z in the Poisson kernel. If $z = re^{i\theta}$ $(r < 1)$ and $w = u + iv$, then we have

$$P_r(\theta - t) = \text{Re}\left[\frac{e^{it} - z}{e^{it} + z}\right]$$

$$= \text{Re}\left[\frac{\dfrac{s - i}{s + i} - \dfrac{w - i}{w + i}}{\dfrac{s - i}{s + i} + \dfrac{w - i}{w + i}}\right]$$

$$= \text{Re}\left[\frac{(s - i)(w + i) - (w - i)(s + i)}{(s - i)(w + i) + (w - i)(s + i)}\right]$$

$$= \frac{(s^2 + 1)v}{(u - s)^2 + v^2}.$$

Thinking of t as a function of s and differentiating each side of (1) we have

$$ie^{it}\frac{dt}{ds} = \frac{(s + i) - (s - i)}{(s + i)^2},$$

or

$$\frac{s - i}{s + i}\frac{dt}{ds} = \frac{2}{(s + i)^2},$$

which simplifies to

$$\frac{dt}{ds} = \frac{2}{s^2 + 1}.$$

We use these calculations to write the Poisson integral of a piecewise continuous function f on the unit circle as an integral over the real line. The function F defined on Im $(w) = 0$ by $F(s) = f[(s - i)/(s + i)]$ is piecewise continuous on Im $(w) = 0$. We have

$$Pf(z) = \frac{1}{2\pi} \int_{[-\pi, \pi]} f(e^{it}) P_r(\theta - t) \, dt$$

$$= \frac{1}{\pi} \int_{]-\infty, \infty[} f\left(\frac{s - i}{s + i}\right) \frac{(s^2 + 1)v}{(s - u)^2 + v} \frac{1}{(1 + s^2)} \, ds$$

$$= \frac{1}{\pi} \int_{]-\infty, \infty[} F(s) \frac{v}{(s - u)^2 + v^2} \, ds = F(w),$$

where we have let the final integral define $F(w)$ for $w = u + iv$ and $z = (w - i)/(w + i)$, $|z| < 1$.

We remarked in (7.1) that the composition of a univalent function with an harmonic function is harmonic. Since the equality $F(w) = f[(w - i)/(w + i)]$ holds and the function of w given by the formula $z = (w - i)/(w + i)$ is univalent, we know that F is harmonic on the half-plane Im $(w) > 0$. This means that F solves the Dirichlet problem in the half-plane Im $(w) > 0$ with boundary values $F(s)$.

Evidently the function F defined by

$$F(w) = \frac{1}{\pi} \int_{[-\infty, \infty]} F(s) \frac{v}{(s - u)^2 + v^2} \, ds$$

solves the Dirichlet problem in the half-plane Im $(w) > 0$ for piecewise continuous functions F on Im $(w) = 0$ whenever we can find a piecewise continuous function f on $\mathbf{T}$ such that $F(s) = f[(s - i)/(s + i)]$. If we have found such an f on T and we happen to know the solution to the Dirichlet problem on the disk corresponding to this f, then we can solve the original problem in the half-plane Im $(w) > 0$ via the formula

$$F(w) = f\left(\frac{w - i}{w + i}\right).$$

The next example illustrates this latter technique.

Example 1.   Consider the function F defined on Im $(w) = 0$ by

$$F(s) = \begin{cases} 1 & \text{if } s > 0 \\ 0 & \text{if } s \le 0. \end{cases}$$

The image of the ray $\{s \le 0\}$ under $(s - i)/(s + i)$ is $\{e^{it} \mid 0 < t \le \pi\}$. Hence the function f is given by

$$f(e^{it}) = \begin{cases} 0 & \text{if } 0 < t \le \pi \\ 1 & \text{if } -\pi < t < 0; \end{cases}$$

it is not defined at 1. We considered the function $1 - f$ in (6.16.ii) and found that

$$P(1 - f)(z) = \frac{1}{2} + \text{Im}\left[\frac{2}{\pi}\sum_{k=1}^{\infty}\frac{z^{2k-1}}{(2k-1)}\right] = \frac{1}{\pi}\text{Arctan}\left(\frac{1-r^2}{-2r\sin\theta}\right)$$

for $z = re^{i\theta}$. It follows that

$$Pf(z) = 1 - \frac{1}{\pi}\text{Arctan}\left(\frac{1-r^2}{-2r\sin\theta}\right).$$

To obtain an expression for $F(w)$ where $z = (w - i)/(w + i)$, write

$$z = \frac{w-i}{w+i} = \frac{w-i}{w+i}\frac{\overline{w}-i}{\overline{w}-i} = \frac{(u^2+v^2-1)-2ui}{u^2+v^2+2v+1}.$$

Hence we have

$$\frac{1-|z|^2}{2\,\text{Im}\,(z)} = \frac{1 - \dfrac{u^2+(v-1)^2}{u^2+(v+1)^2}}{-\dfrac{4u}{u^2+(v+1)^2}} = -\frac{v}{u}.$$

We thus obtain

$$F(w) = 1 - \frac{1}{\pi}\text{Arctan}\left(\frac{v}{u}\right) \quad (w = u + iv).$$

Example 2. Sometimes it is possible to so simplify a problem by a univalent map that the solution becomes obvious. To give an example, consider the Dirichlet problem in the upper half-plane U for the boundary function f defined by

$$f(x) = \begin{cases} 1 & \text{if } x \in [-1, 1] \\ 0 & \text{if } x \notin [-1, 1]. \end{cases}$$

First, we map $[-1, 1]$ to a half-line by a linear fractional transformation which carries the x-axis to itself. The transformation

$$T(z) = \frac{z-1}{z+1}$$

does this. Next, take U to the strip $S = \{0 < \text{Im}\,(z) < \pi\}$ by Log. Thus

$$M(z) = \text{Log}\left(\frac{z-1}{z+1}\right)$$

maps U to S, $]-1, 1]$ to $l_1 = \{\text{Im}(z) = \pi\}$, and $R\backslash]-1, 1]$ to $l_2 = \{\text{Im}(z) = 0\}$. Let $F = f \circ M^{-1}$, so that $F = 0$ on l_2 and $F = 1$ on l_1. The problem

$$\begin{cases} w = F \text{ on } l_1 \cup l_2 \\ \dfrac{\partial^2 w}{\partial x^2} + \dfrac{\partial^2 w}{\partial y^2} = 0 \text{ on } S \end{cases}$$

has the obvious solution $w(z) = (1/\pi) \text{Im}(z)$. Accordingly,

$$u(z) = \frac{1}{\pi} \text{Im}\left[\text{Log}\left(\frac{z-1}{z+1}\right)\right] = \frac{1}{\pi} \text{Arg}\frac{z-1}{z+1}$$

solves the original problem.

EXERCISES

1. Let $F(s)$ be defined on the real line by the formula

$$F(s) = \begin{cases} 1 & s > 0 \\ 0 & s \leq 0. \end{cases}$$

Perform the integration indicated by the formula

$$F(w) = \frac{1}{\pi} \int_{]-\infty, \infty[} F(s)\frac{v}{(u-s)^2 + v^2}\, ds \quad (w = u + iv)$$

to solve the Dirichlet problem on $\text{Im}(w) > 0$. Check your answer against the example in (7.16).

7.17 THE SCHWARZ-CHRISTOFFEL FORMULA

The Schwarz-Christoffel formula is a formula which gives, in terms of an integral, an analytic function F which maps the open upper half-plane $\{z \mid \text{Im } z > 0\}$ onto the interior of a closed polygon. We will present a descriptive derivation of the formula and then examine some examples. Throughout this section, we let $H = \{z \mid \text{Im}(z) > 0\}$ and $\bar{H} = \{z \mid \text{Im}(z) \geq 0\}$.

To set the stage let us have two copies of the complex plane, one called the z-plane and one called the w-plane. The domain of the transformations will be the z-plane and the polygons to which we map will lie in the w-plane. It will be convenient to use two branches of the argument. In the w-plane, we will write

$$w = |w|e^{i\theta} \quad \text{where} \quad \theta = \text{Arg}(w) \quad \text{and} \quad -\pi < \text{Arg}(w) \leq \pi;$$

that is, we use the principal value of arg which we have used throughout the book. In the z-plane we will use $\text{Arg}_{\pi/2}$ so that

$$z = |z|e^{i\theta} \quad \text{where} \quad \theta = \text{Arg}_{\pi/2}(z) \quad \text{and} \quad -\frac{\pi}{2} < \text{Arg}_{\pi/2}(z) \leq \frac{3\pi}{2}.$$

We begin by showing how to construct a continuous map on $\overline{H}$ which is analytic on H and maps the real axis into two half-rays meeting at the origin in the w-plane at a prescribed exterior angle. Recall that the exterior angle ω formed by a triple w_1, w_2, w_3 of distinct complex numbers is defined by

$$e^{i\omega} \frac{w_2 - w_1}{|w_2 - w_1|} = \frac{w_3 - w_2}{|w_3 - w_2|} \quad (-\pi < \omega \leq \pi).$$

See Figure 7.5. The exterior angle which we have in mind can be written $\omega = k\pi$ where k satisfies $-1 < k < 1$.
Let x_1 be a point on the real axis in the z-plane. Let

$$f_1(z) = \frac{1}{(1-k)} e^{(1-k)\,\text{Log}_{\pi/2}\,(z-x_1)} \tag{1}$$

where $\text{Log}_{\pi/2}(z) = \log|z| + i\,\text{Arg}_{\pi/2}(z)$. Note that $\text{Log}_{\pi/2}$ is analytic on $\{z \mid z \neq re^{-i\pi/2}, r \geq 0\}$. The notation in (1) is cumbersome, and we replace it in Sections (7.17)–(7.19) with

$$f_1(z) = \frac{(z-x_1)^{1-k}}{1-k}, \tag{2}$$

where it is understood that the right side of (2) is the same as the right side of (1). Hence f_1 is analytic except on the ray

$$\{z \mid z = re^{-i\pi/2} + x_1, r \geq 0\}.$$

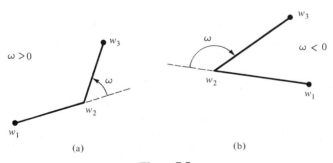

(a)   (b)

Figure 7.5

In particular, f_1 is analytic on an open connected set containing $\overline{H}\setminus\{x_1\}$. On this latter set we have $f_1'(z) = (z - x_1)^{-k}$, where we use the same convention as in (1) and (2). If Im $(z) = 0$ and Re $(z) < x_1$, then we have $\text{Arg}_{\pi/2}\,(z - x_1) = \pi$. If we think of $(z - x_1)^{-k}$ as a point in the w-plane, then we have

$$-\pi < \text{Arg}\,[(z - x_1)^{-k}] = -k\pi < \pi.$$

If Im $(z) = 0$ and Re $(z) > x_1$, then we have

$$\text{Arg}\,(z - x_1) = 0 \quad \text{and} \quad \text{Arg}\,[(z - x_1)^{-k}] = 0.$$

The direction of the tangents to the curves

$$\{f_1(x) \mid x < x_1\} \quad \text{and} \quad \{f_1(x) \mid x > x_1\}$$

at x are given by

$$\text{Arg}\,(f_1'(x)) = \text{Arg}\,(x - x_1)^{-k} = \begin{cases} -k\pi & x < x_1 \\ 0 & x > x_1. \end{cases}$$

Hence the image of each ray $\{x < x_1\}$ and $\{x > x_1\}$ under f_1 is linear. Since $f_1(x_1) = \lim_{x \to x_1} f_1(x) = 0$, the image of $\mathbf{R}$ under the mapping $w = f_1(z)$ thus consists of two rays meeting at the origin in the w-plane with exterior angle $k\pi$. Since $f_1(x) > 0$ if $x > x_1$ we have the interpretation in Figure 7.6.

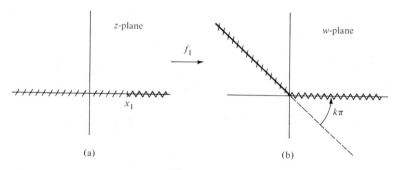

(a)   (b)

Figure 7.6

In summary, for given k ($-1 < k < 1$), we have seen how to find a function which is continuous on $\overline{H}$ so that the mapping is analytic on H, is linear on $\{x < x_1\}$ and $\{x > x_1\}$, and makes an exterior angle $k\pi$ at 0. Our next step is to use this technique to find a mapping which is continuous on $\overline{H}$, analytic on H, and which maps the real axis onto a closed polygon having n sides and prescribed exterior angles.

Let $k_1, k_2, \ldots, k_n$ be numbers such that $-1 < k_j < 1$ and $\sum_{j=1}^{n} k_j = 2$. Let $x_1 < x_2 < \cdots < x_n$ be n points on the real axis in the z-plane. Suppose that f is a function which is analytic on a connected open set containing $\overline{H}\backslash\{x_1, x_2, \ldots, x_n\}$, continuous on $\overline{H}$, and which satisfies

$$f'(z) = (z - x_1)^{-k_1}(z - x_2)^{-k_2} \ldots (z - x_n)^{-k_n} = \prod_{j=1}^{n} f_j'(z) \qquad (3)$$

where

$$f_j(z) = \frac{(z - x_j)^{(1-k_j)}}{1 - k_j}.$$

We use the same conventions as in (1) and (2) to define the exponentiation. Each f_j is the type of function f_1 analyzed in the preceding paragraph. The set $\{f(x) \mid x \text{ is real}\}$ is a curve defined on $\mathbf{R}$. For $x \neq x_j$, the curve has a tangent which makes an angle Arg $(f'(x)) = \sum_{j=1}^{n}$ Arg $(f_j'(x))$ with the real axis. By our analysis of the function f_1, we see that

$$\text{Arg } f_j'(x) = \begin{cases} -k_j\pi & x < x_j \\ 0 & x > x_j. \end{cases}$$

Hence Arg $f'(x)$ is constant on each open interval $]x_{j-1}, x_j[$ $(j = 2, 3, \ldots, n)$. On the ray $\{x < x_1\}$ the angle is -2π and on the ray $\{x > x_n\}$ the angle is zero. Thus if $\tau(x)$ denotes the unit tangent to the image of $\mathbf{R}\backslash\{x_1, \ldots, x_n\}$ under f (that is, $\tau(x) = e^{i \text{ Arg } (f'(x))}$, $x \notin \{x_1, \ldots, x_n\}$) then $\tau(x)$ is constant on the ray $\{x < x_1\}$, on each of the intervals $[x_{j-1}, x_j]$, and on the ray $\{x > x_n\}$. Thus the image of each of these sets under such an f is a line, a ray, or a line segment. If we imagine x moving from left to right along $\mathbf{R}$ we see that the angle that $\tau(x)$ makes with the positive real axis increases by the amount $k_j\pi$ as x crosses the point x_j. Hence $\tau(x)$ sweeps out a full circle in n steps of $k_j\pi$ each as x moves from $-\infty$ to ∞ along $\mathbf{R}$.

In order to obtain a function f with derivative as in (3), let z_0 be a fixed point with Im $(z_0) > 0$, let α be a path in $\overline{H}\backslash\{x_1, x_2, \ldots, x_n\}$ from z_0 to z, and set

$$f(z) = \int_{\alpha} (\xi - x_1)^{-k_1}(\xi - x_2)^{-k_2} \ldots (\xi - x_n)^{-k_n} d\xi. \qquad (4)$$

Since the integrand is analytic on H, the integral is independent of path. We can, for example, take $\alpha = \overrightarrow{[z_0, z]}$. It is clear that f satisfies (3). In order to show that f maps $\mathbf{R}$ onto a closed polygonal path we will show that

(i) f has a continuous extension to $z = x_i$, $i = 1, 2, \ldots, n$; and

(ii) there is a number L such that $\lim_{z \to \infty, \text{ Im } (z) > 0} f(z) = L$.

The continuity of f at the x_j implies that the image of each $[x_j, x_{j+1}]$ is a line segment. The existence of the limit in (ii) implies that the rays or segments $\{f(x) \mid x < x_1\}$ and $\{f(x) \mid x > x_n\}$ meet at L. Hence they are, in fact, segments. Since each of these later segments is parallel to the real axis, together they form the segment from $f(x_1)$ to $f(x_n)$.

To establish (i), let $g_j(z)$ denote

$$\prod_{i \neq j} (z - x_i)^{-k_i},$$

so that $g_j(z)(z - x_j)^{-k_j} = f'(z)$. Then g_j is analytic at $z = x_j$ and there are coefficients $(a_n)_{n=0}^{\infty}$ such that

$$f'(z) = (z - x_j)^{-k_j} g_j(z) = (z - x_j)^{-k_j} \sum_{n=0}^{\infty} a_n (z - x_j)^n$$

$$= a_0 (z - x_j)^{-k_j} + (z - x_j)^{1-k_j} \sum_{n=1}^{\infty} a_n (z - x_j)^{n-1}$$

holds. Let α be a path on $[0, 1]$ from z_0 to x_j such that $\alpha(t) \neq x_i$ if $i \neq j$ and $\text{Im }(\alpha(t)) \geq 0$. Let α_t denote α restricted to $[0, t]$. We have

$$\lim_{t \to 1} a_0 \int_{\alpha_t} (\xi - x_j)^{-k_j} \, d\xi = \frac{a_0}{1 - k_j} \left[\left[\lim_{t \to 0} (\alpha(t) - x_j)^{1-k_j} \right] - (z_0 - x_j)^{1-k_j} \right]$$

$$= \frac{a_0}{k_j - 1} (z_0 - x_j)^{1-k_j}.$$

Since $1 - k_j > 0$, the integral of $(z - x_j)^{1-k_j} \sum_{n=1}^{\infty} a_n (z - x_j)^{n-1}$ over α clearly exists and is the limit of integrals over α_t. Hence $\lim_{\text{Im }(z) \geq 0, \, z \to x_j} f(z)$ exists and if we define $f(x_j)$ as this limit, then f is continuous at $x_j, j = 1, 2, \ldots, n$.

To prove (ii) take $z_0 = i$ in the definition of f and consider first the limit $\lim_{y \to \infty} f(iy)$; that is, consider the limit along the y-axis. Since

$$\left[\left| \prod_{j=1}^{n} (it - x_j)^{-k_j} \right| \right] \leq \frac{1}{t^2},$$

we have

$$\lim_{y \to \infty} \int_{[1, y]} \left| \prod_{j=1}^{n} (it - x_j)^{-k_j} \right| \, dt < \infty \tag{5}$$

and hence the limit

$$\lim_{y \to \infty} \int_{[1, y]} \prod_{j=1}^{n} (it - x_j)^{-k_j} i \, dt = L \tag{6}$$

exists for some complex number L. Hence we have $\lim_{y \to \infty} f(iy) = L$. We will show that $\lim_{z \to \infty, \, \text{Im}(z) \geq 0} f(z) = L$. That is, we will show that for any $\varepsilon > 0$ there is an R such that $|f(z) - L| < \varepsilon$ whenever $|z| > R$ and $\text{Im}(z) \geq 0$. By (5) and (6), there is an R_1 such that

$$\left| \int_{[1, R]} \prod_{j=1}^{n} (iy - x_j)^{-k_j} \, dy - L \right| < \frac{\varepsilon}{2}$$

and

$$\int_{[R, \infty[} \left| \prod_{j=1}^{n} (iy - x_j)^{-k_j} \right| dy < \frac{\varepsilon}{2}$$

hold if $R \geq R_1$.

Let $m = \max \{|x_j| \mid j = 1, 2, \ldots, n\}$. If $\xi = Re^{i\phi}$ then the inequality $|\xi - x_j| \geq R - m$ holds for each j. Since $\sum_{j=1}^{n} k_j = 2$, we conclude that

$$\prod_{j=1}^{n} |\xi - x_j|^{-k_j} \leq \frac{1}{(R - m)^2}$$

if $R = |\xi| > m$. For $R > m$, let γ be the curve given by $\gamma(t) = Re^{it}$, $0 \leq t \leq \pi$. We have

$$\left| \int_{\gamma} \prod_{j=1}^{n} (\xi - x_j)^{-k_j} \, d\xi \right| \leq \frac{\pi R}{(R - m)^2}.$$

Since $\lim_{R \to \infty} \pi R/(R - m)^2 = 0$, there is an $R_2 > m$ so that $\pi R/(R - m)^2 < \varepsilon/2$ if $R \geq R_2$. Let M be the maximum of R_1 and R_2. Let z be such that $\text{Im}(z) \geq 0$ and $|z| > M$. Let $R = |z|$. Let Γ be the subarc of γ which lies in $\{z \mid \text{Im}(z) \geq 0\}$ and connects iR to z. Integrating along $\overrightarrow{[i, iR]}$ and then along Γ to obtain $f(z)$, we see that

$$f(z) = \int_{[1, R]} \prod_{j=1}^{n} (iy - x_j)^{-k_j} i \, dy + \int_{\Gamma} \prod_{j=1}^{n} (\xi - x_j)^{-k_j} \, d\xi.$$

The first integral on the right is $f(iR)$, and so we have

$$|f(z) - L| \leq |f(iR) - L| + \left| \int_{\gamma} \prod_{j=1}^{n} (\xi - x_j)^{-k_j} \, d\xi \right|.$$

Each of the summands on the right is less than $\varepsilon/2$ if $|z| > M$. Hence we have $\lim_{z \to \infty, \, \text{Im}(z) \geq 0} f(z) = L$.

For any positive integer n and any set $\{k_j\}_{j=1}^{n}$ in $[-1, 1]$ such that $\sum_{j=1}^{n} k_j = 2$, we can now find an analytic function f on H which has a continuous extension to $\mathbf{R} = \{z \mid \text{Im}(z) = 0\}$ such that $f(\mathbf{R})$ is a closed n-sided polygon Q^* having exterior angles $(k_j \pi)_{j=1}^{n}$. We will show that $f(H)$ lies in the union of the bounded connected components of $C\backslash Q^*$. Give Q^* the orientation induced by f, and call the resulting path Q. Then Q is essentially parametrized on $\mathbf{R}$ instead of on an interval. For $w_0 \notin Q^*$, we have

$$\text{Ind}_Q (w_0) = \frac{1}{2\pi i} \int_Q \frac{1}{w - w_0}\, dw = \frac{1}{2\pi i} \int_{]-\infty,\, \infty[} \frac{f'(x)}{f(x) - w_0}\, dx. \qquad (7)$$

We will show that $\text{Ind}_Q (w_0)$ is nonzero if and only if w_0 is equal to $f(z')$ for some z' in the upper half-plane. To this end, let $R > 1$ and consider the line $\{z \mid z = x + i/R\}$ and the circle $\{z \mid z = Re^{i\theta}\}$. These curves intersect at the points $-\sqrt{R^2 - 1/R^2} + i/R$ and $\sqrt{R^2 - 1/R^2} + i/R$. Let l_R be the path from $x = -\sqrt{R^2 - 1/R^2}$ to $x = \sqrt{R^2 - 1/R^2}$ along the line $\{z \mid z = x + i/R\}$ and γ_R be the path along the arc of the circle $\{z \mid z = Re^{i\theta}\}$ which lies above this line (see Figure 7.7) parametrized so that $l_R \smile \gamma_R$ is a closed curve. Evidently

$$\lim_{R \to \infty} \int_{\gamma_R} \frac{f'(z)}{f(z) - w_0}\, dz = 0$$

holds, since $\lim_{z \to \infty} f(z) = L$ and $|f'(z)| \le M/|z|^2$ for all large $|z|$ and some fixed M. It is clear that

$$\lim_{R \to \infty} \int_{l_R} \frac{f'(z)}{f(z) - w_0}\, dz = \int_{]-\infty,\, \infty[} \frac{f'(x)}{f(x) - w_0}\, dx.$$

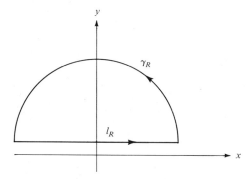

Figure 7.7

Hence from (6) we obtain the equality

$$\text{Ind}_Q (w_0) = \lim_{R \to \infty} \frac{1}{2\pi i} \int_{l_R \smile \gamma_R} \frac{f'(z)}{f(z) - w_0}\, dz \qquad (8)$$

if $w_0 \notin Q^*$.

Suppose first that $f(z') = w_0$ for $z' \in H$. For large enough R, z' is interior to $l_R \smile \gamma_R$. By (5.9) the integral

$$\frac{1}{2\pi i} \int_{l_R \smile \gamma_R} \frac{f'(z)}{f(z) - w_0} \, dz \tag{9}$$

is the number of zeros of $f - w_0$ interior to $l_R \smile \gamma_R$. It is thus nonzero and $\text{Ind}_Q(w_0) \neq 0$ holds.

Conversely if $\text{Ind}_Q(w_0) \neq 0$ holds, then by (7) the integral (8) is nonzero for all sufficiently large R. Hence by (5.9) there are z in H such that $f(z') = w_0$.

In particular, we have

$$\{w \in \mathbf{C} \backslash Q^* \mid \text{Ind}_Q(w) \neq 0\} \subset f(H) \subset Q^* \cup \{w \mid \text{Ind}_Q(w) \neq 0\}. \tag{10}$$

It may happen that points of H map to Q^*. (See Exercise 8, p. 265.) Since $f(H)$ is open, at least we know from (10) that it contains no boundary points of the unbounded component of $\mathbf{C} \backslash Q^*$. By (3), f is conformal on H.

In the case that f is one-to-one on $\mathbf{R}$, the image Q is a simple path. By (4.30) it follows that $\text{Ind}_Q(w_0)$ is 1, -1, or 0. It then follows from (7) above that f is one-to-one on H. Thus in the case that a simple path Q is obtained, f is univalent from H onto the interior of Q.

We summarize what we have proved in the following theorem.

(iii) *Theorem.* The integral in (4) is finite for all $z \in \bar{H}$. The function f which it defines is continuous on $\bar{H}$ and analytic on H. The image $f(\mathbf{R})$ is an n-sided polygon Q having vertices $(f(x_j))_{j=1}^n$. The exterior angle at $f(x_j)$ is $k_j \pi$. The image $f(H)$ satisfies (10) above. If f is one-to-one on $\mathbf{R}$, then f is univalent on H.

There remains an outstanding question. Let P be a given simple closed polygonal path having n sides. Let us number the n vertices of P so that if we started at the first vertex, w_1, and walked around P in such a way that we always kept the interior of P on our left, then we would encounter vertices $w_2, w_3, \ldots, w_n, w_1$ in that order. Let the exterior angle at the vertex w_j be $k_j \pi$. We have

$$e^{ik_j\pi} \frac{(w_j - w_{j-1})}{|w_j - w_{j-1}|} = \frac{(w_{j+1} - w_j)}{|w_{j+1} - w_j|} \quad \text{and} \quad -1 < k_j < 1.$$

Since P is a closed polygonal path, it is intuitively obvious that $\sum_{j=1}^n k_j = 2$. (The reader may want to give a proof by induction.) We can obtain a function f as above for which $f(\mathbf{R})$ is a closed polygonal path Q having the same exterior angles as P. This does not, however, mean that P is a translate of Q, or even that P and Q are similar. (Unless $n = 3$, in which case P and Q must be similar if they have the same exterior angles; see Exercise 8, p. 265.)

Do there exist points $x_1, x_2, \ldots, x_n$ so that f as defined by (4) is one-to-one and so that the Q obtained from f is similar to the given P? The answer is yes, but it is difficult to prove. One can base a proof on the Riemann mapping theorem. Once it is established that a Q similar to P can be obtained, a dilation of Q will yield a polygon congruent to P, and a rotation and translation will give P. Thus for constants A and B, the function $F = Af + B$ is univalent on H, one-to-one and continuous on $\bar{H}$, $F(\mathbf{R}) = P$, and $f(\mathrm{H})$ is the interior of P.

Let f be as in (4). A slightly simpler function having the same image as f can be found by composing f with a linear fractional transformation which maps $\mathbf{R}$ to $\mathbf{R}$, H to H, and ∞ to x_n. One of the vertices of the polygon will then correspond to the point ∞. Such a transformation T is given by

$$T(z) = x_n - \frac{1}{z}, \tag{11}$$

with inverse

$$T^{-1}(z) = \frac{1}{x_n - z}. \tag{12}$$

We have $T(a_j) = x_j$, where $a_j = 1/(x_n - x_j)$. The inequalities $a_j < a_{j+1}$ $(j = 1, 2, \ldots, n - 2)$ hold. We have

$$f(T(z)) = \int_{\overrightarrow{[z_0, \, T(z)]}} \prod_{j=1}^{n} \frac{1}{(\xi - x_j)^{k_j}} \, d\xi$$

$$= \int_{\overrightarrow{[T^{-1}(z_0), \, z]}} \prod_{j=1}^{n} \frac{1}{(T(\xi) - x_j)^{k_j}} \frac{1}{\xi^2} \, d\xi$$

$$= \int_{\overrightarrow{[z_1, z]}} \prod_{j=1}^{n} \frac{1}{(x_n - x_j - 1/\xi)^{k_j}} \frac{1}{\xi^2} \, d\xi$$

$$= B \int_{\overrightarrow{[z_1, \, z]}} \prod_{j=1}^{n-1} \frac{1}{(\xi - a_j)^{k_j}} \, d\xi$$

where $z_1 = T^{-1}(z_0)$ and $B = \prod_{j=1}^{n-1} (x_n - x_j)$. The transformation g given by

$$g(z) = \int_{\overrightarrow{[z_1, z]}} \prod_{j=1}^{n-1} \frac{1}{(\xi - a_j)^{k_j}} \, d\xi \tag{13}$$

satisfies $g = (1/B) f \circ T$ with real B and thus maps $\mathbf{R}$ to a polygon which is similar to $f(\mathbf{R})$.

Conversely, suppose that $(k_j)_{j=1}^{n}$ and $(a_j)_{j=1}^{n-1}$ satisfy $\sum_{j=1}^{n} k_j = 2$ and $a_j < a_{j+1}$ $(1 \le j \le n - 2)$. Define g by (13). Let x_n be any real number larger than $1/a_{n-1}$, define T as in (11), and let $x_j = T(a_j)$. We have $g(z) =$

$B^{-1}f(T(z))$ for f as in (4). Hence g maps **R** to a polygon with exterior angles $(k_j\pi)_{j=1}^n$ and H to the interior of the polygon.

It is generally easier to deal with (13) than with (4). In general, neither is amenable to explicit evaluation.

7.18 EXAMPLES

(i) *Triangles.* Suppose that the exterior angles of a triangle T are πk_1, πk_2, and πk_3. Using (13) of (7.17), we select $a_1 = 0$ and $a_2 = 1$. As we have seen, for any choice of z_0 in H, the integral in (4) from 0 to x_j is independent of path in H. It follows that we may select $z_1 = 0$ in (13). Accordingly, we let

$$g(z) = \int_{\overrightarrow{[0,z]}} \frac{1}{\xi^{k_1}(\xi - 1)^{k_2}} \, d\xi. \tag{1}$$

The vertices of the triangle $T = g(\mathbf{R})$ are thus $g(0) = 0$, $g(1)$, and $g(\infty)$. For $0 < \zeta < 1$, we have $\mathrm{Arg}_{\pi/2}(\zeta) = 0$ and $\mathrm{Arg}_{\pi/2}(\zeta - 1) = \pi$. Thus we have

$$g(t) = e^{-ik_2\pi} \int_{[0,t]} \frac{1}{\zeta^{k_1}(1 - \zeta)^{k_2}} \, d\zeta \qquad (0 \le t \le 1). \tag{2}$$

For $1 < \zeta \le \infty$ we have $\mathrm{Arg}_{\pi/2}(\zeta) = \mathrm{Arg}_{\pi/2}(\zeta - 1) = 0$, and hence

$$g(t) = g(1) + \int_{[1,t]} \frac{1}{\zeta^{k_1}(\zeta - 1)^{k_2}} \, d\zeta \qquad (t > 1) \tag{3}$$

holds. The integral in (3) is positive, so the side $g([1, \infty[)$ of T is parallel to the x-axis and oriented to the right. For $\zeta < 0$, we have $\mathrm{Arg}_{\pi/2}(\zeta) = \mathrm{Arg}_{\pi/2}(\zeta - 1) = \pi$. Hence the equality

$$g(t) = e^{-i(k_1+k_2)\pi} \int_{\overrightarrow{[0,t]}} \frac{1}{(-\zeta)^{k_1}(1 - \zeta)^{k_2}} \, d\zeta \qquad (t < 0) \tag{4}$$

holds. See Figure 7.8.

If $k_2 = \frac{1}{2}$, then a right triangle in the fourth quadrant is obtained. If, in addition, $k_1 = k_3 = \frac{3}{4}$, then the triangle is isosceles. Equating the lengths of the two equal sides, we obtain the equality

$$\int_{[0,1]} \frac{1}{\xi^{3/4}(1 - \xi)^{1/2}} \, d\xi = \int_{[1,\infty[} \frac{1}{\zeta^{3/4}(\zeta - 1)^{1/2}} \, d\zeta.$$

(This equality can also be obtained by a simple change of variables.)

The *Beta function* is a function of (x, y) $(x > 0, y > 0)$ defined by the equality

$$B(x, y) = \int_{[0,1]} \zeta^{x-1}(1 - \zeta)^{y-1} \, d\zeta.$$

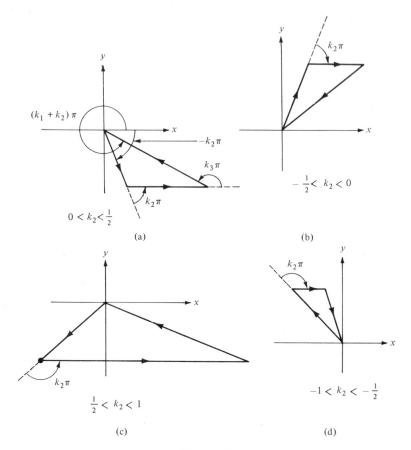

Figure 7.8

In terms of the Beta function, two vertices of the image triangle are 0 and $e^{-ik_2\pi}B(1 - k_1, 1 - k_2)$. The third vertex is of course determined since the angles of the triangle are known. For those readers familiar with the Beta function and with the Gamma function Γ, we mention that the identity

$$B(1 - k_1, 1 - k_2) = \frac{\Gamma(1 - k_1)\Gamma(1 - k_2)}{\Gamma(2 - (k_1 + k_2))} = \frac{\Gamma(1 - k_1)\Gamma(1 - k_2)}{\Gamma(k_3)} \quad (5)$$

provides a means of studying the image triangle in terms of the Gamma function.

(ii) *Rectangles.* In the case of a rectangle, the exterior angles are all $\pi/2$.

We use (13). Let $a_1 = -1$, $a_2 = 0$, and $a_3 = a > 0$ and select $z_1 = -1$. The resulting mapping function is

$$g(z) = \int_{[-1, z]} \frac{1}{\zeta^{1/2}(\zeta + 1)^{1/2}(\zeta - a)^{1/2}} \, d\zeta.$$

For real x, we obtain

$$g(x) = - \int_{[-1, x]} \frac{1}{\sqrt{t(t + 1)(t - a)}} \, dt \qquad (-1 \le x \le 0);$$

$$g(x) = g(0) + e^{-i(\pi/2)} \int_{[0, x]} \frac{1}{\sqrt{t(t + 1)(a - t)}} \, dt \qquad (0 \le x \le a);$$

$$g(x) = g(a) + \int_{[a, x]} \frac{1}{\sqrt{t(t + 1)(t - a)}} \, dt \qquad (a \le x);$$

$$g(x) = e^{-i(3\pi/2)} \int_{[-1, x]} \frac{1}{\sqrt{t(t + 1)(a - t)}} \, dt \qquad (x \le -1).$$

(See also Figure 7.9.) It is easy to deduce from these formulas that g is one-to-one on **R**. Hence by (7.17.iii) g is univalent on H.

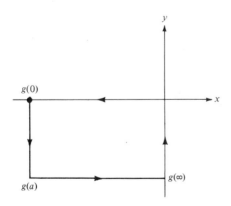

Figure 7.9

The lengths of the sides of the rectangle are

$$\int_{[-1, 0]} \frac{dt}{\sqrt{t(t + 1)(t - a)}} \quad \text{and} \quad \int_{[0, a]} \frac{1}{\sqrt{t(t + 1)(a - t)}} .$$

These are equal if $a = 1$. Hence a univalent mapping of H to a square is given by

$$h(z) = \int_{\overrightarrow{[-1, z]}} \frac{1}{\zeta^{1/2}(\zeta + 1)^{1/2}(\zeta - 1)^{1/2}} \, d\zeta.$$

The formula (4) of (7.17) yields a particularly simple formula in the case of a rectangle. Select the four points to be $-1, -k, k, 1$ where $0 < k < 1$. The mapping function f is thus

$$f(z) = \int_{\overrightarrow{[0, z]}} \frac{1}{(\zeta + 1)^{1/2}(\zeta - 1)^{1/2}(\zeta + k)^{1/2}(\zeta - k)^{1/2}} \, d\zeta.$$

For real x we have

$$f(x) = B(x) \int_{\overrightarrow{[0, x]}} \frac{1}{\sqrt{|\zeta^2 - 1||\zeta^2 - k^2|}} \, d\zeta$$

where $B(x)$ is given by

$$B(x) = \begin{cases} e^{-i(\pi/2)} & k \le x \le 1 \\ -1 & -k \le x \le k \\ e^{-i(3\pi/2)} & -1 \le x \le -k \\ 1 & x \le -1. \end{cases}$$

(iii) *Rectangular strips.* In our general treatment, we restricted the exterior angles to lie strictly between $-\pi$ and π. If we let two exterior angles be $\pi/2$ and the third π, then the figure obtained would appear to be a semi-infinite strip. Using formula (13) of (7.17), it is thus reasonable to suspect that the functions F given by equalities

$$F(z) = A \int_{\overrightarrow{[z_0, z]}} \frac{1}{(\zeta - x_1)^{1/2}(\zeta - x_2)^{1/2}} \, d\zeta + B$$

will map H univalently onto the interior of rectangular strips, the strips being determined by $x_1, x_2, A_1,$ and B_1. Ignoring justifications and calculating formally for a moment, we let $x_1 = -1$, and $x_2 = 1$, and obtain

$$F(z) = A_1 \int_{\overrightarrow{[z_0, z]}} \frac{d\zeta}{(\zeta^2 - 1)^{1/2}} \, d\zeta + B_1.$$

The square root, of course, requires interpretation. We skip this interpretation and notice that it appears that

$$F(z) = -i A \, \mathrm{Sin}^{-1}(z) + B \tag{*}$$

for constants A and B. We now abandon the integrals, treat the above reasoning as heuristic only, and show that functions given by (*) map H

univalently onto strips. This will be clear once we know that sin is univalent from the strip $S = \{w \mid \text{Im } (w) > 0, |\text{Re } (w)| < \pi/2\}$ onto H. To see that it is, let $w = u + iv$ be in S. We have

$$\sin (w) = \sin (u) \cosh (v) + i \cos (u) \sinh (v).$$

If $z = x + iy$ is in $\bar{H}$, then the equations

$$x = \sin (u) \cosh (v)$$
$$y = \cos (u) \sinh (v)$$

have a unique solution for $u + iv$ in $\bar{S}$. Hence Sin^{-1} is well-defined from $\bar{H}$ onto $\bar{S}$. The ray $]-\infty, -1[$ goes to $\{w| \text{ Im } (w) = -\pi/2\}$, the ray $]1, \infty[$ to $\{w \mid \text{Im } (w) = \pi/2\}$ and the interval $[-1, 1]$ to $[-\pi/2, \pi/2]$.

If $b > 0$ and we select $B = 0$ and $A = (2b)/\pi$ in (*), then we find that F maps H univalently onto $W = \{z \mid \text{Im } (z) > 0, |\text{Re } (z)| < b\}$.

7.19 CONCLUDING REMARKS

Our approach to conformal and univalent mappings has been oriented to examples. We presented a number of elementary mappings of familiar geometric regions. We have not presented the most important theorem of the subject, the Riemann mapping theorem. This theorem states that any proper simply connected subset of $\mathbf{C}$ can be mapped univalently onto the open unit disk. The proof of this theorem is difficult and it is not particularly helpful in constructing particular mappings. The theorem, however, is not only of theoretical interest. For example, we know of no way to obtain all the properties of the Schwarz-Christoffel transformations (general form) without appealing to the Riemann mapping theorem. There are extensions of the Riemann mapping theorem which deal with the behaviour of the mapping function on the boundary of the region. These theorems always contain some topological restriction on the boundary. We conclude with exercises which include several more examples of univalent mappings.

EXERCISES

1. Fix θ such that $0 < \theta < \pi$.
   Find a function h on $\{z \mid \text{Im } (z) \geq 0\}$ such that h maps $\{z < 0\}$ to $\{re^{i\theta} \mid r > 0\}$, $\{z > 0\}$ to itself, and $\{z \mid \text{Im } (z) > 0\}$ to $\{re^{iw} \mid 0 < w < \theta\}$.

2. Let f be as in formula (4) of (7.17). Let T be a linear fractional transformation from $\{z \mid |z| < 1\}$ onto the upper half-plane H such that $T(\{|z| = 1\}) = \mathbf{R}$. Show that $f \circ T$ can be written

$$f \circ T(z) = A \int_{\overrightarrow{[0, z]}} \frac{1}{(\zeta - b_1)^{k_1}(\zeta - b_2)^{k_2} \ldots (\zeta - b_n)^{k_n}} \, d\zeta \quad (|z| \leq 1) \ (*)$$

where A is a constant and $b_1, b_2, \ldots, b_n$ are distinct points on the unit circle. In (*), we use the same definition of ξ^{k_j} adopted in (2) of (7.17); that is $\xi^{k_j} = e^{k_j \, \text{Log} \, \xi}$.

Conversely, let $b_1, b_2, \ldots, b_n$ be n distinct points on the unit circle ($|b_i| = 1$) and let $(k_j)_{j=1}^n$ satisfy $-1 < k_j < 1$ and $\sum_{j=1}^n k_j = 2$. Show that the transformation g defined by the integral in (*) maps $\{z \mid |z| = 1\}$ onto a closed polygon having n-sides, vertices $(g(b_j))_{j=1}^n$, and exterior angle at $g(b_j)$ equal to $k_j \pi$. And, g maps $D = \{z \mid |z| < 1\}$ into the interior of the polygon.

3. Consider the triangular function g defined and discussed in (7.18.i). Fix k_2 satisfying $0 < k_2 \leq \frac{1}{2}$.
   (a) Show that $k_1 + k_2 > 1$.
   (b) If $k_2 \leq \frac{1}{2}$, show that

$$\int_{[1, \infty]} \frac{1}{\zeta^{k_1}(- 1)^{k_2}} \, d\zeta$$

is bounded as a function of k_1. What does this boundedness mean for the side of the triangle parallel to the x-axis? Show that

$$\lim_{k_1 \to 1} \int_{[0, 1]} \frac{1}{\zeta^{k_1}(1 - \zeta)^{k_2}} \, d\zeta = \infty.$$

What happens to the triangle as k_1 goes to 1?
   (c) Suppose that $k_2 = \frac{1}{2}$. Show that

$$\lim_{k_1 \to 1/2} \int_{[1, \infty]} \frac{1}{\zeta^{k_1}(\zeta - 1)^{k_2}} \, d\zeta = \infty$$

and that

$$\lim_{k_1 \to 1/2} \int_{[0, 1]} \frac{1}{\zeta^{k_1}(1 - \zeta)^{k_2}} \, d\zeta$$

is finite. What is happening to the triangle as k_1 goes to $\frac{1}{2}$?

4. Let k_1 and k_2 be positive real numbers which satisfy $k_1 + k_2 = 1$. Let $x_1 < x_2$, x_1 and x_2 real numbers. Use a limiting argument or otherwise show that the function g defined by

$$g(z) = \int_{[z_0, z]} \frac{1}{(\zeta - x_1)^{k_1}(\zeta - x_2)^{k_2}} \, d\zeta$$

(Im $(z_0) > 0$) is univalent from the upper half-plane H onto the open region shown in Figure 7.10 and that g is continuous and one-to-one on $\bar{H}$.

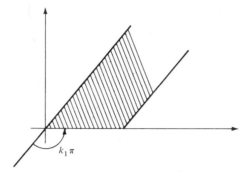

Figure 7.10

5. An n-sided polygon is regular if all its exterior angles are equal. The exterior angle is $2\pi/n$. Use (4) and (13) of (7.17) to write down two mappings of $\bar{H}$ into the interior of an n-sided polygon. Prove that each of the mappings is univalent on H and one-to-one on $\bar{H}$. Give a formula for side length of the polygon.

6. Let g be the transformation defined in Exercise 2, mapping $D = \{z \mid |z| < 1\}$ into the interior of an n-sided polygon. Select $k_j = 2/n$ for each j, so that the resulting polygon is regular. Select b_j so that $(b_j)_{j=1}^n$ are the n n^{th} roots of unity. Show that g can be written

$$g(z) = A \int_{\overrightarrow{[0,\,z]}} \frac{d\xi}{|\xi^n - 1|^{2/n}}$$

where A is a constant of modulus 1.

7. Use (4) or (13) of (7.17) to write a mapping of H onto a region shaped as in Figure 7.11. Show that the mapping is one-to-one on $\mathbf{R}$.

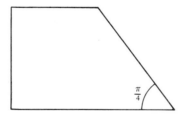

Figure 7.11

8. In formula (4) of (7.17), let $n = 12$ and let k_j equal $\frac{1}{2}$ for eight choices of j and $-\frac{1}{2}$ for the remaining four choices. Sketch various possibilities for $f(\mathbf{R})$. In particular, observe that it is possible for a portion of $f(\mathbf{R})$ to lie in $f(H)$.

9. Let f be as in (4) of (7.17). Use the fact that f' has constant argument on the intervals $[x_j, x_{j+1}]$ to show that

$$\int_{[x_j, x]} |f'(t)|\, dt = \left| \int_{[x_j, x]} f'(t)\, dt \right|$$

for $x_j < x < x_{j+1}$. From this give a proof different from that given in the text that the image of $[x_j, x_{j+1}]$ is a line segment.

Notation

With few exceptions (for example, $x \in A$) all the notation and terminology used in this book are carefully defined on the page where they are first introduced. In this sense this appendix is an index—the page listed for a particular symbol is the page on which that symbol first occurs. The comment column often contains informal remarks intended to supplement the definition given in the body of the text. In case the symbol in question is not defined in the text (for example, $\in$, the symbol of set membership), a definition or explanation is included here.

All symbols are listed in order of appearance in the text.

SYMBOL	PAGE	COMMENT
C	1	The set of complex numbers.
$\{a \mid a \text{ has property } P\}$	1	Logical symbol for defining a set: "The set of all objects a such that a has property P." For example, $\{x + iy \mid x, y \in \mathbf{R}\} = \mathbf{C}$ means $\mathbf{C}$ is the set of all objects of the form $x + iy$ where x and y are real numbers.
i	2	A solution of $x^2 + 1 = 0$.
$\in$	2	Logical symbol for set membership, that is, if A is a set, then $x \in A$ means that x is a member of the set A.
$\|z\|$	2	Modulus of z. If $z = x + iy$, then $\|z\| = (x^2 + y^2)^{1/2}$.
$\bar{z}$	4	The conjugate of z. If $z = x + iy$, then $\bar{z} = x - iy$.

SYMBOL	PAGE	COMMENT
Re (z)	5	The real part of z. If $z = x + iy$, then Re $(z) = x$.
Im (z)	5	The imaginary part of z. If $z = x + iy$, then Im $(z) = y$. Observe that Im (z) is a real number.
Arg (z)	6	The argument of z. $-\pi < $ Arg $(z) \le \pi$ and $z = |z|[\cos (\text{Arg } z) + i\sin (\text{Arg } z)]$.
arg (z)	6	The set of θ such that $z = |z|(\cos \theta + i\sin \theta)$.
T	7	$\{z \mid |z| = 1\}$. Geometrically this is the circle with center 0 and radius 1.
$[z, w]$	10	The line segment connecting the complex numbers z and w and including both z and w. It can be written $$[z, w] = \{z + t(w - z) \mid 0 \le t \le 1\}.$$ If a and b are real numbers, $$[a, b] = \{x \mid a \le x \le b\}.$$
$[z, w[$	10	Those complex number in $[z, w]$ except w; that is, $[z, w]\backslash\{w\}$.
$]z, w]$	10	$[z, w]\backslash\{z\}$.
$]z, w[$	10	$[z, w]\backslash\{z, w\}$.
$\{f(k)\}_{k=0}^M$	11	If $f(k)$ is a complex number for each choice of the integer k, then $\{f(k)\}_{k=0}^M$ is the set $\{f(0), f(1), f(2), \ldots, f(M)\}$.
Z	11	The set of integers.
$\mathbf{Z}^{++}$	11	The set of positive integers.
$\mathbf{Z}^+$	11	The set of nonnegative integers.
R	12	The set of real numbers.
$\Rightarrow$	12	Logical symbol for implication. If A and B are statements, then $A \Rightarrow B$ means that A implies B.
$(z_n)_{n=1}^\infty$	13	A sequence of complex numbers.
$\lim\limits_{n \to \infty} z_n = z$	13	The sequence $(z_n)_{n=1}^\infty$ of complex numbers converges to z.
$\overline{\lim}\limits_{n \to \infty}$; $\underline{\lim}\limits_{n \to a}$	19	Limit superior and limit inferior, respectively.
$\dbinom{n}{k}$	23	$$\binom{n}{k} = \frac{n!}{(n - k)!\, k!}$$ $$= \frac{n(n - 1)(n - 2)\ldots(n - k + 1)}{k!}.$$ Read "the binomial coefficient n over k."
$A\backslash B$	25	If A and B are sets, then $A\backslash B$ is the collection of of objects in A which are not in B.

SYMBOL	PAGE	COMMENT
glb $\{x_n \mid n \geq k\}$	25	glb is the "greatest lower bound." glb $\{x_n \mid n \geq k\}$ is the largest number s_k which satisfies $s_k \leq x_n$ for all $n \geq k$.
lub $\{x_n \mid n \geq k\}$	25	lub is the "least upper bound." lub $\{x_n \mid n \geq k\}$ is the smallest number s_k which satisfies $s_k \geq x_n$ for all $n \geq k$.
$D(r, a)$	26	$\{z \mid \lvert z - a \rvert < r\}$, that is, the collection of complex numbers whose distance from a is less than r. Read "the open disk centered at a." Note that $D(1, 0) = \{z \mid \lvert z \rvert < 1\}$.
$\subset$	26	Logical symbol for containment. If A and B are sets, then $A \subset B$ means every element of A is an element of B.
$\lim_{n \to \infty} \lim_{m \to \infty} a_{nm} = B$	28	This symbol denotes an iterated limit and means first compute $A_n = \lim_{m \to \infty} a_{nm}$ and then compute $\lim_{n \to \infty} A_n = B$.
$\lim_{(n, m) \to \infty} A_{nm}$	28	This symbol denotes a double limit and means n and m are required to be large simultaneously. A precise definition is given on page 28.
$\mathbf{Q}$	30	The set of rational numbers.
$\mathbf{Q}^+$	30	The set of nonnegative rational numbers.
$\mathbf{Q}^{++}$	30	The set of positive rational numbers.
$\mathbf{R}^+$	30	The set of nonnegative real numbers.
$\mathbf{R}^{++}$	30	The set of positive real numbers.
$C(r, a)$	30	The circle of radius r centered at a; $C(r, a) = \{z \mid \lvert z - a \rvert = r\}$.
$f\colon A \to B$	31	Logical symbol for a function. If A and B are sets, then $f\colon A \to B$ means f is a function with domain A and range B.
$\mathbf{R}^2$	31	The Euclidian plane. The collection of ordered pairs of real numbers with vector addition and no scheme for multiplication. $\mathbf{R}^2$ is often written $\mathbf{R} \times \mathbf{R}$.
$\mathbf{C} \times \mathbf{C}$	31	The set of all pairs of complex numbers. $\mathbf{C} \times \mathbf{C}$ is often written $\mathbf{C}^2$.
$f + g$ fg $f/g, \ f \circ g$	36	If f and g are functions, then $(f + g)(z) = f(z) + g(z)$, $(fg)(z) = f(z)g(z)$, $(f/g)(z) = f(z)/g(z)$, and $(f \circ g)(z) = f(g(z))$.
$\lim_{\substack{s \to z \\ s \in S}} f(s) = A$	36	The usual definition of $\lim_{s \to z} f(s) = A$ is restricted to those s which lie in the set S.

SYMBOL	PAGE	COMMENT
$\mathbf{R}^-$	37	The nonpositive real numbers.
$f(S)$	37	The image of the set S under the function f; $f(S) = \{f(s): s \in S\}$.
$\varnothing$	40	The empty set.
T'	41	If T is a set, then T' denotes the complement of T; that is, the collection of those objects under consideration which are not in T.
$\cap$	41	Logical symbol for set intersection. If A and B are sets, then $x \in A \cap B$ if $x \in A$ and $x \in B$.
$\overline{T}$	41	If T is a set, then $\overline{T}$ denotes the closure of T. Roughly speaking, this set consists of T and its boundary. Compare with the notation $\bar{z}$ as defined on page 4.
$\cup$	41	Logical symbol for set union. If A and B are sets, then $x \in A \cup B$ if either $x \in A$ or $x \in B$.
$\bigcup_F F$	41	$\bigcup_F F = \bigcup \{F \mid F \in \mathsf{F}\}$. Both symbols denote the set of elements x such that $x \in F$ for at least one F in the collection of sets F.
$\bigcap_E E$	43	$\bigcap_E E = \bigcap \{E \mid E \in \mathsf{E}\}$. Both symbols denote the set of elements x such that $x \in E$ for all $E \in \mathsf{E}$.
lub	44	Least upper bound. lub $\{f(x): x \in S\}$ is the smallest real number l such that $f(x) \leq l$ for all $x \in S$. See Appendix B, p. 276 for a general definition.
glb	44	Greatest lower bound. glb $\{f(x): x \in S\}$ is the largest real number m such that $f(x) \geq m$ for all $x \in S$. See Appendix B, p. 276.
f'	47	If f is a function on $\mathbf{C}$ or $\mathbf{R}$, then f' denotes the derivative of f. Compare this notation with the notation T' for sets T on page 41.
$\mathbb{C}^1$	53	A $\mathbb{C}^1$ function on a set S is a function defined on S with continuous first derivative.
exp (z)	65	exp $(z) = \sum_{n=0}^{\infty} z^n/n!$. Also written e^z.
log (z)	71	log $(z) = \{w \mid e^w = z\}$.
Log (z)	72	Log $(z) = \log|z| + i$ Arg (z).
Arg$_\theta$ (z)	72	Arg$_\theta$ (z) satisfies $\theta - \pi < $ Arg$_\theta$ $(z) \leq \theta + \pi$ and $z = |z|e^{i\,\text{Arg}_\theta\,(z)}$.
Log$_\theta$ (z)	72	Log$_\theta$ $(z) = \log|z| + i$ Arg$_\theta$ z.
z^α	73	$z^\alpha = \{e^{\alpha w} \mid w \in \log (z)\}$.

SYMBOL	PAGE	COMMENT
$F_\alpha(z)$	74	$F_\alpha(z) = e^{\alpha \operatorname{Log} z}$. F_α is an analytic function with values in the set z^α.
α^*	85	If α is a curve, then α^* is the set of points $\alpha(t)$.
$\overrightarrow{[a, b]}$	86	The oriented line segment from a to b, where a and b are complex numbers.
$\overrightarrow{C(r, a)}$	86	Positively oriented circle with center a and radius r.
$\alpha \smile \beta$	94	α and β are curves. This is read "α cup β." $\alpha \smile \beta$ is a new curve formed by joining the "end" of α to the "beginning" of β.
$v_\alpha(x)$	95	If a is given, then $v_\alpha(x)$ is the variation of α from a to x. This is the same number as the arc length along α^* from $\alpha(a)$ to $\alpha(x)$.
Δ	96	Here Δ is a partition of a real interval into subintervals. In other contexts it may have other meanings.
$\lvert \Delta \rvert$	96	The length of the longest interval in Δ.
$V_a^b(\alpha)$	98	Variation of α from a to b. (The same number as the arc length along α^* from $\alpha(a)$ to $\alpha(b)$, namely $v_\alpha(b)$.
$\operatorname{Ind}_\alpha(z)$	107	For α a closed curve this is read "the index of α with respect to z." This number is the number of times α^* winds around the point z.
$\hat{f}(n)$	109	Fourier coefficient of a function f on $\mathbf{T}$.
$\mathfrak{C}^2(\mathbf{T})$	110	Functions defined on $\mathbf{T}$ having continuous second derivatives as functions on $\mathbf{T}$.
$l(\partial)$	125	If ∂ is a curve, then $l(\partial)$ denotes the length of ∂.
$f^{(n)}(z)$	132	Denotes the nth derivative evaluated at z of the function f.
Cf	143	Cauchy integral of a continuous function.
$\operatorname{Res}(f, s)$	178	The residue of f at s, defined to be a_{-1} in the Laurent expansion of f at s.
$\hat{f}$	194	Fourier transform of a function f defined on R.
$\mathbf{u}$	210	Boldface type indicates a vector.
$P(r, t)$	215	Poisson kernel.
$Q(z, t)$	216	Poisson kernel.
Pf	217	Poisson integral of a function on $\mathbf{T}$.
$P_R f$	219	Poisson integral of a function on a circle $C(r, a)$.

Rational and Real Numbers

In this appendix we review algebraic and analytic properties of the integers **Z**, the rational numbers **Q**, and the real numbers **R**. It is meant as a review and as a convenient reference for the main text. The student unfamiliar with fundamental aspects of real numbers can find the facts here, but will probably find one semester (or quarter) too short a time to study them in any depth and also master some complex analysis. We advise these students to concentrate on the complex analysis and refer to this appendix for information as needed.

B.1 RATIONAL NUMBERS

Mathematics begins with counting; that is, the set of positive integers $\{1, 2, 3, \ldots\}$, which we denote by $\mathbf{Z}^{++}$. Addition in $\mathbf{Z}^{++}$ is deficient from an algebraic point of view because the equation $x + n = m$ for fixed integers n and m may have no solution in $\mathbf{Z}^{++}$; for example, $x + 3 = 1$ has no solution in $\mathbf{Z}^{++}$. The set $\mathbf{Z}^{++}$ is thus augmented by the negative integers and zero to form the integers **Z**:

$$\mathbf{Z} = \{0, \pm 1, \pm 2, \ldots\}.$$

All equations $x + n = m$ for $n, m \in \mathbf{Z}$ have solutions in **Z**. But equations $mx = n, m,$ and n in **Z**, may fail to have solutions in **Z**; for example, $2x = 1$ has no solution in **Z**. The rational numbers, all quotients p/q with p and q in **Z** and $q \neq 0$, are thus formed; we denote the set of rational numbers by **Q**. All equations $mx + n = p$ $(m, n, p \in \mathbf{Q}, m \neq 0)$, have solutions in **Q**. We suppose that the reader is familiar with the arithmetic of **Q**. The basic binary operations are addition $(+)$ and multiplication $(\cdot)$. The fundamental laws governing these operations are listed below; $a, b,$ and c denote elements of **Q**.

(i) *Closure.* $a + b$ and $a \cdot b$ are in **Q**.

(ii) *Commutative laws.* $a + b = b + a$ and $a \cdot b = b \cdot a$.

(iii) *Associative laws.* $(a + b) + c = a + (b + c)$ and
$(a \cdot b) \cdot c = a \cdot (b \cdot c)$.

(iv) *Distributive law.* $a \cdot (b + c) = a \cdot b + a \cdot c$.

(v) *Identities.* $a + 0 = a$ and $a \cdot 1 = a$ (0 is additive identity, 1 is multiplicative identity).

(vi) *Additive inverse.* If $a \in$ **Q**, there is a unique $b \in$ **Q** such that $a + b$
$= 0$; b is called the additive inverse of a and denoted $-a$.

(vii) *Multiplicative inverse.* If $a \in$ **Q**$\backslash\{0\}$, there is a unique $b \in$ **Q** such that
$a \cdot b = 1$; b is called the multiplicative inverse of a and denoted a^{-1}.

Properties (i–vii) mean that the triple (**Q**, $+$, $\cdot$) is a field; any triple
$(K, +, \cdot)$ satisfying these axioms is by definition a field. Subtraction and
division are defined in terms of addition and multiplication:

$$a - b = a + (-b) \quad \text{and} \quad a \div b = a \cdot b^{-1}(b \neq 0).$$

We will often drop the symbol $\cdot$ and write ab for $a \cdot b$. Notice that (i–vi)
are satisfied if **Q** is replaced by **Z**, but that (vii) is not satisfied.

B.2 INEQUALITY AND ABSOLUTE VALUE

A rational number q is *positive* if it can be written $q = m/n$ with m and n
positive integers; it is *nonnegative* if it is positive or zero. The set **Q**$^{++}$ of
positive rationals satisfies (i–iv) below.

(i) If $a, b \in$ **Q**$^{++}$, then $a + b$ and $a \cdot b$ are in **Q**$^{++}$.

(ii) **Q** $= (-$**Q**$^{++}) \cup$ **Q**$^{++} \cup \{0\}$ and
$(-$**Q**$^{++}) \cap$ **Q**$^{++} = \varnothing$ $(-$**Q**$^{++} = \{-q| q \in$ **Q**$^{++}\})$.

(iii) If $q \in$ **Q**$^{++}$, then q^{-1} is also in **Q**$^{++}$.

(iv) If $q \in$ **Q**$^{++}$ and $p \in -$**Q**$^{++}$, then $qp \in -$**Q**$^{++}$.

Properties (i) and (ii) are easily established using the definition of positivity.
To prove (iii), assume that $q \in$ **Q**$^{++}$ but that $q^{-1} \in -$**Q**$^{++}$. Then $-q^{-1}$
is in **Q**$^{++}$. It is easy to see that $-q^{-1} = (-1)q^{-1}$. Hence we have
$-1 = (-1)q^{-1}q = (-q^{-1})q \in$ **Q**$^{++}$, a contradiction. Property (iv) follows
from (iii).

The inequality relation $<$ relates pairs of elements of **Q** by the definition

(v) $a < b$ if $b - a \in Q^{++}$;

and, $a \le b$ means that either $a < b$ or $a = b$. Proofs of the elementary properties of inequalities can be based on the definition (v), using properties (i–iv); examples are given in the exercises. *Absolute value* is defined by

(vi) $|a| \equiv \begin{cases} a & \text{if } a \in \mathbf{Q}^{++} \\ 0 & \text{if } a = 0 \\ -a & \text{if } -a \in \mathbf{Q}^{++}. \end{cases}$

Consideration of cases shows that the equality

(vii) $|ab| = |a| \cdot |b|$, all $a, b \in \mathbf{Q}$,

is valid. Since $|ab| = ab$ if a and b are both positive or both negative, and $ab < 0$ if a and b have opposite parity, $ab \le |ab|$ always holds. We use this inequality to prove the *triangle inequality*:

(viii) $|a + b| \le |a| + |b|$, all $a, b \in \mathbf{Q}$.

The figure below provides a geometric interpretation.

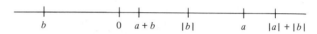

PROOF OF (viii):

$$|a + b|^2 = (a + b)^2 = a^2 + 2ab + b^2$$
$$= |a|^2 + 2ab + |b|^2 \le |a|^2 + 2|a| \cdot |b| + |b|^2 = (|a| + |b|)^2.$$

B.3 PROPERTIES OF Z

The rational numbers $\mathbf{Q}$ have the property that between two distinct rationals there is a third; for example, if $a < b$, then $a < \frac{1}{2}(a + b) < b$. In particular, given a rational a, there is no unique "next" rational. On the other hand, every integer n has a unique successor, namely, $n + 1$. This important property is the basis for mathematical induction, which can be formulated as follows.

(i) *Principal of Mathematical Induction.* If a set $S \subset \mathbf{Z}$ contains the integer n_0, and if the implication "$n \in S$ implies that $n + 1 \in S$" is valid for all $n \ge n_0$, then $\{n \in \mathbf{Z} \mid n \ge n_0\} \subset S$.

The reader will doubtless have encountered many applications of mathematical induction. In an axiomatic development of the integers and rational numbers, it is one of the axioms.

An element $p \in \mathbf{Z}^{++}$ is *prime* if it is divisible only by itself and 1, and $p \ne 1$.

(ii) *Fundamental Theorem of Arithmetic.* For every integer $n \geq 2$ there is a unique set of primes $\{p_i\}_{i=1}^n$ and exponents $\{e_i\}_{i=1}^n$ $(1 \leq e_i)$ such that $n = p_1^{e_1}, p_2^{e_2}, \ldots, p_n^{e_n}$.

B.4 REAL NUMBERS

The equation $x^2 - 2 = 0$ has no solution in $\mathbf{Q}$. This algebraic defect of $\mathbf{Q}$ is also a geometric defect, for the solution should be the length of the hypotenuse of a right isosceles triangle of leg length 1. The difficulty can be remedied by creating an element, called $\sqrt{2}$, with defining property $(\sqrt{2})^2 = 2$, and considering the set $\mathbf{Q}(\sqrt{2}) \equiv \{a + b\sqrt{2} \mid a, b \in \mathbf{Q}\}$. The new set $\mathbf{Q}(\sqrt{2})$ with $(\sqrt{2})^2 = 2$ determining the multiplication law

$$(a + \sqrt{2}b)(c + \sqrt{2}d) = (ac + 2bd) + (bc + ad)\sqrt{2}$$

and addition defined by $(a + b\sqrt{2}) + (c + d\sqrt{2}) = (a + c) + (b + d)\sqrt{2}$ satisfies the axioms in (B.1); that is, $\mathbf{Q}(\sqrt{2})$ is a field. But $\mathbf{Q}(\sqrt{2})$ does not (for example) contain a solution of $x^2 - 5 = 0$; that is, it contains no square root of 5. This algebraic approach to defining number systems (fields) for which polynomial equations have roots can be applied more generally to obtain a field, the field of *algebraic numbers*, in which every polynomial equation

$$a_n x^n + a_{n-1} x^{n-1} + \cdots + a_0 = 0$$

with rational coefficients has a solution. This approach still is not satisfactory for analysis, however, for it can be proved that the sequence $\{(1 + 1/n)^n\}_{n=1}^\infty$ has no limit among the algebraic numbers; that is, e is not a solution of any polynomial equation with rational coefficients. What is needed is an analytic completion of $\mathbf{Q}$ for which every reasonable sequence (for example, every bounded increasing sequence) has a limit. The *real numbers* $\mathbf{R}$ form such a completion. All completion processes yielding $\mathbf{R}$ from $\mathbf{Q}$ are moderately intricate, and we will not give one. Instead we list four fundamental properties of $\mathbf{R}$, and base our subsequent analysis on these.

(i) *Algebraic Property.* The set $\mathbf{R}$ satisfies (i–vii) of (B.1) (that is, $\mathbf{R}$ is a field) and $\mathbf{Q} \subset \mathbf{R}$.

(ii) *Order Property.* There is a subset $\mathbf{R}^{++}$ of $\mathbf{R}$ such that $\mathbf{Q} \cap \mathbf{R}^{++} = \mathbf{Q}^{++}, (-\mathbf{R}^{++}) \cap \mathbf{R}^{++} = \varnothing, (-\mathbf{R}^{++}) \cup \mathbf{R}^{++} \cup \{0\} = \mathbf{R}, \mathbf{R}^{++} + \mathbf{R}^{++} \subset \mathbf{R}^{++}$ and $\mathbf{R}^{++} \cdot \mathbf{R}^{++} \subset \mathbf{R}^{++}$.

Inequality in $\mathbf{R}$ is defined by "$a < b$ if $b - a \in \mathbf{R}^{++}$"; that is, in terms of positivity. Elementary properties of inequalities follow, as for $\mathbf{Q}$. Absolute value is defined as for $\mathbf{Q}$ (B.2.vi), and (B.2.vii–B.2.viii) are proved as for $\mathbf{Q}$.

(iii) *Archimedean Property.*   For any $\varepsilon \in \mathbf{R}^{++}$, there is an $n \in \mathbf{Z}$ such that $1 < n\varepsilon$.

(iv) *Completeness Property.*   If $S \subset \mathbf{R}$ and S is bounded above, there is a unique $s \in \mathbf{R}$ such that

(a) $x \in S$ implies that $x \leq s$

(b) $t < s$ implies that $t < x$ for some $x \in S$.

The unique number s is called the *least upper bound* of S and is written lub S or lub (S). The symbol sup (S) is used for the same number.

B.4.1 GREATEST LOWER BOUND

If $S \subset \mathbf{R}$ and S is bounded below, then $-S$ is bounded above. The number $-\text{lub}\,(-S)$ is denoted glb (S) (the greatest lower bound of S); it is also denoted by inf (S). It is easy to show that glb (S) is the unique real number s satisfying the conditions

(a) $x \in S$ implies that $x \geq s$

(b) $t > s$ implies that $t > x$ for some $x \in S$.

We give one fundamental consequence of (iii), namely:

B.4.2 THEOREM (DENSENESS OF Q IN R)

If $x < y$ for $x, y \in \mathbf{R}$, then there is $q \in \mathbf{Q}$ such that $x < q < y$.

PROOF.   By (ii) and (iii) there is an $N \in \mathbf{Z}$ so that $x + N > 0$. By (iii) there is an $n \in \mathbf{Z}^{++}$ such that $1/n < y - x$. Again using (iii) there are integers k such that $k/n > x + N$; that is, the set $\{k \mid k/n > x + N\}$ is a nonvoid subset of $\mathbf{Z}^{++}$. If m is the smallest element of this set, then we claim that $x + N < m/n < y + N$. The left inequality is clear. By definition of m, $(m - 1)/n \leq x + N$. If $m/n \geq y + N$ held, then we would have

$$\frac{m - 1}{n} = \frac{m}{n} - \frac{1}{n} \geq y + N - \frac{1}{n} > x + N;$$

the preceding contradiction proves that $m/n < y + N$. Since $m/n - N = (m - nN)/n \in \mathbf{Q}$ and $x < (m - nN)/n < y$, the proof is complete.

In the exercises we will outline proofs showing how the above properties are used to prove familiar properties of the real numbers.

B.5 SEQUENCES OF REAL NUMBERS

A function from the positive integers $\mathbf{Z}^{++}$ to $\mathbf{R}$ is called a *sequence in* $\mathbf{R}$; that is, to each integer $n > 0$ there is assigned a real number x_n. The sequence is *bounded above* if there is an $M \in \mathbf{R}$ such that $x_n \leq M$ for all n, *bounded below* if there is an M such that $x_n \geq M$ for all n, and *bounded* if there is an M such that $|x_n| < M$ for all n. The sequence is *increasing* if $x_n \leq x_{n+1}$ for all n; it is *decreasing* if $x_n \geq x_{n+1}$ for all n. The sequence $(x_n)_{n=1}^{\infty}$ *converges* to x if for every positive real number ε there is an n_0 such that the inequality

$$|x_n - x| < \varepsilon$$

holds for all $n \geq n_0$. If $(x_n)_{n=1}^{\infty}$ converges to x, we write $\lim_{n\to\infty} x_n = x$.

B.6 THEOREM

Let $(z_n)_{n=1}^{\infty}$ and $(w_n)_{n=1}^{\infty}$ be sequences in $\mathbf{R}$ such that $\lim_{n\to\infty} z_n = z$ and $\lim_{n\to\infty} w_n = w$. We have

(a) $\lim_{n\to\infty} (z_n + w_n) = z + w$

(b) $\lim_{n\to\infty} (z_n w_n) = zw$

(c) $\lim_{n\to\infty} (z_n/w_n) = z/w$ if $w_n w \neq 0$ for all n.

The proofs of (a), (b), and (c) are elementary and are left to the reader.

B.7 THEOREM

Every bounded increasing sequence of real numbers converges and every bounded decreasing sequence of real numbers converges.

PROOF. If $(x_n)_{n=1}^{\infty}$ is a bounded increasing sequence, then we will show that

(a) $\lim_{n\to\infty} x_n = \text{lub} \{x_n \mid n \in \mathbf{Z}^{++}\}$.

Let $x = \text{lub} \{x_n\}$. If (a) fails, then there is an $\varepsilon > 0$ such that for each n there is an $n_1 \geq n$ such that $x - x_{n_1} \geq \varepsilon$. Since $x_n \leq x_{n_1}$, $x - x_n \geq \varepsilon$ holds also. Thus $x - \varepsilon \geq x_n$ for all n, and $x = \text{lub} \{x_n\}$ cannot hold.

A similar proof works for bounded decreasing sequences.

B.8 CAUCHY SEQUENCES

A sequence $(x_n)_{n=1}^{\infty}$ in $\mathbf{R}$ is called a *Cauchy sequence* if for every $\varepsilon > 0$ there is an integer n_0 such that

$$n \geq n_0 \quad \text{and} \quad m \geq n_0 \quad \text{imply that} \quad |x_n - x_m| < \varepsilon.$$

(i) *Theorem.* If $\{x_n\}_{n=1}^{\infty}$ converges, then it is a Cauchy sequence.

PROOF. Let $\lim_{n\to\infty} x_n = x$, and suppose that $\varepsilon > 0$. Select n_0 such that $|x - x_n| < \varepsilon/2$ if $n \geq n_0$. If $n, m \geq n_0$, then we have

$$|x_n - x_m| \leq |x_n - x| + |x - x_m| < \frac{\varepsilon}{2} + \frac{\varepsilon}{2} = \varepsilon.$$

(iv) *Theorem* (Cauchy criterion). If $\{x_n\}_{n=1}^{\infty}$ is a Cauchy sequence in **R**, then it converges.

PROOF. The set $T = \{x_n \mid n \in \mathbf{Z}^{++}\}$ is bounded, for if n_0 is such that $|x_n - x_m| < 1$ holds for $m, n > n_0$, then $|x_n| < 1 + |x_{n_0}|$ holds for $n \geq n_0$; hence we have $|x_n| \leq M$ for all n, where $M = \{\max |x_1|, |x_2|, \ldots, |x_{n_0-1}|, 1 + |x_{n_0}|\}$. Each set

$$T_k = \{x_n \mid n \geq k\}$$

is a subset of T, so each T_k is bounded. The sequence $\{t_k\}_{k=1}^{\infty}$ defined by

$$t_k = \text{lub}\ (T_k)$$

is decreasing because $T_{k+1} \subset T_k$ for all k. By (B.7), $(t_k)_{k=1}^{\infty}$ converges, say to x. It remains to show that $\lim_{n \to \infty} x_n = x$. To prove this, we use the inequality

(a) $|x - x_n| \leq |x - t_m| + |t_m - x_m| + |x_m - x_n|,$

valid for all n and m. Let $\varepsilon > 0$, and select n_0 such that

(b) $n \geq n_0$ and $k \geq n_0$ imply that $|x_k - x_n| < \varepsilon/3$

and $n \geq n_0$ implies that $|x - t_n| < \varepsilon/3$.

The relations

$$t_m - x_m = \text{lub}\ \{x_k \mid k \geq m\} - x_m \leq \text{lub}\ \{|x_k - x_m| \mid k \geq m\}$$

prove that $|t_m - x_m| \leq \varepsilon/3$ if $m \geq n_0$. To conclude the proof, let $n \geq n_0$ and select $m \geq n_0$ (for example, $m = n_0$); then the right side of (a) is less than ε.

EXERCISES

TYPE I

1. If a and b are real numbers such that for each $\mu > a$ we also have $\mu > b$, prove that $a \geq b$.

2. If $x, y, z \in \mathbf{R}$, prove [using the definitions in (B.4)] that
   (a) $x < y$ implies that $x + z < y + z$
   (b) $0 < x < y$ and $z > 0$ implies that $xz < yz$
   (c) $0 < x < y$ and $z < 0$ implies that $xz > yz$.

3. If $a, b \in \mathbf{R}^{++}$, prove that $a^2 < b^2$ if and only if $a < b$, using (B.4).

4. Use mathematical induction to prove that $\sum_{k=1}^{n} k = [n(n + 1)]/2$.

5. Verify that glb (S) as defined in (B.4.1) is the unique real number satisfying (a) and (b) of (B.4.1).

6. Find $\lim_{n \to \infty} [n^2/(2n^2 + 5n + 1)]$.

7. Find $\lim_{n \to \infty} (\sqrt{n^2 + n} - n)$.

TYPE II

1. Prove that $r^2 \neq 2$ for every rational number r.
   OUTLINE. Suppose $r = m/n$ is in lowest terms and $r^2 = 2$. Since $m^2 = 2n^2$, m^2 is even. Hence m is even, and since $n^2 = m^2/2$, n^2 must be even.

2. Generalize (1), showing that $r^n = m$ cannot hold for rational r unless m is the perfect nth power of an integer.
   [*Hint:* Use (B.3.ii).]

3. Let n be a fixed positive integer, and let $a < b$. Prove that there is a rational number q such that $a < q^n < b$.

4. Let $r \in \mathbf{R}^+$ $(r > 0)$, $n \in \mathbf{Z}^{++}$. Prove that there is exactly one $s \in \mathbf{R}^+$ such that $s^n = r$. (The number s is written $\sqrt[n]{r}$ or $r^{1/n}$.)

   OUTLINE. Prove that the set
   $$S = \{q \in \mathbf{Q} \mid q^n \leq r\}$$
   is bounded above. Use (B.4.iv) to prove that $s = $ lub (S) works.

5. The elementary laws of exponents for *rational* exponents are based on the definition
   $$a^{m/n} = (\sqrt[n]{a})^m, \ a \in \mathbf{R}^{++}, \ n > 0.$$

   The definition can be extended to real x by
   $$a^x = \text{lub } \{a^{m/n} \mid m/n < x\}.$$

   With this definition, prove that

   (a) $x < y, a > 1$ implies that $a^x < a^y$
   (b) $x < y, 0 < a < 1$ implies that $a^x > a^y$
   (c) $a^x a^y = a^{x+y}$, all $x, y \in \mathbf{R}$
   (d) $(a^x)^y = a^{xy}$, all $s, y \in \mathbf{R}$.

6. Show that the set
   $$S = \left\{ \left(1 + \frac{1}{n}\right)^n \ \middle| \ n \in \mathbf{Z}^{++} \right\}$$
   is bounded above. [One definition of e is $e = $ lub(S).]

[*Hint:* In the binomial expansion

$$\sum_{k=0}^{n} \binom{n}{k} \frac{1}{n^k} \quad \text{of} \quad \left(1 + \frac{1}{n}\right)^n,$$

the inequalities

$$\binom{n}{k} \frac{1}{n^k} < \frac{1}{k!} < \frac{1}{2^k}$$

hold if $k \geq 4$. Hence $(1 + 1/n)^n$ is bounded by a constant plus a geometric series.]

7. Prove that $\mathbf{Q}(\sqrt{2})$ contains no solution of the equation $x^2 - 5 = 0$.

Exercises 8–12 can be solved using logarithms and techniques of calculus. They can also be solved using only results and methods presented in Appendix B.

8. Let $a > 0$. Prove that $\lim_{n \to \infty} 1/n^a = 0$.
   [*Remark:* Exponentiation for real exponents is treated in Exercise 5.]

9. For $a > 0$, prove that $\lim_{n \to \infty} \sqrt[n]{a} = 1$.

10. Let α be a positive real number. Prove that $\lim_{n \to \infty} [n/(n + 1)]^\alpha = 1$.
    [*Hint:* Use the laws of exponents for real numbers (see Exercise 5) to show that it suffices to consider only integral α. Then use a binomial expansion and the results of the preceding sections.]

11. Prove that $\lim_{n \to \infty} n^\alpha x^n = 0$ if $|x| < 1$ and $\alpha > 0$ and that $\lim_{n \to \infty} n^{-\alpha} z^n = \infty$ if $\alpha > 0$ and $|x| > 1$.
    [*Hint:* Use Exercise 10.]

12. Prove that $\lim_{n \to \infty} \sqrt[n]{n} = 1$.
    [*Hint:* Show that $(\sqrt[n]{n})_{n=1}^{\infty}$ is eventually decreasing. Let α be the limit of the sequence. If $\alpha > 1$, use the result of Exercise 4 to obtain a contradiction.]

13. Use the continuity of the logarithm (proved in calculus) to show that $\lim_{n \to \infty} (\log n)/n = 0$.
    [*Hint:* Exponentiate and use Exercise 12.]

14. Prove that $\lim_{n \to \infty} (\log n)^\alpha/n = 0$ for any $\alpha > 0$.

15. Let $s_1 = \sqrt{2}$ and let $s_{n+1} = \sqrt{2 + \sqrt{s_n}}$ $(n = 1, 2, 3, \ldots)$. Prove that $s_n < 2$ for all n, that $\alpha = \lim_{n \to \infty} s_n$ exists, and that α satisfies the equality

$$\alpha^4 - 2\alpha^2 - \alpha + 4 = 0.$$

16. The fact that Cauchy sequences in R are convergent is equivalent to completeness. Show this by proving that the Cauchy criterion for real

sequences together with the axioms (i–iii) of (B.4) imply the completeness property (B.4.iv).

17. *Decimal expansion.* Let $q \in \mathbf{Z}^+$, $q \geq 2$. Put $S = \{0, 1, 2, \ldots, q - 1\}$.
    (a) If $(y_n)_{n=1}^\infty$ is a sequence in S, prove that $\sum_{n=1}^\infty y_n/q^n$ converges and is in $[0, 1]$.
    (b) For $0 \leq y \leq 1$, let y_1 be the largest integer in S such that

$$\frac{y_1}{q} \leq y.$$

Inductively, define a sequence $(y_n)_{n=1}^\infty$ in S by letting y_n be the largest integer in S such that

$$\frac{y_1}{q} + \frac{y_2}{q^2} + \cdots + \frac{y_{n-1}}{q^{n-1}} + \frac{y_n}{q^n} \leq y. \qquad (*)$$

Prove that

(i) $\quad y = \sum_{n=1}^\infty \dfrac{y_n}{q^n}$.

    (c) By (a) and (b), every $y \in [0, 1]$ has an expansion of the form (i), and conversely every sequence $(y_n)_{n=1}^\infty$ in S gives an element in $[0, 1]$ by the equality (i). Prove that the expansion is unique subject to the condition that $(*)$ holds for each integer n. Show that there is also a unique expansion (possibly different) subject to the condition $(*)$ with $\leq$ replaced by $<$.
    (d) Show that the expansion in (i) subject to $(*)$ terminates if and only if $y = p/q^k$ for integers p and k.
    (e) Show that the expansion in (i) is periodic if and only if y is rational.
    (f) Let $x \geq 0$. Show that there is a unique $n \in \mathbf{Z}^+$ such that $n \leq x < n + 1$. Let $[x] = n$. ($[x]$ is the greatest integer *less than or equal to* x.) Show that $0 \leq x - [x] < 1$. Thus

(ii) $\quad x = [x] + \sum_{n=1}^\infty \dfrac{x_n}{q^n}$,

x_n in S. The expression (ii) is the q-adic expansion of the real number x.

Index